Brain Repair

ADVANCES IN EXPERIMENTAL MEDICINE AND BIOLOGY

Recent Volumes in this Series

Volume 561
CHEMISTRY AND SAFETY OF ACRYLAMIDE IN FOOD
Edited by Mendel Friedman and Don Mottram

Volume 562
CHOLINERGIC MECHANISMS
Edited by José Gonzalez-Ros

Volume 563
UPDATES IN PATHOLOGY
Edited by David C. Chhieng and Gene P. Siegal

Volume 564
GLYCOBIOLOGY AND MEDICINE: Proceedings of the 7th Jenner Glycobiology and Medicine Symposium
Edited by John S. Axford

Volume 565
SLIDING FILAMENT MECHANISM IN MUSCLE CONTRACTION: Fifty Years of Research
Edited by Haruo Sugi

Volume 566
OXYGEN TRANSPORT TO TISSUE XXVI
Edited by Paul Okunieff, Jacqueline Williams and Yuhchyau Chen

Volume 567
THE GROWTH HORMONE-INSULIN-LIKE GROWTH FACTOR AXIS DURING DEVELOPMENT
Edited by Isabel Varela-Nieto and Julie A. Chowen

Volume 568
HOT TOPICS IN INFECTION AND IMMUNITY IN CHILDREN II
Edited by Andrew J. Pollard and Adam Finn

Volume 569
EARLY NUTRITION AND ITS LATER CONSEQUENCES: New Opportunities
Edited by Berthold Koletzko, Peter Dodds, Hans Akerbloom and Margaret Ashwell

A Continuation Order Plan is available for this series. A continuation order will bring delivery of each new volume immediately upon publication. Volumes are billed only upon actual shipment. For further information please contact the publisher.

Brain Repair

Edited by

Mathias Bähr

Department of Neurology

University of Göttingen

Göttingen, Germany

Springer Science+Business Media

Landes Bioscience / Eurekah.com

Springer Science+Business Media
Eurekah.com / Landes Bioscience

Printed in the U.S.A.

Springer Science+Business Media, 233 Spring Street, New York, New York 10013, U.S.A.
http://www.springeronline.com

Please address all inquiries to the Publishers:
Eurekah.com / Landes Bioscience, 810 South Church Street, Georgetown, Texas, U.S.A. 78626
Phone: 512/ 863 7762; FAX: 512/ 863 0081
http://www.eurekah.com
http://www.landesbioscience.com

Brain Repair edited by Mathias Bähr, Landes Bioscience / Springer Science+Business Media dual imprint / Springer series: Advances in Experimental Medicine and Biology

ISBN: 0-306-47859-5

While the authors, editors and publisher believe that drug selection and dosage and the specifications and usage of equipment and devices, as set forth in this book, are in accord with current recommendations and practice at the time of publication, they make no warranty, expressed or implied, with respect to material described in this book. In view of the ongoing research, equipment development, changes in governmental regulations and the rapid accumulation of information relating to the biomedical sciences, the reader is urged to carefully review and evaluate the information provided herein.

Library of Congress Cataloging-in-Publication Data

Brain repair / [edited by] Mathias Bähr.
p. ; cm. -- (Advances in experimental medicine and biology ; v. 557)
Includes bibliographical references and index.
ISBN 0-306-47859-5
1. Brain--Physiology. 2. Brain--Regeneration. 3. Cell cycle --Regulation. 4. Apoptosis. I. Bähr, Mathias. II. Series.
[DNLM: 1. Nervous System--physiology. 2. Alzheimer Disease. 3. Apoptosis. 4. Cell Cycle. 5. Neurons--physiology. WL 102 B81335 2006]
QP376.B7197 2006
612.8'2--dc22

2005019652

PARTICIPANTS

Young Hwan Ahn
Neuronal Survival Unit
Section for Neuroscience
Department of Experimental Medical Studies
Wallenberg Neuroscience Center
Lund, Sweden

Mathias Bähr
Department of Neurology
University of Göttingen
Göttingen, Germany

Christine E. Bandtlow
Innsbruck Medical University
Biocenter Innsbruck
Division of Neurobiochemistry
Innsbruck, Austria

Patrik Brundin
Neuronal Survival Unit
Section for Neuroscience
Department of Experimental Medical Studies
Wallenberg Neuroscience Center
Lund, Sweden

Fabio Cavaliere
I.R.C.C.S. Santa Lucia Foundation
Rome, Italy

Thomas Deller
Institute of Clinical Neuroanatomy
Johann Wolfgang Goethe-University
Frankfurt am Main, Germany

Jörg Dietrich
Department of Neurology
Massachusetts General Hospital
Harvard Medical School
Boston, Massachusetts, U.S.A.

Alexandre Dobbertin
INSERM U686
Centre Universitaire des Saints-Péres
Paris, France

Andreas Faissner
Department of Cell Morphology and Molecular Neurobiology
Ruhr-University
Bochum, Germany

James W. Fawcett
Centre for Brain Repair
Cambridge University
Cambridge, U.K.

Fulvio Florenzano
I.R.C.C.S. Santa Lucia Foundation
Rome, Italy

Thomas M. Freiman
Department of Neurosurgery
Neurocenter
University Clinic Freiburg
Albert-Ludwigs-University
Freiburg, Germany

Michael Frotscher
Institute of Anatomy and Cell Biology
University of Freiburg
Freiburg, Germany

Yutaka Fukuda
Department of Physiology
Graduate School of Medicine
Osaka University
Osaka, Japan

Jeremy Garwood
Centre de Neurochimie du CNRS
Strasbourg, France

Marcia Gasis
Molecular Neurobiology Laboratory
Department of Neurology
Heinrich-Heine-University
Düsseldorf, Germany

Carola A. Haas
Experimental Epilepsy Group
Neurocenter
University Clinic Freiburg
Albert-Ludwigs-University
Freiburg, Germany

Nicolas Heck
Institute of Physiology
and Pathophysiology
University of Mainz
Mainz, Germany

Susanne Hermanns
NEURAXO BIOTEC GmbH
Düsseldorf, Germany

Mathias Jucker
Department of Cellular Neurology
Hertie-Institute for Clinical Brain
Research
University of Tübingen
Tübingen, Germany

Gerd Kempermann
Max-Delbrück-Center
of Molecular Medicine
Berlin-Buch, Germany
and
Department of Neurology
Charité University Hospital
Humboldt University Berlin
Berlin, Germany

Nicole Klapka
Molecular Neurobiology Laboratory
Department of Neurology
Heinrich-Heine-University
Düsseldorf, Germany

Kerstin Krieglstein
Zentrum Anatomie,
Abt. Neuroanatomie
Universität Göttingen
Göttingen, Germany

Takuji Kurimoto
Department of Ophthalmology
Hyogo College of Medicine
Hyogo, Japan

Jia-Yi Li
Neuronal Survival Unit
Section for Neuroscience
Department of Experimental Medical Studies
Wallenberg Neuroscience Center
Lund, Sweden

Rafael Linden
Instituto de Biofísica da UFRJ
CCS
Cidade Universitária
Rio de Janeiro, Brazil

Paul Lingor
Department of Neurology
University of Göttingen
Göttingen, Germany

Tomomitsu Miyoshi
Department of Physiology
Graduate School of Medicine
Osaka University
Osaka, Japan

Marco Molinari
I.R.C.C.S. Santa Lucia Foundation
Rome, Italy

Hans Werner Müller
Molecular Neurobiology Laboratory
Department of Neurology
Heinrich-Heine-University Düsseldorf
Düsseldorf, Germany

Gesine Paul
Neuronal Survival Unit
Section for Neuroscience
Department of Experimental Medical Studies
Wallenberg Neuroscience Center
Lund, Sweden

Amie Phinney
Department of Neuropathology
Institute of Pathology
University of Basel
Basel, Switzerland

Tommaso Pizzorusso
Institute of Neuroscience CNR and Scuola Normale Superiore
Pisa, Italy

Gian Michele Ratto
Institute of Neuroscience CNR and Scuola Normale Superiore
Pisa, Italy

Mariana S. Silveira
Laboratório de Neurogênese
Instituto de Biofísica da UFRJ
Rio de Janeiro, Brazil

Maria Teresa Viscomi
I.R.C.C.S. Santa Lucia Foundation
Rome, Italy

Cinza Volonté
I.R.C.C.S. Santa Lucia Foundation
Rome, Italy
and
Institute of Neurobiology and Molecular Medicine (CNR)
Rome, Italy

Anne D. Zurn
Department of Experimental Surgery
Lausanne University Hospital
Faculty of Biology and Medicine
Lausanne, Switzerland

CONTENTS

CHAPTER 1

Cell Death in the Nervous System

Kerstin Krieglstein

Programmed cell death is a fundamental and essential process in development and tissue homeostasis of multicellular organisms. About half of all neurons produced during neurogenesis die before the nervous system matures. Failure to kill appropriate cells can lead to severe developmental defects and diseases such as cancer, whereas increased cell death can lead to degenerative diseases.

Programmed Cell Death

The term "programmed cell death" is used for developmentally occurring cell death, where cells are "programmed" to die during normal development.[1,2] As cell death seems to follow an intracellular "program", programmed cell death is sometimes also used as a synonym for apoptosis.

Apoptosis is a cell autonomous genetically defined program, in which cells respond to internal or external signals by actively participating in their suicide and organization of their disposal.[3] Originally, the term "apoptosis" was defined by a characteristic pattern of morphological changes[4,5] and is now increasingly used to describe the underlying molecular mechanisms.[6] Any cell death lacking the features indicative for active cell death is referred to as necrosis. Necrosis presents a passive form of cell death with relatively slow disintegration of the cells.[7]

The morphological changes by which apoptotic cell death can be characterized and identified, occur in a consecutive fashion. Dying cells start to detach from neighboring cells and extracellular matrix and round up. The cells start to show pertrusions from the plasma membrane, referred to as blebs. Many dying cells show nuclear condensation and disintegration of the nucleus into several fragments. Organelles are generally intact, but may be affected at later stages. Mitochondria have been described to either swell or condense, there is dilatation of the ER, release and aggregation of ribosomes and the occurrence of cytoplasmic vacuoles. Whole cells condense and reorganize into so-called apoptotic bodies,[4] which are membrane bound vesicles containing cytosolic elements, organelles and parts of the condensed nuclei in various combinations. Apoptotic cells are rapidly engulfed and digested by neighboring cells,[4,8] which makes it quite difficult to study morphological changes in vivo.

Neuronal Cell Death during Development

A central problem in developmental neurobiology is the understanding of the regulation of neuron survival and death. The neurotrophic theory[9,10] provides a basis for understanding several features of neuronal development, including the question of why, in many populations of developing neurons, only a proportion of the original number of postmitotic cells survives.

Brain Repair, edited by Mathias Bähr. ©2006 Eurekah.com
and Kluwer Academic / Plenum Publishers.

It is well known that in many regions of the nervous system large numbers (approx. 50%) of postmitotic neurons degenerate and die by a process of naturally occurring neuronal death.[1] It has been repeatedly demonstrated that in most populations of neurons this normal cell loss occurs during the period when neurons are establishing synaptic connections with their targets.[11,12] This temporal coincidence, together with the demonstration that manipulations of the availability of putative synaptic targets alters the number of surviving innervating neurons, led to the proposal that neurons compete for a target-derived factor that is supplied in limiting amounts by the targets and thus adjusting neuron numbers so as to provide sufficient innervation for their targets.[13,14] However, there is growing evidence that other mechanisms may be involved in regulating cell death.[15]

Cell Death in Early Neural Development

Cell death in the developing nervous system is already seen prior to neuronal differentiation and synaptogenesis. Early neural cell death is detected as early as neurulation and seems to affect proliferating neural precursor cells as well as young postmitotic cells during and following neurogenesis of the neural tube and spinal cord as well as the neural crest derived peripheral nervous system.[16] The extent of cell death in early neurogenesis is quite large as estimated from the phenotype analyses of caspase3, caspase9 and Apaf1 gene target ablations,[17-20] however, the precise developmental role remains unknown. In general, apoptosis may be used to select for specific parameters, for morphogenesis or for shaping compartments.[21,22]

Glial Cell Death

So far, research on programmed cell death in the nervous system has focused on neurons. However, it is becoming increasingly clear that developmental cell death also occurs in oligodendrocytes and Schwann cells, the glia of the central and the peripheral nervous system, respectively. Establishing quantitative matches between neurons and ensheathing or myelinating glia suggests that the number and appropriate differentiation of oligodendrocytes or Schwann cells during development should be under close regulation. Considerable oligodendrocyte cell death occurs during development,[23] and also after traumatic conditions. Similarly, programmed cell death of Schwann cells has been described during development in response to the presence of the absence of axon-derived signals.[24,25] Programmed cell death of both oligodendrocytes and Schwann cells follow a typical caspases-dependent pathways.[26,27]

Molecular Mechanisms of Programmed Cell Death

Apoptosis is controlled by several pro- and antiapoptotic gene families that are conserved from nematodes through mammals.[3,28] In the nematode C. elegans, neurons destined to die do so according to a stereotyped scheme: 131 of the 1090 somatic cells generated during development are destined to die.[29] Genetic screens for cell death abnormal (ced) mutants have revealed that a cascade involving ced-3 and ced-4 is required to regulate the intrinsic cell death program in C. elegans, whereas ced-9 is negatively regulated by egl-1 in dying cells (Table 1).[3,30] Although the basic components of the vertebrate apoptotic machinery are similar to those seen in nematodes, they are more varied and complex. In addition to the cell intrinsic genetic cascade many death and survival signals lead to the activation of extrinsic or intrinsic pathways being composed of several adaptors, regulators, caspases, inhibitors as well as pro- and antiapoptotic mediators.

The key effector components of apoptosis are caspases.[6,31] A family of currently 15 members has been identified as important regulators of inflammatory response, while 8 of them play important roles during apoptosis. Caspases are cysteine proteases that cleave their substrates after aspartate residues. Caspases are initially produced as inactive zymogens (procaspases). Caspases may be divided in two functional subfamilies: initiator caspases (caspases-2, -8, -9 or

Table 1. Comparison of apoptosis effector mechanisms in nematodes (top) and vertebrates (bottom)

[a]BH3 Only Inhibitor of Regulator	Regulator	Adaptor		Protease			
EGL-1 -|	CED-9 -|	CED-4	➔	CED-3 ➔			death
BAD, BIK etc. -|	Bcl-2 -|	Apaf-1	➔	Casp9 ➔	Casp3 ➔		death

Signals -| inhibiting or ➔activating the cell death pathway

-10), which are involved in upstream regulatory events, and effector caspases (caspases-3, -6, or -7), which are directly responsible for cell disassembly events, i.e., widespread cleavage of multiple substrates ultimately causing cell death.[32] Effector caspases are direct targets of initiator caspases, while activation of initiator caspase precursors is achieved by adaptor proteins that bind to them via shared motifs. For example, caspase 9 is activated after association of the caspase recruitment domain (CARD) in its prodomain with the CARD in another adaptor protein such as Fadd and Apaf1.[33,34] The ability of adaptors to activate the caspases can be regulated by other proteins that appear to directly interact with the adaptors, e.g., members of the Bcl2 family.

Bcl-2 family members are important sensors that receive multiple signals from various signal transduction pathways residing upstream of irreversible cell damage, they play a pivotal role in deciding whether cells will live or die by either blocking or permitting the regulation of downstream cell death effectors (Table 2).[35,36] This decision is carried by various members of the Bcl2 family that have either anti-apoptotic (e.g., Bcl-2, Bcl-xl, Bcl-w, Mcl-1) of pro-apoptotic (e.g., Bax, Bak, Bcl-XS) function. The ratio between the two subsets is thought to be one determinant for the susceptibility to programmed cell death, which may be mediated via formation of heterodimers.[37] Bcl-2 family members are composed of up to four Bcl-2 homology domains (BH), which are highly conserved in anti-apoptotic Bcl-2 family members, whereas they are conserved to a lesser degree in pro-apoptotic family members. There is an emerging subfamily of "BH3-domain-only" members (e.g., Bid, Bad, Bik, Bim), which are all pro-apoptotic to date (Table 1). Many of the Bcl-2 members contain a hydrophobic sequence at their C-terminus that has been shown to target these molecules to membranes such as the outer mitochondrial membrane.[38] Anti-apoptotic members are found as integral membrane proteins in the mitochondria, endoplasmic or nuclear membrane,[39] whereas the pro-apoptotic members localize to cytosol and translocate upon a death signal to the mitochondrial outer membrane.[40,41] A model describing the mechanism of action of pro- and anti-apoptotic molecules at the mitochondria builds on the channel forming capacity of the Bcl-2 family members, whereby the anti-apoptotic molecules are "guarding the mitochondrial gate" while the pro-apoptotic molecules "gain access" following a death signal.[36]

One consequence of a variety of death stimuli is the release of cytochrome c into the cytosol,[42-44] activating Apaf-1, which in turn activates caspase-9 and caspase-3 (Table 2).[45,46]

However, besides the anti-apoptotic Bcl-2-family members there is an additional antiapoptotic acting gene family, the inhibitor of apoptosis (IAP) family of proteins linked to caspases inhibition. Up to now, seven members of the IAP family have been identified (i.e., XIAP, IAP1, IAP2, NAIP, Survivin, Bruce and livin). IAPs seem to archive their anti-apoptotic activity by two separate mechanisms: one involving caspases inhibition, the other requiring TAK1-dependent JNK1 activation.[47] IAPs have been shown to directly bind and inhibit specific members of the caspases family, e.g., XIAP, IAP1, IAP2 and Survivin directly bind and inhibit caspases-3, -7 and –9, but not caspases-8 or –10.[48-50]

Table 2. Mouse models with defective programmed cell death of in the nervous system

Gene	Knockout Phenotype	References
bclx	Embryonic lethality (E13.5); Extensive neuronal phenotype	90
bax	Viable; Neuronal hyperplasia	91
Bak/bax	Perinatal lethality; accumulation of neurons in the nervous system	92
Apaf1	Embryonic lethality (E16.5); exencephaly due to defective apoptosis of neural progenitor cells;	93-95
Caspase 9	Perinatal lethality, neuronal hyperplasia	19, 51, 96
Caspase 3	Perinatal lethality, neuronal hyperplasia	18, 51

Caspase Independent Pathways

The importance of caspases in the apoptotic process is well established, however, there is also strong evidence for a caspase-independent pathway in the nervous system. Evidence for this comes on one hand from the analysis of programmed cell death in caspase null mice, in which in contrast to the striking perturbances in the forebrain, the extent of cell death of brain stem and spinal cord motoneurons as well as ganglia of the peripheral nervous system may be delayed but in essence is unaltered. [51] Cell death may be mediated via other mitochondrially regulated pathways, [52,36] such as via the apoptosis-inducing factor (AIF). [53]

Cell Death Receptors

Programmed cell death is executed by cell intrinsic pathways, yet, the induction of cell death may be the consequence of extracellular signals. Sensors mediating cell death signaling from the outside into the intracellular death machinery are termed death receptors.[54] Death receptors belong to the tumor necrosis factor receptor (TNFR) gene superfamily (Table 3), which is defined by a common extracellular cysteine motif that serves as the ligand binding domain.[55,56] The death receptors also contain an additional common cytoplasmic domain termed the "death domain".[57,58] Upon ligand binding the assembly of a death-inducing signaling complex is initiated. In the case of CD95L, the ligation of the death receptor is induced to form a homotrimeric complex, thereby clustering the death domains of the receptor.[59] A Fas-associated death domain (FADD) binds as an adaptor protein through its own death domain to the clustered death domains of the receptor. FADD in turn binds through its "death effector domain" the zymogen form of caspases 8 (FLICE/MACH).[50,61] The death effector domain is an example of a caspase recruitment domain (CARD) that represents a more global homophilic interaction domain, which is found in several caspases involved in the apoptotic machinery, including caspases-2, -8, -9, -10.[33] Upon interaction with FADD, procaspase 8 oligomerization is activated by self-cleavage and triggers the apoptotic pathway by cleaving appropriate effector caspases.[62] As the death receptors transmit the signals directly from the cell surface to the apoptotic machinery, caspase activation can be achieved within seconds, causing cell death within hours.

The p75 nerve growth factor receptor also contains a death domain,[63] recent evidence supports an involvement of this receptor in neuronal cell death.[64-67] However, in contrast to the cell death receptors discussed above, the proapoptotic function of p75 seems not exclusively to depend on ligand binding, but on the interplay between Trk and p75 that determines neuronal survival. Neurotrophin receptor interacting factor (NRIF) has been identified as an intracellular p75 binding protein transducing the cell death signals during development.[68]

Table 3. Cell death receptors

Death Receptor	Ligand	References
CD95 (Fas, Apo1)	CD96L	55, 56
TNFR1 (p55, CD120a)	TNFα, lymphotoxin α	58, 97
CAR1 (avian)	unknown	98
Death receptor 3 (DR3, Apo3, WSL-1, TRAMP, LARD)	Apo3L (TWEAK)	99-102
DR4	Apo2L (TRAIL)	103-106
DR5 (Apo2, TRAIL-R2, TRICK 2, KILLER)	Apo2L (TRAIL)	107-111
p75	NGF	66, 68

Disruption of the nrif gene leads to a reduction in cell death closely resembling that of p75 (-/-) and ngf (-/-).[68,69]

Extrinsic Mechanisms

Outside the nervous system, triggering of cell death by extrinsic factors plays an important role in normal development. In the immune system, for example, the regulatory elimination of many cells is mediated by signaling through transmembrane receptors of the Fas/TNFR family.[58] In the nervous system, a pro-apoptotic role is emerging for nerve growth factor (NGF) and bone morphogenetic protein 4 (BMP4). While NGF is best known for its trophic functions, recent experiments indicate that it can also cause cell death during development by activating the neurotrophin receptor p75 in the absence of trkA (see above). There is growing evidence that BMPs function in the early nervous system during regional morphogenesis, i.e., of the dorsal telencephalon or of segmented neural crest-derived structures originated of the hindbrain by regulating specific gene expression, cell proliferation and local cell death.[70,71]

Most recently, the pleiotrophic molecule transforming growth factor-β (TGF-β) has been shown to act as a key regulator in the induction of developmental as well as lesion induced cell death.[72,73] TGF-β induced apoptosis affects neurons of the peripheral and central nervous system,[72] Schwann cells[74] in vivo and oligodendrocytes in vitro.[75,76] TGF-β seems to cooperate with other factor to induce cell death, such as NGF,[73] TNF-α,[77] or CD95.[75,78] The molecular mechanism of TGF-β induced cell death is reviewed by Schuster and Krieglstein.[79]

Cell Death during Neurodegenerative Disorders and Aging

Many neurological diseases involve neuronal degeneration and consequently cell death.[80] Acute disorders, occurring within minutes and hours, e.g., brain trauma, or infarction involve injury-induced apoptosis.[81-83] Chronic disorders, such as Parkinson's disease, Alzheimer's disease or amyotrophic lateral sclerosis, involve slow degeneration of the central nervous system, spanning years or decades. There is evidence that the mechanism of neuronal cell death may involve apoptosis in these disorders.[84] Understanding the biochemical signaling events controlling and mediating apoptosis will lead to the identification of potential targets that could be used for developing new therapeutic agents to reduce cell death as a means to promote functional recovery. Along this line multiple experience is now available using caspases as targets in stroke and neurodegenerative diseases.[85] Another successful approach to inhibit injury induced apoptosis involves the application of the X-linked inhibitor of apoptosis (XIAP).[86]

Regulation of apoptosis of inflammatory cells within the injured nervous system does also play an important role in protecting the central nervous system from immune-mediated damage. For further information see e.g., Pender and Rist[87] or Zipp et al.[88]

Whether apoptosis contributes or even explains the aging progress is still under debate. A good overview on the aging hypothesis and apoptosis is presented for example by Higami and Shimokawa.[89]

References

1. Oppenheim RW. Cell death during development of the nervous system. Annu Rev Neurosci 1991; 14:453-501.
2. Vaux DL, Korsmeyer SJ. Cell death in development. Cell 1999; 96:245-254.
3. Horvitz HR. Genetic control of programmed cell death in the nematode Caenorhabditis elegans. Cancer Res 1999; 59:1701s-1706s.
4. Kerr JF, Wyllie AH, Currie AR. Apoptosis: a basis biological phenomen with wide-ranging implications in tissue kinetics. Br J Cancer 1972; 26:239-257.
5. Häcker G. The morphology of apoptosis Cell Tiss Res 2000; 301:5-17.
6. Shi Y. Mechanisms of caspases activation and inhibition during apoptotsis. Mol Cell 2002; 9:459-470.
7. Majino G, Joris I. Apoptosis, oncosis, and necrosis. An overview of cell death. Am J Pathol 1995; 146:3-15.
8. Ellis RE, Yuan JY, Horvitz HR. Mechanisms and functions of cell death. Annu Rev Cell Biol 1991; 7:663-698.
9. Levi-Montalcini, R. The nerve growth factor 35 years later. Science 1987; 237:1154-1162.
10. Thoenen H, Bandtlow C, Heumann R. The physiological function of nerve growth factor in the central nervous system: comparison with the periphery. Rev Physiol Biochem Pharmacol 1987; 109:145-178.
11. Landmesser L, Pilar G. Interactions between neurons and their targets during in vivo synaptogenesis. Fed Proc 1987; 37:2016-2022.
12. Landmesser L. The relationship of intermuscular nerve branching and synaptogenesis to motoneuron survival. J Neurobiol 1992; 23:1131-1139.
13. Sohal GS. The role of target size in neuronal survival. J Neurobiol 1992; 23:1124-1130.
14. Hollyday M, Hamburger V. Reduction of the naturally occurring motor neuron loss by enlargement of the periphery. J Comp Neurol 1976; 170:311 320.
15. Pettmann B, Henderson CE. Neuron cell death. Neuron 1998; 20:633-647.
16. De la Rosa EJ, de Pablo F. Cell death in early neural development: beyond the neurotrophic theory. TINS 2000; 23:454-457.
17. Ranger AM, Malynn BA, Korsmeyer SJ. Mouse models of cell death. Nature Genetics 2001; 28:113-118.
18. Kuida K, Zheng TS, Na TQ et al. Decreased apoptosis in the brain and premature lethality in CPP32-deficient mice. Nature 1996; 384:368-372.
19. Kuida K, Haydar TF, Kuan CY et al. Reduced apoptosis and cytochrome c-mediated caspase activation in mice lacking caspase 9. Cell 1998; 94:325-337.
20. Cecconi F, Gruss P. Apafl in developmental apoptosis and cancer: how many ways to die. CMLS 2001; 58:1688-1697.
21. Glückmann A. Cell death in normal vertebrate ontogeny. Biol Rev 1951; 26:59-86.
22. Jacobson MD, Weil M, Raff M. Programmed cell death in animal development. Cell 1997; 88:347-54.
23. Barres BA, Hart IK, Coles HS et al. Cell death and control of survival in the oligodendrocyte lineage. Cell 1992; 70:31-46.
24. Grinspan JB, Marchionni MA, Reeves M et al. Axonal interactions regulate Schwann cell apoptosis in developing peripheral nerve: neuregulin receptors and the role of neuregulins. J Neurosci 1996; 16:6107-6118.
25. Trachtenberg JT, Thompson WJ. Schwann cell apoptosis at developing neuromuscular junctions is regulated by glial growth factor. Nature 1996; 379:174-177.
26. Gu C, Casaccia-Bonnefil P, Srinivasan A et al. Oligodendrocyte apoptotsis mediated by caspases activation. J Neurosci 1999; 19:3043-3049.

27. Winseck AK, Caldero J, Ciutat D et al. In Vivo analysis of Schwann cell programmed cell death in the embryonic chick: regulation by axons and glial growth factor. J Neurosci 2002; 22:4509-4521.
28. Reed JC. Dysregulation of apoptosis in cancer. J Clin Oncol 1999; 17:2941-1953.
29. Sulston JE, Schierenberg E, White JG et al. The embryonic cell lineage of the nematode Caenorhabditis elegans. Dev Biol 1983; 100:64-119.
30. Metzstein MM, Stanfield GM, Horvitz HR. Genetics of programmed cell death in C. elegans: past, present and future. Trends Genet 1998; 14:410-416.
31. Thornberry NA, Lazebnik Y. Caspases: enemies within. Science 1998; 281:1312-1316.
32. Nicholson DW, Thornberry NA. Caspases—killer proteases. Trends Biochem.. Sci. 1997; 22:299-306.
33. Hofmann K, Bucher P, Tschopp J. The CARD domain—a new apoptotic signaling motif. Trends Biochem Sci 1997; 22:155-156.
34. Chou JJ, Matsuo H, Duan H, Wagner R. Solution structure of the RAIDD CARD and model for CARD/CARD interaction in caspase-2 and caspase-9 recruitment. Cell 1998; 94:171-180.
35. Adams JM, Cory S. The bcl-2 protein family: arbiters of cell survival. Science1998; 281:1322-1326.
36. Gross A, McDonnell JM, Korsmeyer SJ. BCL-2 family members and the mitochondria in apoptosis. Genes Dev 1999; 13:1899-1911.
37. Oltvai ZN, Milliman CL, Korsmeyer SJ. Bcl-2 heterodimerizes in vivowith a conserved homolog, Bax, that accelerates programmed cell death. Cell 1993; 74:609-619.
38. Nguyen M, Millar DG, Yong VW et al. Targeting of Bcl-2 to mitochondrial outer membrane by COOH-terminal signal sequence. J Biol Chem 1993; 268:25265-25268.
39. Hockenberry D, Nunez G, Milliman C et al. Bcl-2 is an inner mitochondrial membrane protein that blocks programmed cell death. Nature 1990; 348:334-336.
40. Hsu YT, Wolter KG, Youle RJ. Cytosol-to-membrane redistribution of Bax and Bcl-XL during apoptosis. Proc Natl Acad Sci USA 1997; 94:3668-3672.
41. Gross A, Jockel J, Wei MC et al. Enforced dimerization of BAX results in its translocation, mitochondrial dysfunction and apoptosis. EMBO J 1998; 17:3878-3885.
42. Liu X, Kim CN, Yang J et al. Induction of apoptotic program in cell-free extreacts: requirement for dATP and cytochrome c. Cell 1996; 86:147-157.
43. Kluck RM, Bossy-Wetzel E, Green DR et al. The release of cytochrome c from mitochondria: a primary site for Bcl-2 regulation of apoptosis. Science 1997; 275:1132-1136.
44. Bossy-Wetzel E, Newmeyer DD, Green DR. Mitochondrial cytochrome c release in apoptosis occurs upstream of DEVD-specific caspases activation and independently of mitochondrial transmembrane depolarization. EMBO J 1998; 17:37-49.
45. Li P, Nijhawan D, Budihardjo I et al. Cytochrome c and dATP-dependent formation of Apaf-1/caspases-9 complex initiates an apoptotic protease cascade. Cell 1997; 91:479-489.
46. Zou H, Henzel WJ, Liu X et al. Apaf-1, a human homolog to C. elegans CED-4 participates in cyrochrome c-dependent activation of caspases-3. Cell 1997; 90:405-413.
47. Sanna MG, da Silva Correia J, Ducrey O et al. IAP suppression of apoptosis involves distinct mechanisms: the TAK1/JNK1 siganling cascade and caspases inhibition. Mol Cell Biol 2002; 22:1754-1766.
48. Roy N, Mahadevan MS, McLean M et al. The gene for neuronal apoptosis inhibitory protein is partially deleted in individuals with spinal muscular atrophy. Cell 1995; 80:167-178.
49. Shin S, Sung BJ, Cho YS et al. An apoptotic protein human surviving is a direct inhibitor of caspases-3 and caspases-7. Biochemistry 2001; 40:1117-1123.
50. Tamm I, Wang Y, Sausville DA et al. IAP-family protein surviving inhibits caspases activity and apoptosis induced by Fas (CD95), Bax, caspases, and anticancer drugs. Cancer Res 1998; 58:5315-5320.
51. Wienseck AK, Caldero J, Ciutat D et al. In vivo analysis of Schwann cell programmed cell death in the embryonic chick: regulation by axons and glial growth factor. J Neurosci 2002; 22:4509-4521.
52. McCarthy NJ, Whyte MK, Gilbert CS, et al. Inhibition of Ced-3/ICE-related proteases does not prevent cell death induced by oncogenes, DNA damage, or the Bcl-2 homologue Bak. J Cell Biol 1997; 136:215-227.
53. Joza N, Susin SA, Daugas E et al. Essential role of mitochondrial apoptosis-inducing factor in programmed cell death. Nature 2001; 410:549-554.

54. Ashkenazi A, Dixit VM. Death receptors: signaling and modulation. Science 1998; 281:1305-1308.
55. Smith CA, Farrah T, Goodwin RG. The TNF receptor superfamily of cellular and viral proteins: activation, costimulation, and death. Cell 1994; 76:959-962.
56. Gruss HJ, Dower SK. Tumor necrosis factor logand superfamily: invovlvement in the pathology of malignant lymphomas. Blood 1995; 85:3378-3404.
57. Tartaglia LA, Ayres TM, Wong GH et al. A novel domain within the 55 kd TNF receptor signals cell death. Cell 1993; 74:845-853.
58. Nagata S. Apoptosis by death factor. Cell 1997; 88:355-365.
59. Huang B, Eberstadt M, Olejniczak ET et al. NMR structure and mutagenesis of the Fas (APO-1/CD95) death domain. Nature 1996; 384:638-641.
60. Boldin MP, Goncharov TM, Goltsev YV et al. Involvement of MACH, a novel MORT1/FADD-interacting protease, in Fas/APO-1- and TNF receptor-induced cell death. Cell 1996; 85:803-815.
61. Muzio M, Chinnaiyan AM, Kischkel FC et al. FLICE, a novel FADD-homologous ICE/CED-3-like protease, is recruited to the CD95 (Fas/APO-1) death-inducing signaling complex. Cell 1996; 85:817-827.
62. Muzio M, Stockwell BR, Stennicke HR et al. An induced proximity model for caspases-8 activation. J Biol Chem 1998; 273:2926-2930.
63. Liepinsh E, Ilag LI, Otting G et al. NMR structure of the death domain of the p75 neurotrophin receptor. EMBO J 1997; 16:4999-5005.
64. Frade JM, Rodriguez-Tebar A, Barde YA. Induction of cell death by endogenous nerve growth factor through its p75 receptor. Nature 1996; 383:166-168.
65. Naumann T, Casademunt E, Hollerbach E et al. Complete deletion of the neurtrophin receptor p75NTR leads to long-lasting increases in the number of basal forebrain cholinergic neurons. J Neurosci 2002; 22:2409-2418.
66. Miller FD, Kaplan DR. Neurotrophin signaling pathways regulating neuronal apoptosis. CMLS 2001; 58:1045-1053.
67. Chao MV, Bothwell M. Neurotrophins: to cleave or not to cleave. Neuron 2002; 33:9-12.
68. Casademunt E, Carter BD, Benzel I et al. The zinc finger protein NRIF interacts with the neurotrophin receptor p75NTR and participates in programmed cell death. EMBO J 1999; 21:6050-6061.
69. Frade JM, Barde YA. Genetic evidence for cell deathmediated by nerve growth factor and the neurotrophin receptor p75 in the developing mouse retina and spinal cord. Development 1999; 126:683-690.
70. Furuta Y, Piston DW, Hogan BL. Bone morphogenetic proteins (BMPs) as regulators of dorsal forebrain development. Development 1997; 124:2203-2212.
71. Graham A, Koentges G, Lumsden A. Neural crest apoptosis and the establishment of craniofacial pattern: an honorable death. Mol Cell Neurosci 1996; 8:76-83.
72. Krieglstein K, Richter S, Farkas L et al. Reduction of endogenous transforming growth factor beta prevents ontogenetic neuron death. Nature Neuroscience 2000; 3:1085-1091.
73. Dünker N, Schuster N, Krieglstein K. Transforming Growth Factor Beta Modulates Programmed Cell Death in the Retina of the Developing Chick Embryo. Development 2001; 128:1933-1942.
74. Parkinson DB, Dong Z, Bunting H et al. Transforming growth factor β (TGF-β) mediates Schwann cell death in vitro and in vivo: examination of c-jun activation, interactions with survival signals, and the relationship of TGF-β-mediated death to Schwann cell differentiation. J Neurosci 2001; 21:8572-8585.
75. Yu C, Takeda M, Soliven B. Regulation of cell cycle proteins by TNF-alpha and TGF-beta in cells of oligodendroglial lineage. J Neuroimmunol 2000; 108:2-10.
76. Schuster N, Bender H, Philippi A et al. TGF-β induces cell death in the oligodendroglial cell line OLI-neu. Glia 2002; 40:95-108.
77. Schuster N, Bender H, Roessler O et al. TGF-β and TNF-α cooperate to induce apoptosis in the oligodendroglial cell line OLI-neu. J Neurosci Res 2003; 73(3):324-33.
78. Hagimoto N, Kuwano K, Inoshima I et al. TGF-β1 as an enhancer of Fas-mediated apoptosis of lung epithelial cells. J Immunol 2002; 168:6470-6478.

79. Schuster N, Krieglstein K. Mechanisms of TGF-beta-mediated apoptosis. Cell Tissue Res 2002; 307:1-14.
80. Honig LS, Rosenberg RN. Apoptosis and neurologic disease. Am J Med 2000; 108:317-330.
81. Mattson MP, Culmsee C, Yu ZF. Apoptotic and antiapoptotic mechanisms in stroke. Cell Tiss Res 2000; 301:173-187.
82. Springer JE, Nottingham SE, McEwen ML et al. Caspase-3 apoptotic signaling following injury to the central nervous system. Clin Chem Lab Med 2001; 39:299-307.
83. Dikranian K, Ishimaru MJ, Tenkova T et al. Apoptosis in the in vivo mammalian forebrain. Neurobiol Dis 2001; 8:359-379.
84. Jellinger KA. Cell death mechanisms in neurodegeneration. J Cell Mol Med 2001; 5:1-17.
85. Schulz JB, Weller M, Moskowitz MA. Caspases as treatment targets in stroke and neurodegenerative diseases. Ann Neurol 1999; 45:421-429.
86. Kugler S, Straten G, Kreppel F. The X-linked inhibitor of apoptosis (XIAP) prevents cell death in axotomized CNS neurons in vivo. Cell Death Differ 2000; 7:815-824.
87. Pender MP, Rist MJ. Apoptosis of inflammatory cells in immune control of the nervous system: role of glia. Glia 2001; 36:137-144.
88. Zipp F, Krammer PH, Weller M. Immune (dys)regulation in multiple sclerosis: role of the CD95-CD95 ligand system. Immunol Today 1999; 20:550-554.
89. Higami Y, Shimokawa I. Apoptosis in the aging process. Cell Tiss Res 2000; 301:125-132.
90. Motoyama N, Wang F, Roth KA et al. Massive cell death of immature hematopoietic cells and neurons in Bcl-x-deficient mice. Science 1995; 267:1506-1510.
91. Knudson CM, Tung KS, Tourtellotte WG et al. Bax-deficient mice with lymphoid hyperplasia and male germ cell death. Science 1995; 270:96-99.
92. Lindsten T, Ross Aj, King A et al. The combined functions of proapoptotic Bcl-2 family members bak and bax are essential for normal development of multiple tissues. Mol Cell 2000; 6:1389-1399.
93. Cecconi F, Alvarezbolado G, Meyer BI et al. Apafl (Ced-4 homolog) regulates programmed cell death in mammalian development. Cell 1998; 94:727-737.
94. Yoshida H, Kong YY, Yoshida R et al. Apafl is required for mitochondrial pathways of apoptosis and barin development. Cell 1998; 94:739-750.
95. Honarpour N, Du C, Richardson JA et al. Adult Apaf-1-deficient mice exhibit male infertility. Dev Biol 2000; 218:248-258.
96. Hakem R, Hakem A, Duncan GS et al. Differential requirement for caspase 9 in apoptotic pathways in vivo. Cell 1998; 94:339-352.
97. Baker SJ, Reddy EP. Modulation of life and death by the TNF receptor superfamily. Oncogene 1998; 17:3261-3270.
98. Brojatsch J, Naughton J, Rolls MM. CAR1, a TNFR-related protein, is a cellular receptor for cytopathic avian leucosis-sarcoma viruses and mediates apoptosis. Cell 1996; 87:845-855.
99. Chinnaiyan AM, O'Rourke K, Yu GL et al. Signal transduction by DR3, a death domain-containing receptor related to TNFR-1 and CD95. Science 1996; 274:990-992.
100. Kitson J, Raven T, Jiang YP et al. A death-domain-containing receptor that mediates apoptosis. Nature 1996; 384:371-375.
101. Chicheportiche Y, Bourdon PR, Xu H et al. TWEAK, a new secreted ligand in the tumor necrosis factor family that weakly induces apoptosis. J Biol Chem 1997; 272:32401-32410.
102. Marsters SA, Sheridan JP, Pitti RM et al. Identification of a ligand for the death-domain-containing receptor Apo3. Curr Biol 1998; 8:525-528.
103. Pan G, Ni J, Wei YF et al. An antagonist decoy receptor and a death domain-containg receptor for TRAIL. Science 1997; 277:815-818.
104. Wiley SR, Schooley K, Smolak PJ et al. Identification and characterization of a new member of the TNF family that induces apoptosis. Immunity 1995; 3:673-682.
105. Pitti RM, Marsters SA, Ruppert S et al. Induction of apoptosis by Apo-2 ligand, a new member of the tumor necrosis factor cytokine family. J Biol Chem 1996; 271:12687-12690.
106. Marsters SA, Pitti RM, Donahue CJ et al. Activation of apoptosis by Apo-2 ligand is independent of FADD but blocked by CrmA. Curr Biol 1996; 6:750-752.
107. Pan G, O'Rourke K, Chinnaiyan AM. The receptor for the cytotoxic ligand TRAIL. Science 1997; 276,111-113.

108. Sheridan JP, Marsters SA, Pitti RM et al. Control of TRAIL-induced apoptosis by a family of signaling and decoy receptors. Science 1997; 277:818-821.
109. Walczak H, Degli-Esposti MA, Johnson RS et al. TRAIL-Ra: a novel apoptosis-mediating receptor for TRAIL. EMBO J 1997; 16:5386-5397.
110. Wu GS, Burns TF, McDonald ER III et al. KILLER/DR5 is a DNA damage-inducible p53-regulated death receptor gene. Nat Genet 1997; 17:141-143.
111. Chaudhary PM, Eby M, Jasmin A et al. Death receptor 5, a new member of the TNFR family, and DR4 induce FADD-dependent apoptosis and activate the NF-kappaB pathway. Innumity 1997; 7:821-830.

Chapter 2

The Glial Response to Injury and Its Role in the Inhibition of CNS Repair

James W. Fawcett

The failure of axon regeneration after CNS injury is due to an inadequate or inappropriate regenerative response from damaged CNS axons and to a CNS environment that inhibits regeneration. This inhibitory environment contains many molecules that promote axon growth as well as molecules that inhibit it, but the balance of activities in the damaged CNS does not favour the regeneration of adult CNS axons. In principle, therefore, axon regeneration could be achieved in three ways: (1) Inhibitory molecules might be destroyed or blocked, (2) the amount of permissive molecules might be increased, or new permissive molecules introduced, (3) Axons might be altered so that they can grow in the inhibitory CNS environment. Some success has been achieved with all three of these approaches. This review addresses the inhibitory properties of the glial scar, a structure which forms wherever the CNS is damaged, and which is one source of axon growth inhibitory molecules in CNS injuries.

What Is the Glial Scar?

The end stage of glial scar formation, which is seen weeks after injury, is a largely astrocytic structure. The astrocytes are hypertrophic, with many tightly interweaving processes, many of them joined with junctional complexes.[1] The appearance is similar in many ways to the astrocytic glia limitans which lies around the surface of the CNS under the meninges. However the scar evolves over the weeks following injury, and other cell types play critical roles. Wherever the CNS is damaged a scarring process is initiated, which involves microglia, oligodendrocyte precursors, astrocytes, meningeal cells and vascular endothelial cells. The first change seen after injury is the appearance at the injury site of blood-borne cells, particularly monocyte/macrophages.[2] Within a few hours the endogenous macrophage lineage cells of the CNS, the microglia, begin to respond. They hypertrophy, begin to divide, and upregulate many molecules including complement receptors, which are the antigens often used to identify the cells and their state of reactivity.[3] Starting at about 24 hours oligodendrocyte precursors (OPCs) become reactive, with a burst of cell division; they hypertrophy and increase in cell surface levels of the chondroitin sulphate proteoglycan (CSPG) NG2 which is the most useful marker of these cells.[4,5] Astrocytes show reactivity after one to two days. They hypertrophy with an increase in size and and length and number of processes, a proportion of the cells divide, and they greatly upregulate the intermediate filament proteins GFAP, vimentin and nestin.[6-8] The GFAP upregulation is widespread, the vimentin and nestin upregulation much more restricted to the site of the lesion. Most lesions are invaded by meningeal cells, either from the meningeal

Brain Repair, edited by Mathias Bähr. ©2006 Eurekah.com and Kluwer Academic / Plenum Publishers.

layer surrounding the CNS, or from the meningeal-like cells that surround major blood vessels. Meningeal cell invasion is a particularly marked feature of spinal cord injuries, where there is often a large plug of these cells filling the lesion cavity.[1] Around CNS injuries capillaries become hypertrophic, with a considerable increase in the amounts of laminin and other matrix molecules, and endothelial cells can be seen invading lesions and forming part of the plug of cells that may fill them. Over time, most of the reactive microglia and oligdendrocyte precursors disappear, leaving behind the reactive astrocytes surrounding the injury and the meningeal cells and vascular endothelial cells filling it.

Control of Glial Scar Formation

The initiation and control of scar formation is an extremely complex subject that needs a review to itself. A recent review summarises current knowledge in this area.[3] The control of scar formation has been most extensively studied in the type of reactive gliosis that occurs when a peripheral nerve is injured. Many of the studies have been performed in the facial nerve nucleus, where there is recruitment and activation of microglia and reactive astrocytosis after nerve crush. However oligodendrocyte precursors are not activated in these lesions. The control of glial reactivity involves a complex interplay between microglia, neurones and astrocytes with signals probably passing in both directions between these three cell types. The molecules involved include IL-6, TGFbeta, FGF-2, MCSF and other cytokines.

The Glial Scar and Axon Regeneration

Numerous in vivo experiments have shown that the glial scar inhibits axon regeneration. There is obviously a correlation between the environment in which axon regeneration fails and scar formation, since a scar will form where ever axons are cut. Even very small lesions, which are insufficient to excite an visible disruption of glial architecture can cause changes in the CNS environment sufficient to block axon regrowth.[9] In order to show more than a correlation between glia scarring and inhibition of axon growth various transplant experiments have been performed. CNS tissue, even immature CNS tissue containing largely astrocytes and few oligodendrocytes blocks axon regeneration when it is transplanted to peripheral nerves, and tissue removed from scarred areas is very inhibitory.[10-13] Until recently the question of whether all CNS tissue is equally inhibitory to axon regeneration, or whether scar tissue is particularly inhibitory was not resolved. Two experiments from Davies and Silver show convincingly that the glial scar is much more inhibitory than the rest of the CNS tissue. The first experiment was to transplant adult sensory neurones into the corpus callosum of adult rats, using an atraumatic microtransplantation technique that does not usually initiate scar formation. In most animals there was extensive regeneration of axons from the transplanted neurones through the corpus callosum. However in a few cases regeneration failed, and it was in these cases that the transplant had been more traumatic, and had excited reactive gliosis around the transplant.[14] In the second experiment adult sensory neurones were transplanted into the spinal cord and a distance from a spinal cord injury. Again axons regenerated from the transplanted neurones through the spinal cord white matter, but growth was blocked when the axons approached the scar tissue surrounding the injury site.[15] The general conclusion is that all CNS tissue is inhospitable to axon regeneration, but glial scar tissue is much more inhibitory than undamaged tissue.

Inhibitory Glial Boundaries

In addition to the inhibition of axon regeneration by normal and damaged CNS tissue, there are situations where axon growth stops at places where the glial environment changes sharply from one type of glial cell to another; these are called glial boundaries. The most studied of these boundaries is that found between peripheral nerve tissue and CNS tissue, where the Schwann cell environment of peripheral nerve changes to the astrocytic environment of the CNS, as is seen at the dorsal root entry zone (DREZ). Axons will regenerate within the dorsal root following a crush, particularly if the regenerative response is increased by a concomitant crush of the peripheral nerve attached to the same dorsal root ganglia. However the axons stop precisely at the DREZ, being unable to transit from a Schwann cell environment to an astrocyte/oligodendrocyte environment.[16-19] A similar situation occurs when a peripheral nerve or a Schwann cell transplant is placed in the CNS. These transplants attract axons successfully, so axons experience no difficulty in passing from astrocytes to Schwann cells. However when the axons reach the other end of the graft most are blocked at the Schwann cell/astrocyte boundary that exists there.[20] Meningeal cells also appear to have boundary-forming properties. Most CNS injuries become lined with invading meningeal cells within a few days. Some axons are able to regenerate to the interface between astrocytes and meningeal cells, but regeneration into the meningeal cell plug is very rarely seen. For instance in optic nerve crush lesions a largely astrocyte-free zone is seen at the crush site which is invaded by meningeal cells, microglia and oligodendrocyte precursors (OPCs).[21] Axons can be seen that have grown to the edge of the astrocyte zone, but they do not penetrate into the crush lesion unless they accompany astrocytic processes, some of which eventually invade these injuries.

Inhibitory Molecules in the Damaged CNS

There are many molecules that have inhibitory activity towards axon growth in the adult CNS. These can be divided into three categories: (1) Molecules present in myelin produced by oligodendrocytes which are present in the normal and damaged CNS, (2) Molecules upregulated in reactive glial cells around areas of injury, which are mostly chondroitin sulphate proteoglycans, (3) Axon guidance molecules which play a role in the develpment of the CNS, and which are also present in CNS injuries.

Inhibitory Molecules Produced by Oligodendrocytes

Oligodendrocytes produce several molecules which are extremely inhibitory to axon growth, and which play an important part in the inhibition of axon growth in the CNS. The subject of this review is the glial scar-related molecules, so only a brief description is given here. NogoA is a molecule of the reticulon family, expressed in oligodendrocytes and also some classes of neuron, which is inhibitory to all classes of axon except some embryonic axons.[22] Much work has been done with a blocking antibody, IN-1 and more recently with other antibodies with blocking activity. These antibodies have been shown to stimulate axon regeneration in a variety of lesion models in the spinal cord and elsewhere.[23] A receptor molecule has recently been identified.[24,25] MAG is expressed in myelinating oligodendrocytes, and is also released from the cell surface to diffuse more widely. It is inhibitory to many types of axon, although inhibition varies with axonal type and age.[26-28] A MAG knockout showed little regeneration in the CNS, demonstrating that other inhibitory mechanisms are able to block regeneration even in the absence of MAG, although the properties of MAG in vitro make it clear that it must be an important inhibitor of axon regeneration in vivo. Tenascin-R is a member of the tenascin family with axon growth inhibitory properties expressed at particularly high level in white matter, produced by oligodendrocytes and oligodendrocyte precursors. Also produced by oligodendrocytes is the CSPG versican, which is described below.

Inhibitory Molecules Produced in the Glial Scar

The first investigations into glial scar inhibitory molecules were performed in astrocytes, since they are the main component of mature scar tissue. The first in vitro experiments did not reveal inhibitory properties, since axons grow fairly well on monolayers of astrocytes. However a glial scar is a three-dimensional tissue of tightly interwoven astrocytes, not a monolayer, and when astrocytes are grown as three-dimensional tissues axons grow through them very poorly.[29] In order to try and identify the types of inhibitory molecules produced by astrocytes a variety of astrocytic cell lines were produced, some of which were inhibitory to axon regeneration, some of which were permissive. Comparisons between these cells showed that the ability of the different cell lines to support axon regeneration was mirrored closely by the growth-promoting properties of their extracellular matrix. Yet the matrix of even the most inhibitory lines contained large amounts of laminin and other growth-promoting molecules. Clearly these cells were producing extracellular matrix molecules with the ability to block the growth-promoting effects of laminin. Various lines of evidence showed that these inhibitory molecules were CSPGs, and that much of the inhibition due to these CSPGs could be abrogated either by digesting away their glycosaminoglycan chains (GAGs) with chondroitinase ABC, or by preventing the sulfation of the GAGs with chlorate, or by preventing the GAGs from attaching to their protein cores by treatment with beta-D-xylosides.[30,31] For one particular inhibitory astrocyte line the main inhibitory molecule was the CSPG NG2.[32] That CSPGs are important inhibitory molecules for primary astrocyte cultures was shown by treating three-dimensional cultures with chlorate to block GAG sulfation, resulting in astrocytes that were much more permissive for axon regeneration than untreated cells.[30]

There were also indications from in vivo work that CSPGs might be important inhibitory molecules in glial scars. Using an antibody, CS56, that binds to sulfated forms of the chondroitin sulfate (CS) GAG chain, it was shown that in many types of CNS injury CS is upregulated within a few days of injury, and remains increased for a month or more afterwards.[33] Two types of experiment had also shown that these molecules are inhibitory. In the first, filter material was implanted into a CNS lesion, then removed several days later covered in scar astrocytes. These provided an inhibitory environment for axon growth, but when digested with chondroitinase the cultures were more permissive.[11,34] A second approach was to dissect scar tissue from around a CNS lesion and extract inhibitory material from it. This investigation showed that the inhibition produced by these extracts was sensitive to chondroitinase and heparitinase.[35,36]

More recently infusions of chondroitinase ABC to the damaged CNS have shown that digesting away GAG chains from CSPGs can make the glial scar less inhibitory to axon regeneration. In the first experiment chondroitinase was infused into a knife cut lesion of the nigro-striatal tract. Since the turnover of matrix in the CNS is fairly slow, these injections were given as bolus injections on alternate days, ensuring that the enzyme was present at a high enough concentration to be effective. In animals treated in this way around 2000 out of a complement of 45,000 axons were able to regenerate back to the striatum, compared with no regeneration in controls.[37] In a second set of experiments chondroitinase ABC has been infused into lesions of the dorsal columns of the rat spinal cord at level C4. These animals have shown regeneration of both sensory and corticospinal axons, with return of postsynaptic potentials resulting from cortical stimulation below the level of the lesion. The animals were assessed behaviourally, and showed a dramatic and rapid return of almost normal function in beam and grid walking tasks, but no improvement in a pure sensory task that would require axons to regrow back to the dorsal column nuclei.[38]

When astrocytes participate in the glial scar they greatly upregulate two cytoskeletal proteins, GFAP and vimentin. There has always, therefore, been an question as to whether this process is responsible for some of the inhibition seen in the glial scar. This has recently been tested in animals in which the GFAP and vimentin genes have been knocked out. Animals with a GFAP knockout showed increased axon regeneration after spinal cord injuries, but there was no effect after vimentin knockout.[39]

Individual CSPGs

Neurocan

Neurocan belongs to the family of lecticans, which also includes versican, brevican and aggecan. These molecules are all secreted molecules, which have a hyaluronate-binding motif. In the adult CNS much of the neurocan is in a processed form, in which the molecule is cleaved by proteolytic action into an N and C terminal fragment, known as neurocan N and neurocan C.[40] When the CNS is extracted into detergent-free saline much of the intact neurocan and neurocan C is removed, indicating that they are not attached to the cell surface. However a proportion of the intact neurocan and about half the neurocan N can only be extracted if the brain homogenate is treated with hyaluronidase, indicating that this neurocan is hyaluronate-bound.[41] When production of neurocan is investigated in purified glial cell populations in vitro it is produced by astrocytes and by OPCs, but not by mature oligodendrocytes. Neurocan is greatly upregulated following CNS damage. Immunostaining of normal brain showns neurocan to be present in white matter and at the glia limitans and also in perineuronal nets around many neurones. After cortical damage intense immunoreactivity is seen in grey and white matter surrounding the lesion. Western blots show that much of this upregulation is of the intact form of neurocan.[41,42] When proteoglycans are run on western blots they are generally present as a long diffuse smear rather than as a single discrete band. This is because of the variable quantity of GAG attached to the molecules, giving a range of molecular weights. However if the extract is treated with chondroitinase the GAGs are removed and the core protein runs as a single band. From this the amount of GAG (glycanation) can be estimated. Estimates of this type for neurocan show that glycantion is increased following CNS injury, each neurocan molecule carrying more GAG. Neurocan is inhibitory for axon growth: in a stripe assay in which stripes of L1 alternated with L1 plus neurocan axons from cerebellum chose to grow on the neurocan-free stripes.[41] Neurocan has been shown to interact with N-CAM, Ng-CAM/Ll, TAG-l/ axonin-l, and tenascin, and to inhibit axon growth mediated by Ng-CAM/Ll.[43,44] The difference in inhibitory properties between monolayers and three-dimensonal cultures of astrocytes is probably due to the way in which neurocan associates with astrocytes. In monolayer culture neurocan does not associate with the astrocyte surface, although it sticks to the dish surrounding the cells. Thus in astrocyte monolayers most of the neurocan is secreted into the medium, is diluted away and therefore is unable to inhibit axon growth on the astrocyte surface. In three-dimensional cultures, however, the neurocan cannot diffuse away and is trapped in between the cells, It is therefore present at high concentration in the environment in which the axons are attempting to grow, and can therefore block axon regeneration. Neurocan production by astrocytes is upregulated by TGFalpha, TGFbeta and FGF-2.

Versican

Versican can exist in four splice variants. The major form in the normal CNS is the smallest form, V2.[45] In purified glial cultures versican is not produced by astrocytes, but the V2 form is produced in a differentiation-related fashion by cells of the OP lineage.[46,47] Undifferentiated bipolar OPCs do not make versican, but mulitpolar precusors and pre-oligodendrocytes do. There is less production by fully mature oligodendrocytes with myelin-like sheets. Versican is also made by meningeal cells in vitro, but these cells produce the larger V0 and V1 forms of the molecule. Extracting versican with saline, and following hyaluronidase digestion showns that about half the versican in the brain is bound to hyaluronate, about half is in a form which can be extracted by detergent-free saline. The versican from brain appears to be less glycanated than neurocan, since even without chondroitinase treatment it runs on western blots as a single band. However this band shifts slightly after chondroitinase digestion showing that there is some glycanation. Versican produced in vitro, however, is much more glycanated, and much of it will not enter a gel unless chondroitinase is used. In the normal brain versican is present in the white matter, and after injury it is upregulated in and around the injury in white and grey matter.[46] This upregulation is presumably due to production by the OPCs that are recruited in large numbers to CNS injuries, although the secreted versican diffuses too readily to be identified around individual OPCs in immunostains. Versican has been shown to be inhibitory to axon growth in stripe and other assays.[46,48] In addition medium conditioned by oligodendrocyte cultures has large amounts of versican and brevican in it, and is inhibitory to axon growth. When the medium is depleted of versican it is less inhibitory.[47] Versican production by OPCs in culture is increased by TGFbeta, Il-1 and CNTF.[41]

Brevican

Brevican is the smallest of the lecticans found in the CNS. It can be produced by astrocytes, is upregulated following injury and has axon growth inhibitory properties.[49,50]

NG2

NG2 is produced as a membrane spanning molecule. It is a part time proteoglycan in that it can be produced both with and without GAG chains attached. NG2 can be cleaved from the cell surface by the action of an unidentified metalloproteinase, so that in the normal CNS around half of the NG2 is membrane associated, and half can be extracted without the use of detergents.[51] In the CNS NG2 is seen on three types of cell, oligodendrocyte precursors, vascular endothelial cells and meningeal cells, and these cell types also produce NG2 in vitro. NG2 positive OPCs are present throughout the CNS in both grey and white matter, where they may contact nodes of Ranvier and synapses.[4] After CNS injury there is rapid proliferation of OPCs within 2mm of the injury site, and the OPCs hypertrophy and greatly increase the around of NG2 on the cell surface. Western blot analysis shows a large increase in levels of NG2 after injury, starting at 24 hours and peaking around 7 days, after which levels decline over the following two or three weeks.[52] Comparisons between chondroitinase digested and undigested lanes show that the glycanation of NG2 is increased after injury. NG2 is strongly inhibitory to axon regeneration, and a blocking antibody to NG2 has been shown to promote axon growth on an inhibitory astrocyte cell line that produces large amounts of the proteoglycan (32). NG2 has other functions. It associates with the PDGFalpha receptor and acts as a necessary cofactor to its ability to transduce effects from this growth factor.[53,54] Since a large proportion of the NG2 in the CNS is in the released form, it is possible that there is some functional competition between released and cell surface NG2, but this issue has not yet been addressed experimentally.

Phosphacan

Phosphacan and its mouse homologue DSD-1 belong to a family of alternatively spliced molecules. Phosphacan, which is a secreted molecule, is the extracellular domain of the receptor tyrosine phosphatase RPTPbeta/zeta. RPTPbeta/zeta itself is found in two forms, a short receptor form and the full length form. In vitro phosphacan is expressed by astrocytes and by OPCs. Both cell types also make RPTPbeta/zeta, with OPCs making predominantly the short receptor form.[42,55,56] Phosphacan is inhibitory to some axons, but promotes growth in others.[43,57]

Glycanation of CSPGs

Various experiments described above demonstrate that the inhibitory properties of CSPGs are partly dependent on the sulphated GAGs attached to the protein cores. The biology of this form of inhibition is not established. To some extent the inhibitory properties of CSPGs can be reproduced by GAGs by themselves, if presented at high concentration.[58] However in general the GAGs and proteoglycan core proteins must be attached to one another for them to show their normal inhibitory properties. Exactly how the inhibition works is not established. Many of the CSPG core proteins bind to other matrix molecules such as laminin and tenascin, or to cell surface adhesion molecules such as L1.[44] It seems probable that if the highly charged sulfated GAGs are localised to a region of a protein by the binding of the CSPG protein core, there might be masking of epitopes that promote axon growth, or a change in the tertiary structure of the protein sufficient to alter its function. Some experiments have examined whether the position in which the GAGs are sulfated affects their inhibitory function. These experiments suggest that the 6-sulfated GAGs are particularly inhibitory, while GAGs sulfated in the 6 position and the 2 position on the glucuronic acids are more permissive to axon growth.[59]

We have examined glycanation of NG2, neurocan and versican in the normal and injured brain. This can be done on western blots by comparing lanes in which the extract has or has not been digested with chondroitinase ABC. In undigested lanes GAGs of various lengths cause the CSPGs to run as a smear, which resolves to a single band after digestion. Densitomitry of the smear region gives an estimate of glycanation. These studies show that for NG2 and neurocan both core protein and the extent of glycanation are increased following cortical injury. Versican, however, is less glycanated in vivo than the other two CSPGs, and this does not change greatly after injury. When CSPGs from glia grown in vitro are examined in the same way, the degree of glycanation is greater than is seen in vivo, particularly for versican which is so highly glycanated when produced by oligodendrocyte lineage cells in vitro that it will not enter a gel at all unless digested with chondroitinase ABC.[41,46,51,60]

Chondroitin Sulphate Proteoglycans and Regeneration in the CNS

Since CSPGs are expressed in large amounts in CNS injuries, and since most of them are inhibitory to axon growth they must play some part in blocking axon regeneration after injury. Some of the evidence that injury-related CSPGs are a significant factor in CNS regeneration is discussed above. The most direct evidence are the experiments in which the enzyme chondroitinase has been infused into brain and spinal cord injuries, also described above.

If CSPGs are inhibitory to axon regeneration, what strategies might be used to counteract their effects and so promote regeneration after CNS injury? The first consideration in planning a strategy is to decide where CSPGs exert their effect. The molecules are expressed to some extent in the normal CNS, particularly in white matter, but are upregulated with an increase in their glycanation around injuries. If the molecules only have to be counteracted in the immediate vicinity of an injury, the task will be easier than trying to clear inhibition from the whole region where axon regeneration is needed. There is little evidence on this issue. Three experiments from Davies and Raisman and Davies and Silver, mentioned previously, suggest that inhibition due to CSPGs is much greater around injuries than in normal CNS.[9,14,15] The first experiment examined axon regeneration after very small lesions in the spinal cord which did not disturb glial architecture: following this cut axons sprouted away from the lesion through undamaged tissue but not through it, suggesting that only the lesion area is very inhibitory. The second and third experiments were to transplant adult sensory neurones into the rat CNS using an atraumatic technique. Axon growth from these transplants was only blocked where they encountered glial scar tissue, with high levels of CSPG. The conclusion is that normal CNS is probably moderately inhibitory to axon growth, but scar tissue is highly inhibitory. If axons can be enabled to pass through the scar area they therefore stand a chance of being able to regenerate on through the undamaged tissue. What treatment might be applied? The problem is that there are many CSPGs in the injured CNS, all of them inhibitory. It will be difficult to design treatments for each CSPG individually. However all CSPGs possess GAG chains, and much of their inhibitory activity relies on them, making them an attractive therapeutic target. Moreover the experiments in which chondroitinase has been infused into the CNS to remove GAG chains have been successful in promoting regeneration. In principal inhibition of GAG synthesis would be a good way of preventing the buildup of inhibition in CNS injuries. However at present on sodium chlorate and beta-D-xylosides are able to do this, and both promote axon regeneration in in vitro models, but there are problems with both these agents.

Glial Boundaries in the CNS

The inhibitory mechanisms due to CSPGs and myelin-related molecules can be called surround inhibition, since the inhibitory molecules are diffusely expressed within the glial scar. There are places where axons meet boundaries between different glial cell types, and at some of these boundaries axon growth is stopped abruptly. For CNS repair two types of boundary are significant, Schwann cell/astrocyte and astrocyte/meningeal cell boundaries.

Schwann Cell/Astrocyte Boundaries

Boundaries between Schwann cells and astrocytes occur wherever peripheral nerves contact the CNS, for instance at the dorsal and ventral roots. Schwann cells and astrocytes show no ability to mix, leading to sharp boundaries between the cell types and dorsal root and ventral root entry zones.[20,61-63] The form of myelination changes at these entry zones, so that motor and sensory axons as they enter or leave the CNS have a node of Ranvier with an oligodendrocyte on one side, a Schwann cell on the other. If motor axons are damaged within the spinal cord, they can usually regenerate out of the CNS across the ventral root entry zone and into peripheral nerve, indicating that these axons can cross from astrocytes onto Schwann cells.[16] However if the dorsal root is crushed the axons will regenerate back towards the spinal cord, particularly if the peripheral nerve is crushed at the same time to increase the vigour of regeneration, but when the axons encounter astrocytes at the dorsal root entry zone their growth is blocked.[17,64] The stopped axon growth cones undergo morphological changes that indicate that the axon has received a stop signal as if it has reached an appropriate target.[65,66] These axons are unable, therefore to cross from Schwann cells to astrocytes. Similar boundary behaviours are seen when a peripheral nerve or Schwann cells are transplanted into the CNS in an attempt to bridge across an injury. Many types of CNS axons will grow into these grafts quite readily, indicating that they can cross from astrocytes to Schwann cells. However when axons attempt to leave the grafts back into an astrocytic environment their growth is blocked, as they are unable to cross from a Schwann cell environment back into an astrocytic one.[20,67,68]

Just as Schwann cells and astrocytes seldom mix in vivo, so they mix poorly in vitro. Therefore when Schwann cells and astrocytes are placed together in the same cultures they tend to separate out into separate patches.[61] This makes it possible to examine the ability of axons to grow across boundaries between Schwann cells and astrocytes in vitro. We have examined the growth of sensory axons in such cultures. Their behaviour is very similar to that seen in vivo. Where axons growing on astrocytes encounter a patch of Schwann cells almost all of them will cross onto the Schwann cells. However when axons growing on Schwann cells encounter a boundary with astrocytes only between 10% and 30% are able to cross.[69] One of the reasons why axons do not like to leave a Schwann cell environment for an astrocyte one may involve the adhesion molecule L1, a potent promoter of axon growth that is present on Schwann cells but not astrocytes. Thus when exogenous L1 in the form of L1-Fc is applied to the cultures, or when L1 is blocked by a functional blocking antibody the proportion of axons able to cross boundaries onto astrocytes is greatly increased. We were not able to find evidence of inhibition by astrocyte-produced CSPGs as a factor in preventing boundary crossing, probably because the main astrocyte CSPG, neurocan, is not attaced to the astrocyte cell surface but released into the culture medium. The recognition of Schwann cell astrocyte boundaries that leads to the blockage of growth appears to involve a signalling pathway that has been shown to affect growth in other models. cAMP levels in the growth cone has been shown to have a strong effect on the ability of axons to grow on inhibitory molecules such as MAG. We find that raising cAMP levels in axons allows more of them to cross from Schwann cells onto astrocytes. Developmental changes are also significant. Axons growing from embryonic sensory neurones or embryonic retina can cross from Schwann cells to astrocytes in much larger numbers than axons from postnatal neurones (Adcock, Shewan, Czvitkovich, Fawcett unpublished results).

Boundary crossing behaviour has also been studied in a different in vitro model, using frozen sections of the dorsal root entry zones a the culture surface. The behaviour of axons in this model is similar to that seen in vivo, and to that seen in the culture model described above. Axons cross readily from the CNS to peripheral nerve, but not in the other direction. Developmental age has a critical influence. Thus when the substrate is taken from embryonic spinal cord the boundary is crossed by many more axons, and embryonic axons can cross adult boundaries. Calcium signalling plays a part in axonal behaviour at these boundaries, since blocking calcium release from internal stores with dantrolene increases the number of axons crossing dorsal root boundaries.[18,19]

Astrocyte/Meningeal Cell Boundaries

Meningeal cells are specialised fibroblast-like cells that surround the CNS and its major blood vessels, and are responsible for co-operating with astrocytes in the setting up of the glia limitans.[70-72] This is a layer of hypertrophic astrocyte processes running parallel to the surface of the brain and spinal cord, with a layer of basal lamina on top in between the astrocytes and meningeal cells. Following CNS injuries, particularly those that penetrate the meninges, the meningeal cells divide, and migrate into the injury cavity. Within a few days they line the entire injury cavity, and if the injury has reduced the density of astrocytes and created space within the tissue meningeal cells may invade more diffusely.[71,73] As the invading meningeal cells come into contact with astrocytes, they induce the same changes that are seen at the glia limitans. This leads eventually to the formation of a new glia limitans and often more general reactive astrocytic changes. Just as axons do not grow out of the brain or spinal cord through the normal glia limitans, the new glia limitans that forms after CNS injury appears to present a barrier to regenerating axons. In various regeneration studies axons can be seen to have approached the astrocyte-meningeal cell boundary and stopped at that point.[74,75] For instance Beattie et al[76] have reported that corticospinal axons retract from the injury site after spinal cord injury, but will then show some regenerative growth leading them to the meningeal boundary, where they stop. Davies and Silver performed transplants of adult sensory axons into the injured spinal cord, placed so that axons would grow through the dorsal columns until they encounter the injury. When the transplants were performed some time after the cord injury the CSPG reaction had abated sufficiently that a proportion of the axons were able to penetrate the glial scar tissue to reach the meningeal boundary, and there they stopped.[15] In all the experiments in which regeneration has been induced in the spinal cord, by blocking NogoA, chondroitinase and other treatments, the regenerating axons have gone around the meningeal boundary, not through it. However, by inhibiting the buildup of collagen and extracellular matrix in the injury it has been possible to induce axon regeneration across the injury site.[77,78]

We have been able to model axon behaviour at astrocyte/meningeal cell boundaries in vitro. As with Schwann cells and astrocytes, meningeal cells and astrocytes separate out in vitro into separate territories, the result being islands of meningeal cells in a lawn of astrocytes. We have plated sensory and other neuronal types onto these cultures to observe their behaviour at boundaries. Axons will grow on both astrocytes and meningeal cells, although axons growing on meningeal cells tend to be fasciculated and tortuous. When axons growing on meningeal cells reach a boundary of astrocytes around 90% will cross, with the remainder following around the interface between the cell types. However, when axons growing on astrocytes meet meningeal cells only 15-30% will cross, many of the axons turning to follow the interface between the cell types.[79] Why are meningeal cells inhibitory? In principle this could be because they express inhibitory molecules, or because they lack the growth-promoting molecules present on astrocytes. Meningeal cells express at least three inhibitory molecules; the CSPGs NG2 and versican, and the axon guidance molecule Semaphorin 3A/ collapsin1. We have applied blocking antibodies to both NG2 and to the semaphorin receptors plexin 1 and plexin 2, and we find that the NG2 and plexin 2 antibodies both ncrease the number of axons that can cross onto meningeal cells from arond 20 to around 40%. There is no specific way of blocking inhibition due to versican. The main growth-promting molecule that is present on astrocytes but absent on meningeal cells is N-Cadherin, but we find that blocking this molecule with HAV peptides does not enhance boundary crossing onto meningeal cells. As with Schwann cell/astrocyte boundaries, it is possible to promote boundary crossing by manipulating growth cone signalling. Thus increasing cAMP levels and blocking the GTPase Rho both increase the number of axons crossing boundaries.[79,80]

Glial Boundaries and Repairing the CNS

To date the only treatment designed specifically to deal with glial boundaries is the inhibition of collagen synthesis in the lesion, which has succeeded in promoting some regeneration.[77,78] In addition two of the treatments that affect axonal boundary crossing as well as growth on inhibitory substrates, namely increasing cAMP levels and blocking Rho have been applied. The Rho ribosylating toxin, C3 from clostridium botulinum, has been applied to optic nerve and spinal cord injuries. In both cases axon regeneration was promoted.[81,82] In the optic nerve experiments it is likely that the axons encountered a meningeal boundary, since these cells invade optic nerve crushes within 24 hours, and the crushed axons retracted from the injury and probably did not grow through it for at least 24 hours after crush. There has only been one experiment to manipulate cAMP levels in vivo, by injecting it directly into dorsal root ganglia. This promoted some regeneration in the spinal cord, although it is not clear that any glial boundaries were involved. However the actions of neurotrophins are at least in part though cAMP, and these molecules do promote boundary crossing. Many experiments have attempted to promote axon growth across the Schwann cell/astrocyte boundary of the dorsal root entry zone. The most successful techniques have used neurotrophns. NT3 infused into the region of the entry zone allowed many axons to cross the boundary into the CNS.[83,84] Neurotrophins have also been expressed in the cord by adenoviral infection, and NGF and FGF-2 have attracted axons across the boundary.[85,86] The technique has been used with a degree of success to attract axons out of Schwann cell grafts to the spinal cord back into cord tissue. An infusion of BDNF and NT-3 allowed a modest number of axons to grow across the boundary.[67]

Strategies for Repairing the CNS

The events that remodel the CNS glial environment after injury are complex, leading to a terrain with several types of inhibitory obstacle. It would be daunting to devise a treatment that modifies all these inhibitory mechanisms simultaneously. However, it may not be necessary to deal with every inhibitory molecule. Axon growth cones integrate together all the various signals that they receive both growth promoting and growth inhibitory. In principle, therefore, axon regeneration could be achieved by increasing the amount of growth promoting activity or by reducing the amount of inhibition. Various different treatments to block inhibition in the damaged CNS have been applied, particularly anti NogoA antibodies, chondroitinase and demyelination all promote growth to a comparable degree. It will be important to find out whether these treatments have additive effects, or whether axon growth is the same once a certain level of permissiveness in the environment has been achieved. The issue of whether treatments are additive also applies to the other ways in which axon growth can be promoted, by manipulating signalling pathways and with neurotrophins. If all these treatments converge on the same growth cone mechanisms it is unlikely that they will be additive. All the treatments that have been identified to date are able to promote growth of a fairly modest number of axons over distances of 1-2 cm. This is just at the margin of what would be useful to a spinal injury patient, where bringing the effective level of an injury down by even one spinal level would be of great benefit to patients with cervical level injuries. However, achieving robust growth of large numbers of axons over long distances is still an elusive aim. It is likely that there are axon growth control mechanisms that are yet to be discovered which we will need to manipulate to achieve this aim.

References

1. Reier PJ, Stensaas LJ, Guth L. The astrocytic scar as an impediment to regeneration in the central nervous system. In: Kao CC, Bunge RP, Reier PJ, eds. Spinal cord reconstruction. New York: Raven Press, 1983:163-195.
2. Perry VH, Andersson PB, Gordon S. Macrophages and inflammation in the central-nervous-system. Trends Neurosci 1993; 16:268-273.
3 Raivich G, Bohatschek M, Kloss CU et al. Neuroglial activation repertiore in the injured brain: graded response, molecular mechanisms and cues to physiological function. Brain Res Rev 1999; 30:77-105.
4. Levine JM, Reynolds R, Fawcett JW. The oligodendrocyte precursor cell in health and disease. Trends Neurosci 2001; 24(1):39-47.
5. Levine JM, Reynolds R. Activation and proliferation of endogenous oligodendrocyte precursor cells during ethidium bromide-induced demyelination. Exp Neurol 1999; 160(2):333-347.
6. Wu VW, Schwartz JP. Cell culture models for reactive gliosis: New perspectives. J Neurosci Res 1998; 51:675-681.
7. Eddleston M, Mucke L. Molecular profile of reactive astrocytes—implications for their role in neurologic disease. Neuroscience 1993; 54:15-36.
8. Ridet JL, Malhotra SK, Privat A et al. Reactive astrocytes: cellular and molecular cues to biological function. Trends Neurosci 1997; 20:570-577.
9. Davies SJA, Field PM, Raisman G. Regeneration of cut adult axons fails even in the presence of continuous aligned glial pathways. Exp Neurol 1996; 142:203-216.
10. Bahr M. Adult-rat retinal glia in vitro—effects of invivo crush- activation on glia proliferation and permissiveness for regenerating retinal ganglion-cell axons. Exp Neurol 1991; 111:65-73.
11. McKeon RJ, Schreiber RC, Rudge JS et al. Reduction of neurite outgrowth in a model of glial scarring following CNS injury is correlated with the expression of inhibitory molecules on reactive astrocytes. J Neurosci 1991; 11:3398-3411.
12. Anders JJ, Hurlock JA. Transplanted glial scar impedes olfactory bulb reinnervation. Exp Neurol 1996; 142:144-150.
13. Giftochristos N, David S. Immature optic nerve glia of rat do not promote axonal regeneration when transplanted into a peripheral nerve. Brain Res 1988; 467:149-153.
14. Davies SJA, Fitch MT, Memberg SP et al. Regeneration of adult axons in white matter tracts of the central nervous system. Nature 1997; 390:680-684.
15. Davies SJA, Goucher DR, Doller C et al. Robust regeneration of adult sensory axons in degenerating white matter of the adult rat spinal cord. J Neurosci 1999; 19(14):5810-5822.
16. Carlstedt T, Cullheim S, Risling M et al. Nerve fibre regeneration across the PNS-CNS interface at the root-spinal cord junction. Brain Res Bull 1989; 22:93-102.
17. Chong MS, Woolf CJ, Turmaine M et al. Intrinsic versus extrinsic factors in determining the regeneration of the central processes of rat dorsal root ganglion neurons: the influence of a periperal nerve graft. J Comp Neurol 1996; 370:97-104.
18. Golding JP, Bird C, McMahon S et al. Behaviour of DRG sensory neurites at the intact and injured adult rat dorsal root entry zone: postnatal neurites become paralysed, whilst injury improves the growth of embryonic neurites. Glia 1999; 26(4):309-323.

19. Golding J, Shewan D, Cohen J. Maturation of the mammalian dorsal root entry zone—From entry to no entry. Trends Neurosci 1997; 20:303-308.
20. Xu XM, Guénard V, Kleitman N et al. Axonal regeneration into Schwann cell-seeded guidance channels grafted into transected adult rat spinal cord. J Comp Neurol 1995; 351:145-160.
21. Chierzi S, Strettoi E, Cenni MC et al. Optic nerve crush: Axonal responses in wild-type and bcl-2 transgenic mice. J Neurosci 1999; 19(19):8367-8376.
22. Bandtlow CE, Schwab ME. NI-35/250/nogo-a: a neurite growth inhibitor restricting structural plasticity and regeneration of nerve fibers in the adult vertebrate CNS. Glia 2000; 29(2):175-181.
23. Schwab ME, Bartholdi D. Degeneration and regeneration of axons in the lesioned spinal cord. Physiol Rev 1996; 76:319-370.
24. Grandpre T, Strittmatter SM. Nogo: a molecular determinant of axonal growth and regeneration. Neuroscientist 2001; 7(5):377-386.
25. Grandpre T, Nakamura F, Vartanian T et al. Identification of the Nogo inhibitor of axon regeneration as a Reticulon protein. Nature 2000; 403(6768):439-444.
26. Cai D, Qiu J, Cao Z et al. Neuronal cyclic amp controls the developmental loss in ability of axons to regenerate. J Neurosci 2001; 21(13):4731-4739.
27. Tang S, Woodhall RW, Shen YJ et al. Soluble myelin-associated glycoprotein (MAG) found in vivo inhibits axonal regeneration. Mol Cell Neurosci 1997; 9(5-6):333-346.
28. Mukhopadhyay G, Doherty P, Walsh FS et al. A novel role for myelin-associated glycoprotein as an inhibitor of axonal regeneration. Neuron 1994; 13:757-767.
29. Fawcett JW, Housden E, Smith-Thomas L et al. The growth of axons in three dimensional astrocyte cultures. Dev Biol 1989; 135:449-458.
30. Smith-Thomas L, Stevens J, Fok-Seang J et al. Increased axon regeneration in astrocytes grown in the presence of proteoglycan synthesis inhibitors. J Cell Sci 1995; 108(3):1307-1315.
31. Smith-Thomas L, Fok-Seang J, Stevens J et al. An inhibitor of neurite outgrowth produced by astrocytes. J Cell Sci 1994; 107:1687-1695.
32. Fidler PS, Schuette K, Asher RA et al. Comparing astocytic cells lines that are inhibitory or permissive for axon growth: the major axon -inhibitory proteoglycan is NG2. J Neurosci 1999; 19:8778-8788.
33. Fitch MT, Silver J. Glial cell extracellular matrix: Boundaries for axon growth in development and regeneration. Cell Tissue Res 1997; 290:379-384.
34. McKeon RJ, Höke A, Silver J. Injury-induced proteoglycans inhibit the potential for laminin-mediated axon growth on astrocytic scars. Exp Neurol 1995; 136:32-43.
35. Fernaud-Espinosa I, Nieto-Sampedro M, Bovolenta P. A neurite outgrowth-inhibitory proteoglycan expressed during development is similar to that isolated from adult brain after isomorphic injury. Journal Of Neurobiology 1998; 36(1):16-29.
36. Bovolenta P, Wandosell F, Nieto-Sampedro M. Characterization of a neurite outgrowth inhibitor expressed after CNS injury. Eur J Neurosci 1993; 5:454-465.
37. Moon LDF, Asher RA, Rhodes KE et al. Regeneration of CNS axons back to their original target following treatment of adult rat brain with chondroitinase ABC. Nat Neurosci 2001; 4:465-466.
38. Bradbury EJ, Moon LDF, Popat RJ et al. Chondroitinase ABC promotes axon regeneration and functional recovery following spinal cord injury. Nature 2002; in press.
39. Menet V, Ribotta M, Chauvet N et al. Inactivation of the glial fibrillary acidic protein gene, but not that of vimentin, improves neuronal survival and neurite growth by modifying adhesion molecule expression. J Neurosci 2001; 21(16):6147-6158.
40. Matsui F, Watanabe E, Oohira A. Immunological identification of two proteoglycan fragments derived from neurocan, and brain specific chondroitin sulphate proteoglycan. Neurochem Int 1994; 25:425-431.
41. Asher RA, Fidler PS, Morgenstern DA et al. Neurocan is upregulated in injured brain and in cytokine-treated astrocytes. J Neurosci 2000; 20:2427-2438.
42. McKeon RJ, Jurynec MJ, Buck CR. The chondroitin sulfate proteoglycans neurocan and phosphacan are expressed by reactive astrocytes in the chronic CNS glial scar. Journal Of Neuroscience 1999; 19(24):10778-10788.
43. Margolis RU, Margolis RK. Chondroitin sulfate proteoglycans as mediators of axon growth and pathfinding. Cell Tissue Res 1997; 290:343-348.

44. Grumet M, Friedlander DR, Sakurai T. Functions of brain chondroitin sulfate proteoglycans during developments: interactions with adhesion molecules. Perspect Dev Neurobiol 1996; 3(4):319-330.
45. Schmalfeldt M, Dours ZM, Winterhalter KH et al. Versican V2 is a major extracellular matrix component of the mature bovine brain. J Biol Chem 1998; 273(25):15758-15764.
46. Asher RA, Morgenstern DA, Shearer MC et al. Versican is upregulated following CNS injury and is a product of oligodendrocyte lineage cells. J Neurosci 2002; 22:2225-2236.
47. Niederost BP, Zimmermann DR, Schwab ME et al. Bovine CNS myelin contains neurite growth-inhibitory activity associated with chondroitin sulfate proteoglycans. J Neurosci 1999; 19:8979-8989.
48. Schmalfeldt M, Bandtlow CE, Dours-Zimmermann MT et al. Brain derived versican V2 is a potent inhibitor of axonal growth. J Cell Sci 2000; 113(5):807-816.
49. Jaworski DM, Kelly GM, Hockfield S. Intracranial injury acutely induces the expression of the secreted isoform of the CNS-specific hyaluronan-binding protein BEHAB brevican. Experimental Neurology 1999; 157(2):327-337.
50. Yamada H, Fredette B, Shitara K et al. The brain chondroitin sulfate proteoglycan brevican associates with astrocytes ensheathing cerebellar glomeruli and inhibits neurite outgrowth from granule neurons. J Neurosci 1997; 17(20):7784-7795.
51. Asher RA, Morgenstern DA, Moon LD et al. Chondroitin sulphate proteoglycans: inhibitory components of the glial scar. Prog Brain Res 2001; 132:611-619.
52. Levine JM. Increased expression of the ng2 chondroitin-sulfate proteoglycan after brain injury. Journal Of Neuroscience 1994; 14:4716-4730.
53. Goretzki L, Burg MA, Grako KA et al. High-affinity binding of basic fibroblast growth factor and platelet- derived growth factor-AA to the core protein of the NG2 proteoglycan. J Biol Chem 1999; 274(24):16831-16837.
54. Grako KA, Ochiya T, Barritt D et al. PDGF (alpha)-receptor is unresponsive to PDGF-AA in aortic smooth muscle cells from the NG2 knockout mouse. J Cell Sci 1999; 112 (Pt 6):905-915.
55. Garwood J, Schnadelbach O, Clement A et al. DSD-1-proteoglycan is the mouse homolog of phosphacan and displas aopposing effects on neurite outrgrowth dependent on neuronal lineage. J Neurosci 1999; 19(10):3888-99.
56. Margolis RK, Rauch U, Maurel P et al. Neurocan and phosphacan—2 major nervous tissue-specific chondroitin sulfate proteoglycans. Perspect Devel Neurobiol 1996; 3:273-290.
57. Garwood J, Schnädelbach O, Clement A et al. DSD-1-proteoglycan is the mouse homolog of phosphacan and displays opposing effects on neurite outgrowth dependent on neuronal lineage. J Neurosci 1999; 19(10):3888-3899.
58. Snow DM, Brown EM, Letourneau PC. Growth cone behavior in the presence of soluble chondroitin sulfate proteoglycan (CSPG), compared to behavior on CSPG bound to laminin or fibronectin. Int J Dev Neurosci 1996; 14(3):331-349.
59. Clement A, Sugahara K, Faissner A. Chondroitin sulfate E promotes neurite outgrowth of embryonic day 18 hippocampal neurons. Neurosci Lett 1999; 269(3):125-8.
60. Morgenstern DA, Asher RA, Fawcett JW. Chondroitin sulphate proteoglycans in the CNS injury response. Progress In Brain Research 2002; Prog Brain Res 2002; 137:313-32.
61. Wilby MJ, Fok-Seang J, Blaschuk OW et al. N-Cadherin inhibits Schwann cell migration on astrocytes. Mol Cell Neurosci 1999; 14:66-84.
62. Iwashita Y, Fawcett JW, Crang AJ et al. Schwann cell transplanted into normal and irriadiated adult white matter do not migrate extensively and show poor long-term survival. Exp Neurol 2000; 164:192-202.
63. Shields SA, Blakemore WF, Franklin RJM. Schwann cell remyelination is restricted to astrocyte-deficient areas after transplantation into demyelinated adult rat brain. J Neurosci Res 2000; 60(5):571-578.
64. Carlstedt T. Nerve fibre regeneration across the peripheral-central transitional zone. J Anat 1997; 190:51-56.
65. Carlstedt T. Regenerating axons form nerve terminals on astrocytes. Brain Res 1985; 347:188-191.
66. Liuzzi FJ, Lasek RJ. Astrocytes block axonal regeneration in mammals by activating the physiological stop pathway. Science 1987; 237:642-645.

67. Xu XM, Guénard V, Kleitman N et al. Combination of BDNF and NT-3 promotes supraspinal axonal regeneration into schwann cell grafts in adult rat thoracic spinal cord. Exp Neurol 1995; 134:261-272.
68. David S, Aguayo AJ. Axonal elongation into peripheral nervous system bridges after central nervous system injury in adult rats. Science 1981; 241:931-933.
69. Adcock KH, Brown DJ, Shearer MC et al. Axon behaviour at Schwann cell astrocyte boundaries. Manipulation of axon signalling pathways and glia can enable axons to cross. Eur J Neurosci 2003; 20:1425-1435.
70. Pehlemann FW, Sievers J, Berry M. Meningeal cells are involved in foliation, lamination, and neurogenesis of the cerebellum: evidence from 6-hydroxydopamine induced destruction of meningeal cells. Dev Biol 1985; 110:136-146.
71. Krueger S, Sievers J, Hansen C et al. Three morphologically distinct types of interface develop between adult host and fetal brain transplants: implications for scar formation in the adult central nervous system. J Comp Neurol 1986; 249:103-116.
72. Sievers J, Pehlemann FW, Gude S et al. Meningeal cells organize the superficial glia limitans of the cerebellum and produce components of both the interstitial matrix and the basement membrane. J Neurocytol 1994; 23:135-149.
73. Mathewson AJ, Berry M. Observations on the astrocyte response to a cerebral stab wound in adult rats. Brain Res 1985; 327:61-69.
74. Stichel CC, Muller HW. The CNS lesion scar: new vistas on an old regeneration barrier. Cell Tissue Res 1998; 294(1):1-9.
75. Stichel CC, Muller HW. Relationship between injury-induced astrogliosis, laminin expression and axonal sprouting in the adult-rat brain. J Neurocytol 1994; 23:615-630.
76. Beattie MS, Bresnahan JC, Komon J et al. Endogenous repair after spinal cord contusion injuries in the rat. Exp Neurol 1997; 148(2):453-463.
77. Stichel CC, Hermanns S, Luhmann HJ et al. Inhibition of collagen IV deposition promotes regeneration of injured CNS axons. Eur J Neurosci 1999; 11(2):632-646.
78. Stichel CC, Niermann H, D'Urso D et al. Basal membrane-depleted scar in lesioned CNS: characteristics and relationships with regenerating axons. Neuroscience 1999; 93(1):321-333.
79. Shearer MC, Niclou SP, Brown D et al. The astrocyte/meningeal cell interface is a barrier to neurtie outgrowth which can be overcome by manipulation of inhibitory molecules or axonal signalling pathways. Mol Cell Neurosci 2003; 24:913-925.
80. Shearer MC, Fawcett JW. The astrocyte/meningeal cell interface: A barrier to successful nerve regeneration? Cell Tiss Res 2001; 305:267-273.
81. McKerracher L. Spinal cord repair: strategies to promote axon regeneration. Neurobiol Dis 2001; 8(1):11-18.
82. Lehmann M, Fournier A, Selles-Navarro I et al. Inactivation of Rho signaling pathway promotes CNS axon regeneration. J Neurosci 1999; 19(17):7537-7547.
83. Ramer MS, Duraisingam I, Priestley JV et al. Two-Tiered Inhibition of Axon Regeneration at the Dorsal Root Entry Zone. J Neurosci 2001; 21(8):2651-2660.
84. Ramer MS, Priestley JV, McMahon SB. Functional regeneration of sensory axons into the adult spinal cord. Nature 2000; 403:312-316.
85. Romero MI, Rangappa N, Garry MG et al. Functional regeneration of chronically injured sensory afferents into adult spinal cord after neurotrophin gene therapy. J Neurosci 2001; 21(21):8408-8416.
86. Romero MI, Rangappa N, Li L et al. Extensive sprouting of sensory afferents and hyperalgesia induced by conditional expression of nerve growth factor in the adult spinal cord. Journal Of Neuroscience 2000; 20(12):4435-4445.

Chapter 3

DSD-1-Proteoglycan/Phosphacan and Receptor Protein Tyrosine Phosphatase-Beta Isoforms during Development and Regeneration of Neural Tissues

Andreas Faissner, Nicolas Heck, Alexandre Dobbertin and Jeremy Garwood

Summary

Interactions between neurons and glial cells play important roles in regulating key events of development and regeneration of the CNS. Thus, migrating neurons are partly guided by radial glia to their target, and glial scaffolds direct the growth and directional choice of advancing axons, e.g., at the midline. In the adult, reactive astrocytes and myelin components play a pivotal role in the inhibition of regeneration. The past years have shown that astrocytic functions are mediated on the molecular level by extracellular matrix components, which include various glycoproteins and proteoglycans. One important, developmentally regulated chondroitin sulfate proteoglycan is DSD-1-PG/phosphacan, a glial derived proteoglycan which represents a splice variant of the receptor protein tyrosine phosphatase (RPTP)-beta (also known as PTP-zeta). Current evidence suggests that this proteoglycan influences axon growth in development and regeneration, displaying inhibitory or stimulatory effects dependent on the mode of presentation, and the neuronal lineage. These effects seem to be mediated by neuronal receptors of the Ig-CAM superfamily.

General Introduction

The Extracellular Matrix in Neural Development and Regeneration

The development of the central nervous system (CNS) evolves as a series of specifiable, separate morphogenetic steps which include cell proliferation, migration of neuronal precursors to their respective locations, differentiation of neurons, process extension, axon outgrowth, guidance, synaptogenesis and, finally, selective neuronal death due to limiting amounts of growth factors,[1] for review).

Brain Repair, edited by Mathias Bähr. ©2006 Eurekah.com and Kluwer Academic / Plenum Publishers.

In all of these events, interactions of the cell with its environment play an important regulatory role. Cells can interact both directly with other cell surfaces or with the complex of secreted proteins and carbohydrate polymers which make up the extracellular matrix (ECM).

Cell-cell interactions are mediated by molecules such as the IgCAM-superfamily,[2,3] the cadherins,[4,5] and the Eph-tyrosine kinases and their ephrin-A and ephrin-B ligands,[6-8] which are implicated in mechanisms such as the regulation of axon growth by fasciculation (N-CAM) or the establishment of functional boundaries within the tissue (interrhombomeric boundary, ephrin).[9]

Many potentially important interactions also occur within between cells and the neural extracellular matrix (ECM). Although the organization of the ECM in the vertebrate CNS is not well understood it can be considered as a complex and dynamic association of extracellular molecules that is relatively rich in chondroitin sulphate proteoglycans (CSPGs) and hyaluronan, but poor in fibrous elements (reviewed in ref. 10).

ECM molecules are thought to contribute to the regulation of CNS histogenesis by a variety of functional properties. Thus, the ECM has been implicated in the control of cellular migration, the storage of soluble growth factors, the promotion of neurite outgrowth, or the inhibition of neuritogenesis by the formation of tissue boundaries.[11-15]

Cell interactions with the ECM are mediated by receptors such as the growing family of integrin heterodimers[16,17] and proteins from the Ig-CAM family.[18]

CSPGs in the ECM

Chondroitin sulfate proteoglycans (CSPGs), a heterogeneous set of proteins bearing glycosaminoglycans (GAGs) of the chondroitin sulfate (CS) class, account for most of the "soluble" proteoglycans in the brain.[19] Tissue fractionation studies performed with rat brain have revealed that whereas most heparan sulfate proteoglycans are tightly associated with cell membranes, the CSPGs, which constitute the major population of proteoglycans (PGs) in the CNS, are recovered in detergent-free salt extracts.[20-22] It is known that soluble CSPG preparations from postnatal rat brain contain at least eight core glycoproteins which are differentially expressed during rat CNS development.[21]

ECM Proteins Control Axon Growth by Promoting and Inhibiting

During development, axon growth and growth cone guidance are regulated by numerous proteins of the ECM[23] including the semaphorin gene family and their receptors,[24-26] and the Nogo proteins.[27-29] These proteins are expressed in differents parts of the CNS and their expression is regulated during time. Depending on the cell type with which they interact, they can exert promotory or inhibitory effects, and display attractive or repulsive properties.

The CSPGs are also ECM molecules involved in the regulation of axon growth as has been demonstrated by in vitro studies on CSPGs such as NG2,[30] neurocan and phosphacan.[31] By virtue of inhibitory or stimulatory influences on neurite growth, CSPGs could be involved in the regulation of sprouting, as has been proposed for CAT 301.[32-34]

CSPGs Are Present in Functional Boundaries in the CNS

During development, strong immunostaining for CS often localizes to territories thought to act as barriers to migrating neurons or extending axons such as the roof plate and midline dorsal tectum,[35,36] the posterior sclerotome,[37,38] the glomeruli of the olfactory bulb,[39,40] the somatosensory barrel field[41-44] and the dorsal root entry zone and dorsal columns in the spinal cord.[45,46] Thus, the CSPGs could contribute to the regulation of the establishement of axonal highways by its spatial distribution in boundaries associated with its neurite outgrowth inhibitory properties. These boundaries seem to share common properties with the glial scar, that is, the expression of CSPGs and tenascin-C at lesion sites is associated with inhibition of axonal regrowth, as has also been proposed for the dorsal root entry zone.[46]

ECM Proteins Expressed during Development Are Also Upregulated in Lesions: Upregulated CSPGs Could Inhibit Regeneration in Lesions

Most of the expression of the proteoglycans occurs during the embryonic and postnatal period and decreases to a low basal level of expression in the adult. Interestingly, it has been shown that the expression of these CSPGs is upregulated at sites of damage to the CNS. For example, NG2,[47] neurocan,[48-50] decorin,[51] versican,[52] and brevican[53,54] are all upregulated in lesions. It is meanwhile well established that the inability of the CNS to regenerate is at least in part due to inhibitory factors released by glial cells into the lesion environment.[55-58] The astrocytes react to lesion of the CNS to form a gliotic scar.[57,59-61] These reactive astrocytes and the ECM that they produce have been shown to inhibit neurite outgrowth in vitro[62,63] and in vivo.[45,64,65] The demonstration that CSPGs can modulate neurite outgrowth in in vitro studies has led to the suggestion that the CSPGs may be contributing to the failure of axonal regeneration in lesions.

Numerous studies have emphasized that the reactive astrocytes up-regulate their expression of CSPGs and keratan sulfate epitopes, and have shown that these components are sufficient to override the beneficial influence of laminin-1, thus impeding the axon growth process.[46,66-71] It has recently been proposed that neurons implanted by a non-traumatic technique into the corpus callosum are able to regenerate long fibers in the presence of intact myelin, provided that the fibers can "escape" from the environment of a ring of tenascin-C and CSPG-expressing reactive astrocytes which emerge around the implantation site.[72,73] In recent in vivo experiments, treatment of the tissue with the enzyme, chondroitinase ABC, which cleaves CS-glycosaminoglycan chains, leads to a substantial regrowth of axons through the lesion.[74-77]

Using Cell Lines and Dissecting the Reactive Astrocyte: Confirmation of a Role for CSPGs

These results have strongly stimulated further investigations into the molecular identity of the CSPGs expressed by reactive astrocytes. An astrocytic cell line selected for its properties similar to those of reactive astrocytes has been shown to express CSPGs.[78] In another study, using a cell line model system of reactive astrocytes, it could be shown that inhibitory activities are associated with the proteoglycan fraction and that these are sensitive to the digestion of GAGs and the blocking of sulfation of polymeric sugars.[79-82] The CSPG, NG2, was identified as one important component in the inhibitory pathway.[83] NG2 is expressed by a subclass of glial cells, up-regulated in lesions, and inhibits neurite outgrowth in several in vitro assays.[30,47,84-87]

DSD-1-PG/Phosphacan/RPTPbeta

Introduction

Thus, proteoglycans have attracted increasing attention because recent investigations suggest that this biochemically heterogeneous group of molecules is functionally significant both during development and regeneration of neural structures.[88,89] Among them, phosphacan/RPTPbeta has been shown to be expressed during the embryonic and postnatal period where it could, according to various in vitro studies, play a role in the regulation of axon growth,[90] cell migration[91] and myelination.[92] The expression of this molecule is regulated in the site of lesion in the adult central nervous system[48] (Dobbertin et al, in preparation). Hence, these CSPGs could be playing a critical role in the regulation of axonal regeneration.

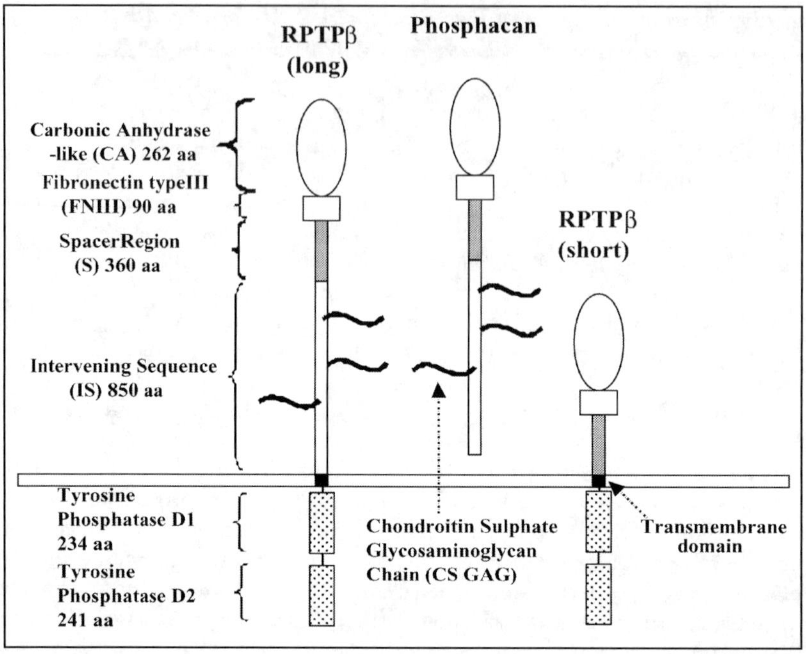

Figure 1. The structure of the phosphacan/RPTP-beta isoforms. The symbols are as indicated in the text. The phosphacan isoform misses the transmembrane domain and is released into the cellular environment.

Structure and Expression of DSD-1-PG/Phosphacan, a Splice Variant of RPTP-Beta, and Related Isoforms

Previously, it has been shown that DSD-1-PG/phosphacan, one of the more abundant of the soluble CSPGs in the postnatal mouse brain, is the mouse homologue of the rat CSPG, phosphacan,[90,93] which closely resembles the cytotactin-binding-proteoglycan CTBP. [94,95] The GAG-composition of DSD-1-PG/phosphacan is characterised by CS-A and a preponderance of CS-C chains. In addition, the CS-D-dependent DSD-1-epitope has been identified with the help of the MAb 473HD, and a keratan sulfate moiety has been detected with the MAb 3H1 which is expressed at least by a subclass of the DSD-1-PG/phosphacan molecules.[90,93,96] As shown in Figure 1, the secreted proteoglycan, DSD-1-PG/phosphacan, is the product of a splice variant of the transmembrane receptor protein tyrosine phosphatase, RPTP-beta, and it corresponds to the entire extracellular region of the largest isoform of RPTP-beta, which is also heavily glycosylated with chondroitin sulphate glycosaminoglycan chains (CS GAGs).[97-100] In the short RPTP-beta isoform, there is a deletion of 850 amino acids between the S domain and the transmembrane domain, corresponding to the GAG attachment sites.[90,101] The expression profile of DSD-1-PG/phosphacan matches that reported for rat phosphacan, with a rapid increase in its concentration during the late embryonic and early postnatal period, the levels remaining high in adult brain.[102] Recently, evidence has been obtained for the existence of a novel isoform of RPTP-beta, corresponding to the N-terminal 600 residues of the extracellular domain common to the three known isoforms of this protein. The novel cDNA transcript was isolated by expression screening of a mouse brain cDNA library using a polyclonal antibody raised against the DSD-1-PG/phosphacan isoform, with which the other isoforms of RPTP-beta had been previously cloned[90] (Garwood et al, submitted). Data collected in amphibians open the possibility that an even wider range of isoforms exists.[103] Whether this is also true for the mammalian nervous system remains to be seen.

The spatiotemporal expression pattern of the RPTP-beta isoforms during the development, maintenance and pathology of the CNS has been correlated with a range of developmental processes which involve cell-cell and cell-ECM signalling, including cellular proliferation, migration, differentiation, circuit formation, synaptogenesis, synaptic function and tissue regeneration,[104,105] reviewed in reference 106. Based on the essentially glial expression of DSD-1-PG/phosphacan/RPTP-beta, the effects on neuronal behaviour of extracellular signals presented by RPTP-beta have been considered, whether as protein sequences/domains or associated with the extensive carbohydrate molecules, especially CS-GAG chains, with which they are modified.

The systematic study of primary cerebellar cultures of distinct developmental stages showed that DSD-1-PG/phosphacan is surface-expressed on a subclass of immature glial cells. Analysis by immunoprecipitation of biosynthetically labelled primary cultures and cell lines and by biochemical techniques such as ion exchange and size-exclusion chromatography supports the conclusion that the proteoglycan synthesized in vitro corresponds to the molecule obtained from postnatal mouse brain.[93] In particular, at later stages, and in secondary cultures enriched for astrocytes, a substantial overlap of DSD-1-PG/phosphacan with O4-positive and O1-negative oligodendrocytes was found.[93,107,108] In mixed glial cell cultures from neonatal rat, antibodies directed against DSD-1-PG/phosphacan stained bipolar precursors and more mature oligodendrocytes, presumably O4-positive (Heck N, Perrault M, Faissner A, unpublished observations) (Fig. 2). With regard to lineage-related restriction of expression, the large RPTP-beta and the released form, DSD-1-PG/phosphacan, seem to be strongly expressed by oligodendrocyte precursors. By comparison, mature astrocytes preferentially express the short RPTP-beta.[101,109] Glial precursor cells, radial glia, Golgi cells, and astrocytes from different developmental stages and parts of the CNS have all been shown to express RPTP-beta isoforms.[100,101]

Studies of the distribution of both the phosphacan mRNA[110] and of the expressed protein[105] show the phosphacan mRNA at E13-E16 largely confined to areas of active cell proliferation such as the ventricular zone of the brain and the ependymal layer surrounding the central canal of the spinal cord. Also, although the mRNA is mostly in the neuroepithelium of the embryonic brain and spinal cord, the protein is widely distributed in these tissues, presumably as a consequence of transport in or along glial processes, local secretion and/or redistribution as a consequence of cell migration.[105,110] In P7 mouse cerebellum, DSD-1-PG/phosphacan staining is strongest in the prospective white matter and is also present in the granule layer and molecular layers, whereas the external granule layer is unstained, with the exception of the Bergmann glia fibres (Garwood J, Heck N, Faissner A, unpublished observations) (Fig. 3), matching a similar distribution of phosphacan in P7 rat cerebellum.[111] The distribution of DSD-1-PG/phosphacan during development corresponds to regions implicated in the formation of axonal trajectories. In this respect, it might play either a neurite promoting role as in the interrhomberic boundaries in chick,[112] or an inhibitory role which would correspond to its presence in glial barrel field boundaries in the developing somatosensory cortex of mouse.[41-44]

In the adult rat brain, it has been shown that DSD-1-PG/phosphacan occurs in the circumference of a selected subpopulation of neurons which expressed the calcium-binding protein, parvalbumin, occupying the extracellular space in close vicinity to the cell body, surrounding axon terminals and glial end feet, but not the synaptic clefts.[113] It has been suggested that CSPGs associate with hyaluronic acid in such perineuronal nets or pericellular matrices to form a neuronal ECM structure analogous to that found in connective tissue.[114,115] Different neuronal subsets have different complements of CSPGs[116] such that perineuronal CSPGs could regulate the extracellular milieu of neurons in cell type-specific ways (see Fig. 4A). For example, late in development, the mature ECM may be an important element in limiting synaptic plasticity.[32,33] Another intriguing expression in the adult brain has been observed around the

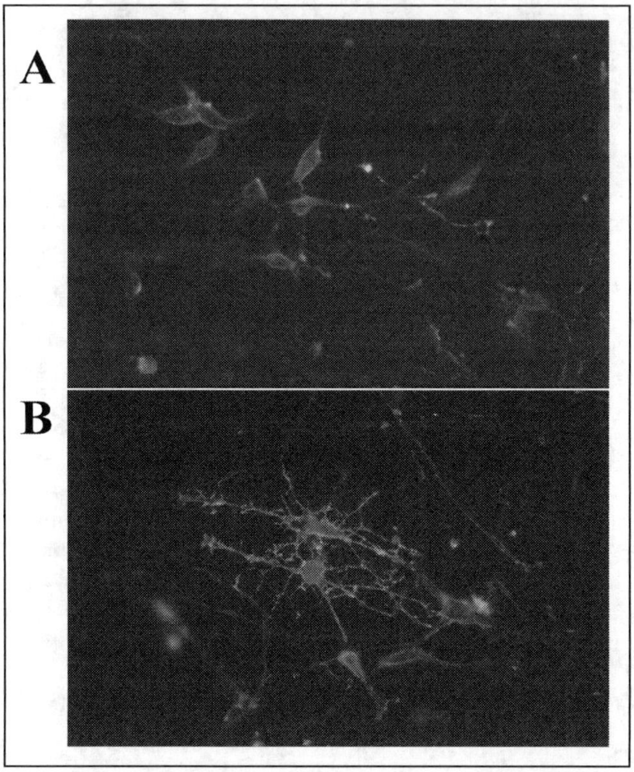

Figure 2. Expression of phosphacan/RPTPbeta by oligodendrocytes. A) The mAb 473HD which recognizes the DSD-1 epitope stains precursors of oligodendrocytes in mixed glial cultures from neonatal rat. B) In cerebral cell cultures prepared from embryonic mice, the polyclonal antibody KAF13 which recognizes all isoforms of phosphacan/RPTPbeta shows a staining of the plasma membrane of mature oligodendrocytes.

ventricular zone, where cell proliferation occurs (Fig. 4B). At the dorsal edge of the lateral ventricle, the presence of phosphacan/RPTPbeta seems to be corrrelated with the pathway of the migrating neurons of the olfactory bulb, which are permanently regenerated.[117-119]

There have been conflicting reports concerning the expression pattern of RPTP-beta isoforms. Although most studies agree that the majority of the protein is of glial origin, there have been several reports of neuronal expression. On E16 cortical neurons, there is immunostaining with anti-6B4-PG/phosphacan of cell bodies, neurites, the rims of growth cones, and filopodial processes.[104,120] However, using an antibody against the tyrosine phosphatase D2 domain on cultured cortical neurons from E17, it seems that at least some of this expression on the cellular surface corresponds to receptor forms of RPTP-beta.[91] On the other hand, the protein expression by neurons, including migrating neurons in the cerebrum and cerebellar Purkinje cells,[121] has been difficult to interpret since DSD-1-PG/phosphacan is present in the ECM surrounding certain subsets of neurons[113,122] and such an extracellular distribution, although apparently associated with the cell surface of neurons, could be glial in origin. A better analysis of the cellular origin of RPTP-beta isoform transcripts has been elaborated by in situ hybridisation analysis, which has confirmed a restricted neuronal expression pattern[123] and, more recently, a heterozygote mouse has been generated in which a lacZ reporter gene has been placed under the control of the RPTP-beta promoter.[120] This latter has shown that, in addition to expression by GFAP-positive astrocytes in the P7 cerebral cortex, there are subsets

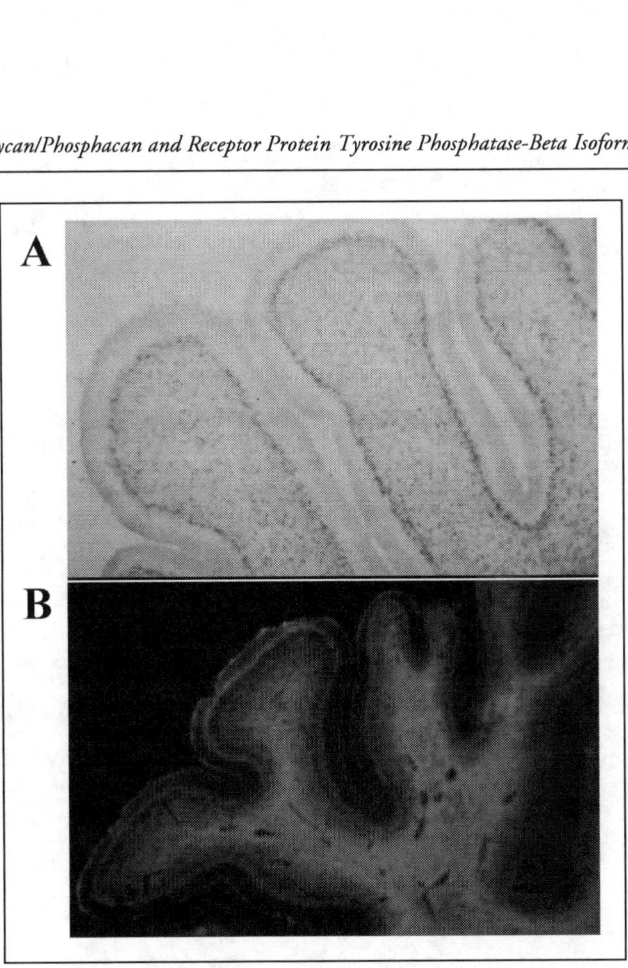

Figure 3. Phosphacan/RPTP-beta expression in P7 cerebellum. A) In situ hybridization on rat cerebellum at postnatal day 7. The riboprobe used recognizes all known isoforms of phosphacan/RPTPbeta. An expression is seen in the Purkinje cell layer, in the inner granule cell layer and in the white matter. B) Immunodetection of the all isoforms of phosphacan/RPTPbeta with the polyclonal antibody KAF13 in the rat cerebellum at postnatal day 7.

of expressing neurons, including pyramidal neurons, found mostly in layers II, III, and V. Again, the use of antibodies against the intracellular tyrosine phosphatase domains of RPTP-beta have indicated that some of the neuronal expression described corresponds to the receptor forms of RPTP-beta.[91] However, in the postnatal cerebral cortex, our in situ hybridisation studies limit expression of the receptor transcripts to the ventricular zone in accordance with other reports[100,117,123] (Garwood et al, submitted).

CSPGs and DSD-1-PG/Phosphacan in Neurite Outgrowth Inhibition

Several studies have reported the enhanced expression of growth inhibiting CSPGs in the context of CNS lesions[46,67,68,70,124] and DSD-1-epitope is also upregulated upon wounding in the CNS (Fig. 5).[48,123,125-128] Such an upregulation of the DSD-1-epitope in the wound reaction might be due to the action of TGF-beta based on studies of Oli-neu, an oligodendrocyte precursor cell line[108] (see below).

Chondroitin sulfate proteoglycans (CSPGs), such as DSD-1-PG/phosphacan and neurocan[111,129] have in several cases been regarded as barriers for neurite outgrowth and credited with inhibitory properties with respect to cell adhesion.[30,35,71,86,130-132] In vitro studies show that CSPGs can inhibit neurite outgrowth and elongation effects[30,71,104,130,132-134] which

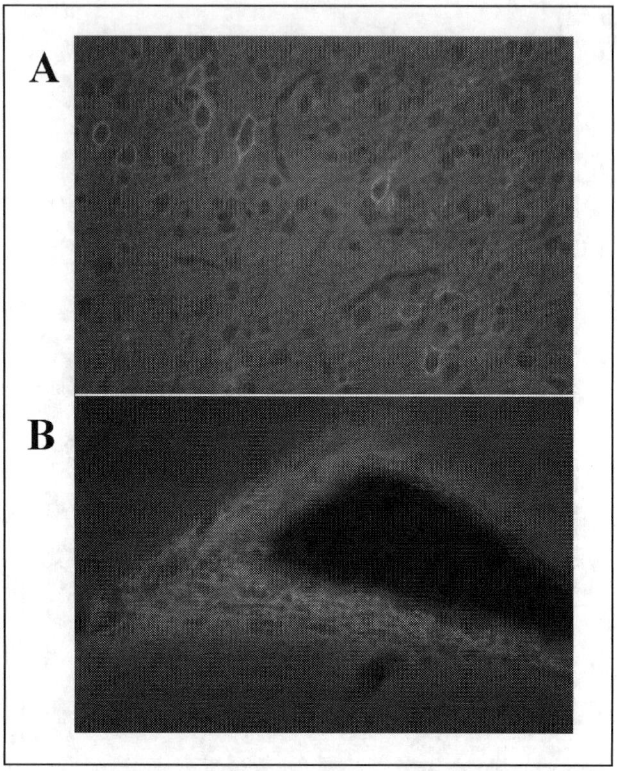

Figure 4. Adult expression of phosphacan/RPTP-beta. A) A subset of adult mouse cortical neurons are surrounded by perineuronal nets, a specialised extracellular matrix closely related with the plasmamembrane which can be visualised by immunohistochemistry with the polyclonal antibody KAF13, which recognizes all phosphacan/RPTP-beta isoforms. B) Strong phosphacan/RPTP-beta expression around the ventricle of the adult rat revealed with the polyclonal antibody KAF13.

can be associated with either the whole CSPG, the chondroitin sulfate (CS) GAGs[28,71,79] or the protein cores[130,135]

In particular the expression of a KS/CSPG in the dorsal roof plate is noteworthy because these regions contain laminated bands of specialized glial cells which are believed to subdivide the dorsal neural tube during development and to prevent the crossing of axons to the opposite side. Another example of KS-rich zones has been reported for neuroanatomical subdivisions between the cortex and thalamic nuclei.[136] Glial cells are a plausible source of KS/CSPG and it has been hypothesized that the KS/CSPG(s) expressed in these regions might mediate a boundary function of glia in this case.[13] Consistent with this hypothesis neurites growing from chicken embryonic DRG explants avoid KS/CSPG spots applied to laminin or N-CAM substrates designed to support DRG process outgrowth in vitro[35,38,71] The MAb AH10 can be used to purifiy the KSPG claustrin from embryonic chicken brain. In contrast to KS antibodies, AH10 does not react with tissues like bone and muscle also known to express KS-GAGs, hinting at GAG microheterogeneities with tissue-specific distribution.[13] Claustrin inhibits the growth of E12 optic lobe neurites from laminin-containing into laminin/claustrin coated areas of a patterned culture substrate, an effect which is abolished by keratanase treatment or incubation of the cultures with an antibody recognizing the KS chains, suggesting that the GAG is the

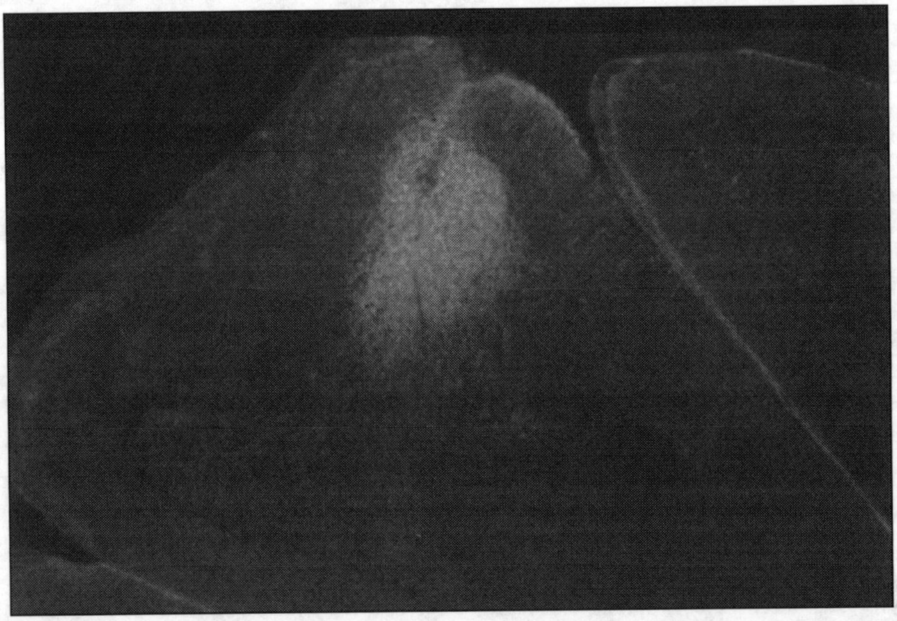

Figure 5. Upregulation of DSD-1-PG in a stab wound. Immunodetection of the CS GAG-epitope DSD-1 with the mAb 473 HD in adult cerebellum into which a stab wound has been made. The expression of the DSD-1 epitope on DSD-1-PG /Phosphacan/RPTP-beta, is strongly upregulated in the region around the lesion site. Figure by courtesy of Dr. Steindler.

functionally active component[71,137] These effects of KS are reminiscent of inhibitory influences of CS GAGs or CSPGs on neurite growth reported by other workers. For example, embryonic chick DRG neurites do not grow out when maintained in a gel that contains heparin, CS and hyaluronic acid (HA),[138] and avoid dermatan sulfate (DS), CS and HA in combination with collagen type I.[139] Further, a CSPG expressed in embryonic chick epidermis prevents embryonic chick DRG neurite ingrowth into epidermal explants in an in vitro co-culture system,[139] and antibodies to CS GAGs neutralize this inhibitory property.[134] Along these lines, a schwannoma-derived inhibitor of the neurite outgrowth promoting activity of laminin-1 contains a KS/CSPG as a prominent component.[140] Interestingly, DSD-1-PG/phosphacan also expresses a KS-epitope recognized by MAb 3H1 and inhibits neurite outgrowth from postnatal day 1 DRGs in a laminin-1-rich environment (Fig. 6).[90] A summary of in vitro neurite outgrowth studies employing DSD-1-PG/phosphacan is presented in Table 1. With regard to its localisation, DSD-1-PG/phosphacan has been observed in the extracellular matrix spaces between glial cells in the roof plate of the developing spinal cord as well as in the DRGs, the dorsal root entry zone and the ventral roots[90,105] and in various boundary-like structures.[141,142] Based on this distribution pattern, it has been suggested that DSD-1-PG/phosphacan may be the CSPG which serves as the glial barrier to axonal extension in the roof plate and that it may restrain axonal growth and movement in these glial-bordered extracellular spaces of the spinal cord.[105] In this perspective, DSD-1-PG/phosphacan could act by either preventing invasion of elongating axons into the roof plate from the floor plate and ventral commissure, or by guiding axonal extension around the floor plate region. Supporting this interpretation, inhibitory effects of 6B4/phosphacan have been found for neurite outgrowth from E17 cerebellar neurons,[91] and P6-P8 retinal ganglion cells.[143]

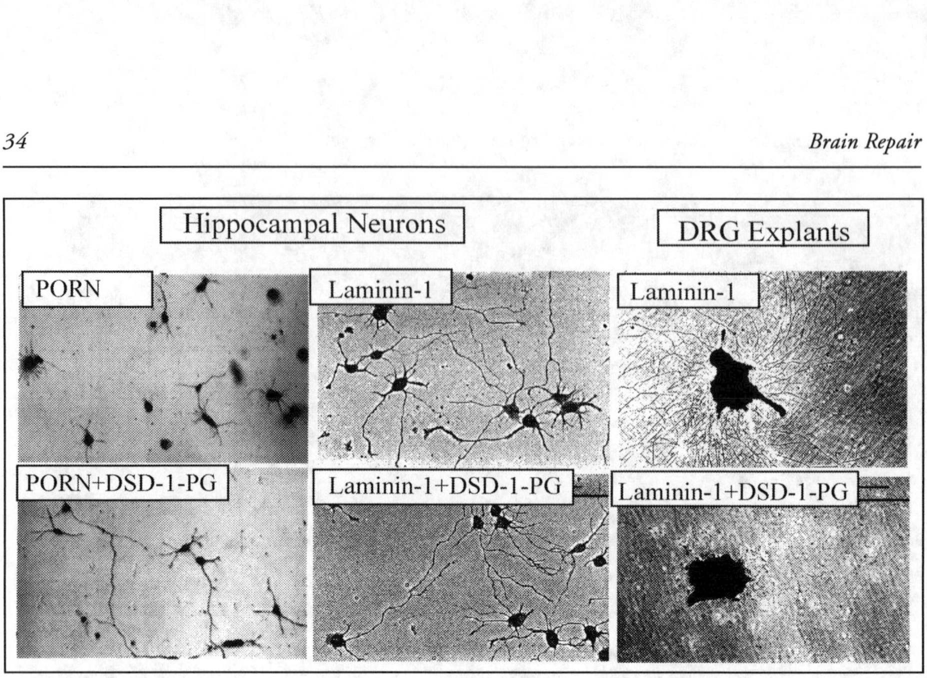

Figure 6. Contrasting effects of DSD-1-PG on neurite outgrowth. Left: outgrowth from hippocampal neurons plated on a minimal substrate (PORN) is promoted by addition of DSD-1-PG. MidFigure 6: Contrasting effects of DSD-1-PG on neurite outgrowth. Left: outgrowth from hippocampal neurons plated on a minimal substrate (PORN) is promoted by addition of DSD-1-PG. Middle: the strong promotion of outgrowth from hippocampal neurons plated on laminin-1 is not affected by addition of DSD-1-PG. Right: addition of DSD-1-PG to a laminin-1 substrate results in inhibition of outgrowth from dorsal root ganglion explants.dle: the strong promotion of outgrowth from hippocampal neurons plated on laminin-1 is not affected by addition of DSD-1-PG. Right: addition of DSD-1-PG to a laminin-1 substrate results in inhibition of outgrowth from dorsal root ganglion explants.

Finally, the neuronal CSPG, neurocan, a member of the aggrecan family of PGs, binds directly to CAMs of the Ig-superfamily, inhibits homophilic L1- or N-CAM-mediated cell adhesion and interferes with both neuron adhesion to and neurite outgrowth on substrates consisting of combinations of cell adhesion molecules or monoclonal antibodies[132,135] The inhibitory properties of neurocan partially reside in the core protein, consistent with a recent report which has attributed inhibition of neurite outgrowth by CSPGs to the core glycoproteins rather than the GAG moieties.[130] Thus, several examples illustrate that chondroitin sulfate proteoglycans may exert inhibitory influences on axon outgrowth in the context of astroglial scar formation, or the construction of transient glial boundaries of neural tissues.[13] Injection of chondroitin sufate-degrading enzymes has resulted in the modification of axon growth trajectories in vivo and application of these enzymes to choice paradigms of axon growth in vitro has indicated a reduction of inhibitory properties of chondroitin sulfate-expressing structures. In light of these experiments, chondroitin sulfate proteoglycans have been viewed as constituents of glial scars with neurite outgrowth inhibiting properties.[76,133,144,145] A study of the role of CSPGs in the outgrowth and adhesion of thalamic neurons plated onto living slices of the mouse embryonic neocortex[146] demonstrated that CS digestion could affect both the permissive environment of the sub-cortical plate and the neurite-repellent properties of the cortical plate. However, rather than being associated with the presence of different CSPGs, they suggest that the opposing activities of these different zones is due to differentially localised CS-binding factors.

Growth Promoting Effects of CSPGs and DSD-1-PG/Phosphacan

On the other hand, it is undisputed that there are also regions in which CSPGs have been found, such as the neocortex and retinal neurons in which CSPGs cannot be regarded as a

Table 1. A summary of in vitro neurite outgrowth studies using whole DSD-1PG/ phosphacan (also known as 6B4-PG) and recombinant protein domains as coated substrates for different neuronal types

Coated Substrates	Neuronal Types	Animal/Age	Effect on Neurite Growth	Ref.
DSD1-PG	Hippocampic neurons	Rat E18	Promotion (due to CS-GAG)	93
DSD1-PG	Mesencephalic neurons	Rat E14	Promotion (due to CS-GAG)	93
DSD1-PG + LN	Hippocampic neurons	Rat E18	No effect	90
DSD1-PG + LN	Dorsal Root Ganglion	Rat P0	Inhibition (due to core protein)	90
Purified Phosphacan	Retinal Ganglion Cells	Rat P6-8	Inhibition (due to core protein)	143
Purified Phosphacan + LN	Retinal Ganglion Cells	Rat P6-8	Inhibition (due to core protein)	143
6B4-PG	Cortical neurons	Rat E16	Promotion (due to core protein)	104
6B4-PG + FN	Cortical neurons	Rat E16	Promotion	104
6B4-PG + TN	Cortical neurons	Rat E16	Promotion	104
6B4-PG	Thalamic neurons	Rat E16	No effect	104
NgCAM + Phosphacan	Neurons	Chick E9	Inhibition (due to core protein)	131
C6 monolayer (Astrocytic cell line producing Pcan/RPTPb) + Ab vs 6B4-PG	Cerebellar neurons	Mouse P5	Inhibition	239
C6 monolayer (Astrocytic cell line producing Pcan/RPTPb + Ab vs F3	Cerebellar neurons	Mouse P5	Inhibition	239
bF, bS	Tectal neurons	Chick	No effect	155
bC, bCF	Tectal neurons	Chick	Promotion	156
bCFS	Tectal neurons	Chick	Maximal promotion (against bC and bCF)	155
bCFS + Ab vs Ng-CAM	Tectal neurons	Chick	Abolition of the	156
bCFS + Ab vs Contactin			promotion	155

barrier to axonal outgrowth. In these cases, tissues that express CSPGs do not exclude the entry of axons, and CSPG staining partially coincides with developing axon pathways.[147-151] In support of this, several in vitro studies suggest that CSPGs[93,152] and isolated core proteins[153] favour rather than inhibit neurite outgrowth. Indeed, initial studies using a mixture of at least three soluble postnatal rat brain CSPGs showed that CSPGs promote neurite outgrowth from E16 rat embryonic neurons. The promoting property was assigned to the core proteins while the CS-GAGs, mostly chondroitin 4-sulfate (CS-A, 80%) and chondroitin 6-sulfate (CS-C, 20%), prepared from the CSPGs were without effect.[153,154] A summary of in vitro neurite outgrowth studies employing DSD-1-PG/phosphacan is presented in Table 1. The 6B4-PG/ phosphacan homologue of DSD-1-PG/phosphacan was found to stimulate outgrowth from E16 cortical neurons in a situation where the substrate was neutral for outgrowth from E16 thalamic neurons.[104] This study using 6B4/phosphacan coated on poly-L-lysine displayed an increased percentage of neurite-bearing cells for both E16 cortical and E16 thalamic neurons, but an enhanced length of the resulting neurites for cortical neurons only. Bacterially and eukaryotically expressed protein domain constructs corresponding to different parts of the core protein of phosphacan/RPTPbeta have been tested in neurite outgrowth assays[155-157] These indicated that outgrowth from chick tectal neurons could be supported by the carbonic anhydrase domain, an effect potentiated by addition of the 'S' domain. Another example relates to cerebellar granule cell neurons which migrate postnatally along the processes of Bergmann glia cells into the prospective internal granular layer. During this migration, the CSPG astrochondrin is expressed at contact sites between granule cells and Bergmann glia in vivo, and the monoclonal antibody L5 directed against a Lewis-X-carbohydrate epitope present on astrochondrin reduces the migration of granule cells in the early postnatal mouse cerebellar cortex in a living explant system, and process formation of astrocytes on laminin and collagen IV in vitro[152,158]

Although these studies partly operate with mixtures of different neuronal subtypes, they are not inconsistent with the contrasting observations relating to DSD-1-PG/phosphacan. On the one hand, this large neural CSPG promotes neurite growth of rat embryonic day 14 (E14) mesencephalic and E18 hippocampal neurones when coated as substrate on poly-DL-ornithine-conditioned plastic (Fig. 6).[93] Yet on the other hand, a striking inhibition by DSD-1-PG/phosphacan of neurite outgrowth from DRGs plated on a mixed laminin-1/CSPG substrate was also observed (Fig. 6).[90] The inhibitory effects of DSD-1-PG/phosphacan are not removed by chondroitinase ABC digestion indicating that they are associated with the core glycoprotein. Interestingly, digestion of the CS-GAG chains does not affect either the inhibitory nor the promotory effects of DSD-1-PG/phosphacan in related approaches.[104.131] Although it has been found that the digestion of the CS-GAG chains does not alleviate the inhibitory effects of DSD-1-PG on DRGs, the neurite outgrowth promoting effect of DSD-1-PG/phosphacan on hippocampal neurons is mediated by the particular CS-GAG structure, DSD-1.[93] In fact, the neurite outgrowth promoting effect can be neutralized by enzymatic digestion with chondroitinase ABC or addition of the blocking monoclonal antibody, 473 HD, which is specific for the DSD-1-epitope contained in the chondroitin sulfate chains. This is reminiscent of earlier reports that HS-, DS- and CS-GAGs potentiate neurite outgrowth by PC12 cells to varying extents upon exposure to aFGF, bFGF and NGF, which might reflect an improvement of cytokine effects.[159]

Similarly, some in vitro studies indicate that CS-GAG chains and heparan sulfate can promote neurite outgrowth.[93,104,153,160,161] A summary of the in vitro neurite outgrowth studies of the different CS GAG types is presented in Table 2. For example, purified HS-, DS- and CS-GAGs induce neurite outgrowth by E14 mesencephalic neurons.[162,163] In the light of these results, the DSD-1-epitope had originally been considered as a putative chondroitin sulfate/dermatan sulfate hybrid GAG structure, based on its differential sensitivity to the endoglycosidase, chondroitinase ACI, and its resistance to digestion by the exoglycosidase, chondroitinase ACII (DSD-1: for "dermatan-sulfate-dependent".[93] Later investigations have, however, established, that the epitope depends on sulfation and is associated with the chondoitin sulfate type D glycosaminoglycan (and should hence read "D-type chondroitin sulfate dependent".[96,164] In a recent study, the effects of diverse chondroitin sulfate GAGs on E18 hippocampal neurite outgrowth were systematically investigated. Consistent with the results of the earlier studies, CS-D and CS-E slightly enhance the fraction of neurite bearing neurons when applied as a substrate in comparison to CS-A, CS-B and CS-C.[96,164] It was observed that neither CS-A, CS-B nor CS-C significantly increased neurite length, CS-D displayed a mild promotory effect and CS-E exerted the strongest effect on neurite extension. The addition of MAb 473HD which is known to block neurite outgrowth promotion enacted by CS-D, did not neutralize the CS-E effect.[164,165] Consistent with this observation, the DSD-1-epitope

Table 2. Summary of in vitro neurite outgrowth studies using different types of glycosaminoglycans as both coated substrates and in soluble form

GAG Types *	Mode of Application	Coated Substrate	Neuronal Type	Animal/Age	Effect on Neurite Growth	Refs.
CS-4, CS-6, DS	coated	LN, LN, L1	Cerebellar neurons	Rat P5-P6	Inhibition	30
CS-4, CS-6, DS	coated	L1,L1,LN	Cerebellar neurons	Rat P5-P6	No effect	30
CS-4	coated	L1	Dorsal root ganglions neurons	Rat E15-E16	Promotion	30
CS-4, CS-6	coated	LN, LN	Dorsal root ganglions neurons	Rat E15-E16	Inhibition	30
CS-6, DS, DS	coated	L1,LN, L1	Dorsal root ganglions neurons	Rat E15-E16	No effect	30
CS-A, CS-C, DS	soluble		NGF induced PC12 cells		No effect	130
CS	soluble		Thalamic neurons	Rat E15	Promotion	161
CS, DS, DS	soluble	LN, LN	Thalamic neurons	Rat E18	No effect	161
CS, CS, DS, DS	soluble	LN, LN	Hippocampal	Rat E18 neurons	No effect	161
CS-A, CS-B, CS-C	coated		Hippocampal neurons	Rat E18	No effect	164
CS-D, CS-E	coated		Hippocampal neurons	Rat E18	Promotion	164
CS-A, CS-B, CS-C	soluble	FN,FN,FN	Dorsal root ganglions neurons	Chick E8-E11	No effect	240

* CS-A: chondroitin sulfate A (C4-S); CS-B: chondroitin sulfate B; CS-C: chondroitin sulfate C (C6-S); DS:dermatan sulfate

which is contained in commercially available CS-D preparations could not be revealed in CS-E preparations.[164,165] CS-E carries a similar charge to CS-D, but displays a different sulfation pattern. Hence, while CS-D is sulfated in the C6-position of N-acetylgalactosamine and the C2-position of hexuronic acid, the constitutive carbohydrate dimer of CS-E carries sulfate groups in the C4- and C6-positions of N-acetylgalactosamine. Recently, the importance of sulfate group distributions on carbohydrate polymers has been highlighted by the observation, that the DSD-1-epitope requires sulfation of a tetra-deca-saccharide.[96] These findings suggest that both the sequence of disaccharide motifs and the attachment of sulfate groups contribute to the formation of chondroitin sulfate domains with functional properties and that a novel, hitherto unknown structural domain with neurite outgrowth promoting properties is contained in CS-E. These observations contrast with the inhibitory properties of GAGs discussed before. The seemingly conflicting reports might be reconciled because the repulsive effects of GAGs have in many cases been observed in choice situations where neurites were confronted with favourable substrates, e.g., collagens, laminin and N-CAM on the one side, and areas of the culture dish containing GAGs and/or PGs in addition on the other. There are precedents that molecules displaying repulsive or inhibitory activities in choice situations may permit, or even enhance neurite growth or cell migration as homogeneous substrata, for example tenascin-C, that carries distinct functional domains.[14,106,166-168] Choice experiments on patterned substrates confront growth cones and/or cell bodies with a 100% step gradient of GAGs and/or

PGs and reveal repulsive properties which may not appear under conditions of uniform concentration.[169] Alternatively, these opposite results could reflect a structural heterogeneity of the complex chondroitin sulfate polymers. Furthermore, a lineage-dependence of neuronal responses to chondroitin sulfates might contribute to the seemingly contradictory reports from different studies.

Multiple Ligands of DSD-1-PG/Phosphacan

The effects on process outgrowth described above seem to be mediated by both the core glycoprotein and the CS GAGs (Table 1), and likely neuronal receptors for these interactions are members of the IgCAM superfamily. In effect, DSD-1-PG/phosphacan has been shown to possess three levels at which it can interact with other molecules either in the ECM or on cell membranes. These are the GAG chains, the other N- and O-linked oligosaccharides, and finally the regions of the protein core which are not covered by carbohydrate modifications. With such a range of possible interactions, it is not surprising that DSD-1-PG/phosphacan is implicated in many developmental processes such as migration and neurite outgrowth. In addition to variations in the presentation of such sites of interaction on the CSPG, the amplitude of its effects is likely to be dependent upon localised combinatory variations, both quantitative and qualitative, of promoting and inhibitory factors which recognise these sites. Cell-type specific differences in the cell-surface receptors (mostly IgCAMs) when confronted with a variety of potential ligands in the ECM, and the relative responsiveness of their intracellular signalling mechanisms to such factors in either cis- or trans- with other cell surface receptor molecules could then account for the differential cellular behaviour observed.[155,170] Models proposed to explain the relationship between the secreted DSD-1-PG/phosphacan isoform and the transmembrane receptor forms of RPTP-beta suggest that extracellular interaction sites common to the different isoforms could compete for extracellular ligands and thereby modulate the nature of the tyrosine phosphatase signal.[171] Other intracellular targets for RPTP-beta also imply roles in the regulation of G-protein coupled receptor signaling cascades (GIT1/Cat-1),[172] and synaptic organisation (PSD-95/SAP90).[173]

A plausible interpretation for the variable effects of phosphacan/RPTP-beta on neurite outgrowth discussed above might reflect variations in the repertoire of CAM receptors present on the different neuronal types (Fig. 7). In this context, the aminoterminal CA- and FNIII-domains and parts of the S-domain seem to be involved in the interactions which have been described for phosphacan/RPTP-beta. Members of the neuronal IgCAMs[170] found to interact with phosphacan/RPTP-beta include L1/NgCAM,[135] F3/contactin[2,156,174-176] TAG-1/axonin-1, and NrCAM.[155] It has been shown that the GPI-anchored F3/contactin binds to the CA-domain,[156] whereas the related GPI-anchored IgCAM, TAG-1/Axonin-1, binds via the CS GAG chains.[177] Similarly, interactions have been demonstrated between the S-domain of phosphacan/RPTP-beta and NCAM, NgCAM/L1, and NrCAM.[155] Several of these binding interactions have been shown to result in promotion of neurite outgrowth, effects which could be blocked by specific antibodies directed against the receptor molecules.[155,156]

In addition to the CS GAGs, DSD-1-PG/phosphacan is highly glycosylated with other carbohydrate modifications as illustrated by the presence of the 3H1, 3F8, L2/HNK-1 and L5/Lewis-X epitopes.[90] The latter is an N-linked carbohydrate which bears similarities with the forse antigen, a topographically restricted epitope of the developing nervous system.[158,178-180] It appears likely that the sulfation, carbohydrate composition and oligosaccharide structure of DSD-1-PG/phosphacan is developmentally regulated and that at least some of these carbohydrate modifications could alter its affinity for other proteins, for example, the N-linked sugars on the carbonic anhydrase and FNIII domains of DSD-1-PG/phosphacan which mediate its interactions with NgCAM, NCAM and TN-C.[181]

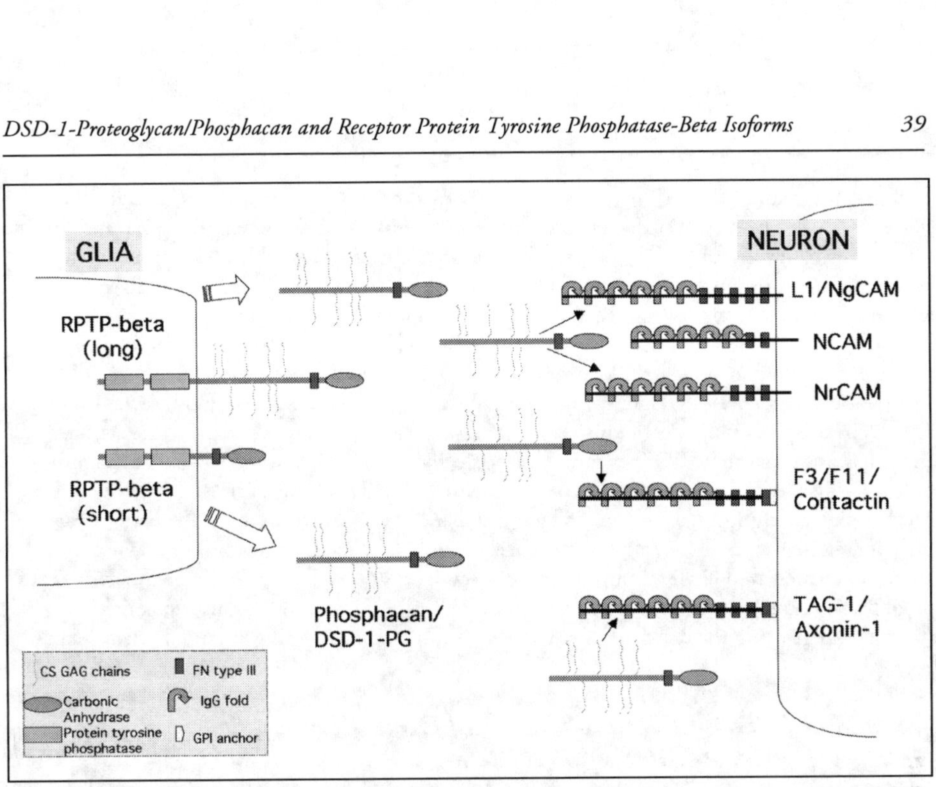

Figure 7. Neuron-glia interactions mediated by phosphacan/RPTP-beta. The Ig CAMs represented as neuronal receptors can all interact with extracellular sites on phosphacan/RPTP-beta: F3/F11/contactin via the CA domain; TAG-1/Axonin via the CS GAGs; the others via the S region.

Recombinant Fc-fusion proteins with the different domains have been used to characterise a range of molecular interactions underlying changes in cell behaviour. Hence, the CA-domain has been shown to support neuronal adhesion and neurite outgrowth by binding to F3/contactin on the surface of neurons.[156] This interaction with F3/contactin also implies a transmembrane receptor, Caspr[157] in a neuronal signaling complex which has been localised in vivo at paranodal junctions between axons and paranodal loops of myelinating glia.[182] It has recently been shown that the CA-domain can block the localisation to the paranodes of the F3/contactin-Caspr complex,[183] suggesting that this complex may be targeted via ECM interactions with RPTP-beta expressed by myelinating glia. Studies on mice deficient for RPTP-beta indicate that, although there is no gross abnormality in the overall brain architecture, there is a fragility in the myelin sheaths of CNS neurons,[92] and a recent report indicates dysfunctioning of the central dopaminergic system associated with locomotion defects.[184]

It has also recently been shown that the CA domain can interact with the extracellular beta-1 domain of the voltage-gated sodium channel,[185] an interaction which could modulate the action potential of the neuronal membrane. This study also demonstrated that the cytoplasmic tyrosine phosphatase domains of short RPTP-beta could interact with, and dephosphorylate, the pore-forming alpha-subunit. Hence, it is possible that competitive binding of different RPTP-beta isoforms could modulate sodium channel activity in a biphasic mechanism: the CA-domain binding to the extracellular domain in contrast to tyrosine phosphatase interactions with the cytoplasmic domain.

In vitro functional studies have provided evidence for roles in the development and maintenance of the CNS, and biochemical studies have demonstrated a number of potential binding partners in the ECM, such as the tenascins, TN-C and TN-R.[186] Meanwhile the RPTP-beta FNIII domain has been shown to bind to a distinct ligand on the surface of glial cells, including astrocytes, which has been recently identified as TN-C.[171] It has been suggested that this interaction is a primary adhesion receptor system to the ECM for glial tumour cells. The presence of TN-C on the cell surface is intriguing since TN-C is a large secreted glycoprotein with many isoform variants which is mostly present in the ECM.[187,188] However, it seems that there are glial receptors for TN-C which can retain it on the cell surface in a configuration which permits interaction with the RPTP-beta FNIII domain. Furthermore one of these glial receptors for TN-C may well be the short RPTP-beta itself, adding a further level of complexity to the potential range of cell-ECM interactions, since it has also been shown that the adhesion of glioblastoma cells to TN-C via the short RPTP-beta is modulated by the secretion of DSD-1-PG/phosphacan.[109,189] In this context it is also interesting to note that it has been recently demonstrated that RPTP-beta was intrinsically active in unstimulated cells in a study of signalling by the heparin-binding cytokine, pleiotrophin/HB-GAM/HBNF.[190] As such, it seems that pleiotrophin signals through a ligand-dependent inactivation of RPTP-beta, consequently increasing levels of tyrosine phosphorylation of beta-catenin to initiate downstream signalling and associated changes in process outgrowth and cell migration. Hence the competition of phosphacan, the released soluble form of RPTP-beta, for common ligands might also play a role in regulating the enzymatic activity of RPTP-beta itself and subsequent changes in cell behaviour.

Regulation of DSD-1-PG/Phosphacan and Other CSPGs by Cytokines in Lesions: Role of TGF-Beta

Using chondroitin sulfate markers, CSPGs have been shown to be markedly upregulated at CNS injury sites.[67-69,72,73,126,191] With the use of antibodies directed against the core proteins, several studies have reported the enhanced expression of growth inhibitory CSPGs including lecticans such as neurocan, brevican and versican,[49,50,52-54] the CSPG NG2[47] and decorin.[51] In contrast, DSD-1-PG/phosphacan core glycoprotein levels have been shown to be decreased in cerebral cortex lesions during the first week following injury[241] and also in kainate-induced seizures models and Ihara's epileptic rats.[193,194] The phosphacan diminution during the first week postlesion in cerebral cortex lesions could result, at least in part, from a reduced production since it was shown that the phosphacan mRNA level is downregulated at 2 days postlesion (dpl).[241] However, extracellular degradation might also contribute to the rapid decrease in phosphacan protein levels as the decrease of phosphacan in the hippocampus after kainate-induced seizure is largely due to its proteolysis by plasmin.[193] After 7 dpl the phosphacan mRNA levels are slightly increased in cerebral cortex lesions as observed following deafferentation of the hippocampus.[123] Following peripheral nerve crush, phosphacan mRNA is significantly induced in the distal segments of the sciatic nerve.[192] In contrast to the core glycoprotein, the phosphacan glycosaminoglycan epitope DSD-1, recognized by the 473mAb, has been found upregulated upon stab wounding in the CNS,[128,195,241] revealing differential modulations of the phosphacan core glycoprotein and DSD-1 epitope after injury.

The factors responsible for the regulation of proteoglycans in neural tissues are poorly understood. In lesion zones of the CNS, activated microglial cells and astrocytes express a variety of modulatory cytokines and growth factors[196-199] and some of these are known to modulate proteoglycan expression in non-neural test systems. For example, platelet-derived growth factor (PDGF) and transforming growth factor (TGF)-beta1 both increase the amount of a versican-like CSPG on arterial smooth muscle cells,[200] and TGF-beta1 enhances the rate of proteoglycan synthesis in rabbit articular chondrocytes.[201] Likewise, TGF-beta1 selectively induces the expression of biglycan, but not decorin, on human embryonic lung fibroblasts.[202] Cytokines such as interleukin-1 can directly inhibit proteoglycan synthesis.[203] In the nervous system, TGFbeta1 was shown to increase the expression of neurocan by astrocytes[50] and of versican by oligodendrocyte precursors.[52] The astrocytic expression of neurocan was also enhanced by epidermal growth factor (EGF) and in a higher magnitude than by TGFbeta1, whereas it was reduced by IFNgamma and PDGF.[50] Recently, it was shown that the astrocytic expression of phosphacan core glycoprotein is strongly augmentated by EGF and is decreased by tumor necrosis factor alpha (TNFalpha), interferon gamma (IFNgamma) and TGFbeta1.[241]

In order to further understanding of DSD-1-PG/phosphacan expression in oligodendrocyte precursors, the effects of soluble mediators known to be implicated in inflammation or wound reaction were systematically examined on the regulation of DSD-1-PG/phosphacan in the cell line Oli-neu. This cell line has been generated by immortalizing cultures enriched for oligodendrocytes with the oncogene t-neu,[204] and it expresses the antigen O4, a marker of immature oligodendrocytes, and can be driven to express myelin associated glycoprotein (MAG), a marker of mature oligodendrocytes, by continuous db-cAMP treatment.[204] In this regard Oli-neu behaves like differentiating primary oligodendrocytes in vitro. It is interesting to compare these observations with earlier studies concerning CSPG expression by glial cells. Thus, the detection of chondroitin sulfate epitopes on immature oligodendrocytes has been reported,[66] and at least one defined CSPG has been detected on oligodendrocyte surfaces, namely NG2.[30,84,85,87] The DSD-1-epitope and DSD-1-PG/phosphacan are known to be expressed by oligodendrocyte precursors (OPCs) in vitro and no other cell line or primary culture system was found which would present comparable advantages in terms of cell numbers and homogeneity of expression. Finally, the ability of Oli-neu to migrate on laminin and the cellular response upon contact with either astrocytes or meningeal cells is similar to that of an enriched oligodendrocyte precursor cell preparation.[205] For these reasons, Oli-neu was chosen as a model system for the study of DSD-1-PG/phosphacan regulation by oligodendrocyte precursors. To this end, an enzyme-linked immunosorbent assay using the cell line Oli-neu in conjunction with MAb 473HD was developed and combined with biosynthetic labeling and immunoprecipitation techniques. Using these approaches, it was found that, of all of the compounds tested so far, the growth factors TGF-beta1-3 induced a significant upregulation of the CSPG, both in the supernatant and in the detergent extract of cultured Oli-neu cells. The roles attributed to TGF-beta in the CNS to date are manifold and include the regulation of migratory properties of neural crest cells, the control of neural progenitor cell proliferation, an influence on neuronal survival and differentiation, the increase of neurite growth of dorsal root ganglia, and the modification of adhesive properties of radial glia and cells migrating in close contact with radial glia[206-208] for review, see references 209, 210.

TGF-beta 2 and 3 isoforms are known to be expressed in most cells present in the uninjured CNS such as astrocytes, neurons and oligodendrocytes while TGFbeta 1 is mainly synthesized in lesioned tissue.[206,208] Upon CNS injury, TGF-beta1 is upregulated by macrophages and astrocytes in the vicinity of adult CNS wounds[212] and contributes to the formation of the glial scar, thus impairing regeneration.[209,213] TGF-betas could influence core protein expression, GAG chain length, -number and -composition, or a combination of these parameters. For example, TGFbetas enlarged the size of individual GAG chains by approximately 25 kDa on a versican-like molecule expressed by smooth muscle cells, and enhance the level of mRNA that was detectable by a versican-cDNA probe 3-fold over unstimulated controls.[200] It also increased the number and length of chondroitin sulfate chains attached to the proteoglycan syndecan in mouse mammary epithelial cells, while the core protein levels remained unchanged.[211] TGF-beta1 apparently increased the expression of the GAG-chain associated DSD-1-epitope and, to a lesser degree, of the DSD-1/phosphacan-proteoglycan core protein. Indeed, the immunoprecipitation of DSD-1-PG from TGF-beta1 treated Oli-neu cells showed that there was both an increase of core protein expression and an enhanced incorporation of sulfate into the GAG chains. This might result from an increased length and/or number of chondroitin sulfate chains, structural rearrangements or varied sulfation, concomittant with an enhanced density of DSD-1-epitope expression on individual cell surfaces. Interestingly, a comparable enhanced sulfation of GAG-chains and an increase of GAG-chain attachment has also been described for NG2 as synthesized in the cell line Neu-7, a model of reactive astrocytes in vitro.[79,83]

DSD-1-PG/Phosphacan and RPTP-Beta Isoforms in the Regulation of Cell-Substrate Interactions and Cell Motility

The importance of ECM molecules for cell motility could be demonstrated in several culture models of cellular migration. For example, neural crest cells migrate through the embryo to differentiate into peripheral nervous system, facial skeleton or pigment cells using migration pathways comprising collagens, laminin-1 or fibronectin.[214] Directional restraints may be imposed by substrates unfavourable for migration, e.g., CSPGs.[37,215,216] An example of extensive migration in the CNS relates to the O2A (oligodendrocyte-type 2 astrocyte) precursor cells that are generated outside the optic nerve and immigrate from its chiasmal side to the retina.[217] Likewise, oligodendrocyte precursors in the developing spinal cord are generated in the subventricular zone and subsequently migrate away to populate the spinal cord.[218] In the adult CNS, a progenitor cell type with limited migratory potential, termed O2A-adult, is retained in the postnatal brain.[219] Migration of oligodendrocyte precursors (OPCs) hence seems important for the determination of myelination territories during development and the remyelination of axons after lesion events. Cell-cell and cell-substratum interactions of oligodendrocytes and their precursors are likely to influence proliferation, migration to sites of differentiation[217,220] re-immigration to places of demyelinating lesions, and remyelination. It has been proposed that the polysialylated form of the neural cell adhesion molecule N-CAM is involved in the control of the motility of O2A (oligodendrocyte-type 2 astrocyte) glial progenitors.[221-223] The corresponding characteristic carbohydrate polymer alpha 2-8 linked polysialic acid has been detected on the surface of O2A-progenitor cells from the hypophysis of newborn rats. Enzymatic digestion of the carbohydrate structure by endoneuraminidase and blockade of the NMDA-receptor activation-dependent exposure of PSA on the cell surface completely blocked the motility of the OPCs.[222,223] The mechanisms involved in the migration of OPCs have been elucidated in some detail. Growth factors such as platelet-derived growth factor (PDGF) stimulate migration presumably by chemotactic mechanisms[224] or by keeping oligodendrocyte precursors in an undifferentiated state.[225] Furthermore, the importance of extracellular matrix molecules as migration substrates has been increasingly appreciated. Oligodendrocyte precursors can migrate on substrates of laminin-1[205] and fibronectin glycoproteins, while tenascin-C as a cosubstrate to fibronectin reduces their migratory ability.[226] Along these lines, integrins as cell-surface receptors for several ECM glycoproteins (Hynes, 1992) have been shown to play an important role in this context (Milner et al, 1996). Integrins [12,16,227] mediate adhesion to laminin-1,[228] merosin (laminin-2),[229] fibronectin,[230] tenascin-C[231,232] in part by an RGD-dependent mechanism. TGF-beta differentially regulates the expression of various integrin genes depending on the cell type and subunit composition of the heterodimers. For example, TGF-beta stimulates adhesion of human mononuclear phagocytes to laminin and fibronectin by increasing integrin beta 2 and alpha 5 mRNA and protein synthesis,[233] and a comparable differential regulation of distinct integrin chains has been described for rat alveolar epithelial cells,[234] This might also be relevant in the context of tumour invasion and metastasis, because the invasion of U-138MG glioma cells was facilitated due to an increased adhesion to ECM after alpha-5 integrin upregulation by TGF-beta.[235] While in most cases studied so far TGF-beta strengthens adhesion to various ECM substrates, the data relating to DSD-1-PG/phosphacan indicate that the Oli-neu cell line shows decreased attachment to laminin upon TGF-beta-treatment. Antibody perturbation studies with mono- and polyclonal antibodies suggest that this can at least partially be attributed to DSD-1-PG/phosphacan. The precise role of DSD-1-PG/phosphacan in this process is, however, currently unclear.

Interestingly, it has been reported that a CSPG isolated from human melanoma cells directly binds an amino acid sequence motif contained in a fibronectin fragment and thereby contributes to the adhesion of melanoma cells to fibronectin.[236] It is conceivable that DSD-1-PG/phosphacan influences the interaction between laminins and their receptors, potentially by interfering with the activation state of appropriate integrins. Because differential expression of integrins is believed to regulate oligodendrocyte precursor cell (OPC) motility, this might impinge on the movement of OPCs.[221,224,237] Alternatively, a fine-tuning of ligand binding by cis-interactions of DSD-1-PG/phosphacan with receptor-integrins in the cellular membrane might be assumed. This might primarily concern the transmembrane isoforms of RPTP-beta, because the largest variant is expressed by oligodendrocyte precursors.[101] It contains the tyrosine-phosphatase domains and could, therefore, potentially anatagonise the activation of integrins, which may involve the tyrosine kinase "focal adhesion kinase" (FAK).[17] The tuning of integrins in the cell membrane presumably affects both the substrate attachment and the motility behaviour of the cell concerned, e.g., the OPC in this case.[238] In conclusion, it seems reasonable to conclude that the levels of DSD-1-PG/phosphacan in the cell membrane modulate the adhesive interactions of Oli-neu with the ECM. The modulation of cell-substrate interactions presumably translates in the tuning of cell motility of Oli-neu cells, and the OPC equivalent (Schnädelbach and Faissner, in preparation).

References

1. Jacobson M. Developmental neurobiology. 3rd ed. New York and London: Plenum Press; 1991.
2. Brummendorf T, Rathjen FG. Axonal glycoproteins with immunoglobulin- and fibronectin type III-related domains in vertebrates: Structural features, binding activities, and signal transduction. J Neurochem 1993; 61(4):1207-1219.
3. Brummendorf T, Rathjen FG. Structure/function relationships of axon-associated adhesion receptors of the immunoglobulin superfamily. Curr Opin Neurobiol 1996; 6(5):584-593.
4. Takeichi M. Morphogenetic roles of classic cadherins. Curr Opin Cell Biol 1995; 7(5):619-627.
5. Uemura T. The cadherin superfamily at the synapse: More members, more missions. Cell 1998; 93(7):1095-1098.
6. Tessier-Lavigne M. Eph receptor tyrosine kinases, axon repulsion, and the development of topographic maps. Cell 1995; 82(3):345-348.
7. Drescher U, Bonhoeffer F, Muller BK. The Eph family in retinal axon guidance. Curr Opin Neurobiol 1997; 7(1):75-80.
8. Klein R. Bidirectional signals establish boundaries. Curr Biol 1999; 9(18):R691-694.
9. Xu Q, Mellitzer G, Robinson V et al. In vivo cell sorting in complementary segmental domains mediated by Eph receptors and ephrins. Nature 1999; 399(6733):267-271.
10. Garwood J, Heck N, Rigato F et al. The extracellular matrix in neural development, plasticity and regeneration. In: Walz W, ed. The neuronal microenvironment. New Jersey, USA: Humana Press; 2002:109-158.
11. Sanes JR. Extracellular matrix molecules that influence neural development. Annu Rev Neurosci 1989; 12:491-516.
12. Hynes RO, Lander AD. Contact and adhesive specificities in the associations, migrations, and targeting of cells and axons. Cell 1992; 68(2):303-322.
13. Faissner A, Steindler D. Boundaries and inhibitory molecules in developing neural tissues. Glia 1995; 13(4):233-254.
14. Faissner A. The tenascin gene family in axon growth and guidance. Cell Tissue Res 1997; 290(2):331-341.
15. Miao HQ, Ishai-Michaeli R, Atzmon R et al. Sulfate moieties in the subendothelial extracellular matrix are involved in basic fibroblast growth factor sequestration, dimerization, and stimulation of cell proliferation. J Biol Chem 1996; 271(9):4879-4886.
16. Hynes RO. Integrins: Versatility, modulation, and signaling in cell adhesion. Cell 1992; 69(1):11-25.
17. Dedhar S, Hannigan GE. Integrin cytoplasmic interactions and bidirectional transmembrane signalling. Curr Opin Cell Biol 1996; 8(5):657-669.

18. Fields RD, Itoh K. Neural cell adhesion molecules in activity-dependent development and synaptic plasticity. Trends Neurosci 1996; 19(11):473-480.
19. Lander AD. Proteoglycans: Master regulators of molecular encounter? Matrix Biol 1998; 17(7):465-472.
20. Kiang WL, Margolis RU, Margolis RK. Fractionation and properties of a chondroitin sulfate proteoglycan and the soluble glycoproteins of brain. J Biol Chem 1981; 256(20):10529-10537.
21. Herndon ME, Lander AD. A diverse set of developmentally regulated proteoglycans is expressed in the rat central nervous system. Neuron 1990; 4(6):949-961.
22. Klinger MM, Margolis RU, Margolis RK. Isolation and characterization of the heparan sulfate proteoglycans of brain. Use of affinity chromatography on lipoprotein lipase-agarose. J Biol Chem 1985; 260(7):4082-4090.
23. Tessier-Lavigne M, Goodman CS. The molecular biology of axon guidance. Science 1996; 274(5290):1123-1133.
24. Luo Y, Raible D, Raper JA. Collapsin: A protein in brain that induces the collapse and paralysis of neuronal growth cones. Cell 1993; 75(2):217-227.
25. Goodman CS. Mechanisms and molecules that control growth cone guidance. Annu Rev Neurosci 1996; 19:341-377.
26. Luo Y, Raper JA. Inhibitory factors controlling growth cone motility and guidance. Curr Opin Neurobiol 1994; 4(5):648-654.
27. Fournier AE, Strittmatter SM. Repulsive factors and axon regeneration in the CNS. Curr Opin Neurobiol 2001; 11(1):89-94.
28. Schwab ME, Kapfhammer JP, Bandtlow CE. Inhibitors of neurite growth. Annu Rev Neurosci 1993; 16:565-595.
29. Huber AB, Schwab ME. Nogo-A, a potent inhibitor of neurite outgrowth and regeneration. Biol Chem 2000; 381(5-6):407-419.
30. Dou CL, Levine JM. Inhibition of neurite growth by the NG2 chondroitin sulfate proteoglycan. J Neurosci 1994; 14(12):7616-7628.
31. Margolis RK, Rauch U, Maurel P et al. Neurocan and phosphacan: Two major nervous tissue-specific chondroitin sulfate proteoglycans. Perspect Dev Neurobiol 1996; 3(4):273-290.
32. Zaremba S, Guimaraes A, Kalb RG et al. Characterization of an activity-dependent, neuronal surface proteoglycan identified with monoclonal antibody Cat-301. Neuron 1989; 2(3):1207-1219.
33. Hockfield S, Tootell RB, Zaremba S. Molecular differences among neurons reveal an organization of human visual cortex. Proc Natl Acad Sci USA 1990; 87(8):3027-3031.
34. Kalb RG, Hockfield S. Induction of a neuronal proteoglycan by the NMDA receptor in the developing spinal cord. Science 1990; 250(4978):294-296.
35. Snow DM, Steindler DA, Silver J. Molecular and cellular characterization of the glial roof plate of the spinal cord and optic tectum: A possible role for a proteoglycan in the development of an axon barrier. Dev Biol 1990; 138(2):359-376.
36. Katoh-Semba R, Matsuda M, Kato K et al. Chondroitin sulphate proteoglycans in the rat brain: Candidates for axon barriers of sensory neurons and the possible modification by laminin of their actions. Eur J Neurosci 1995; 7(4):613-621.
37. Landolt RM, Vaughan L, Winterhalter KH et al. Versican is selectively expressed in embryonic tissues that act as barriers to neural crest cell migration and axon outgrowth. Development 1995; 121(8):2303-2312.
38. Oakley RA, Tosney KW. Peanut agglutinin and chondroitin-6-sulfate are molecular markers for tissues that act as barriers to axon advance in the avian embryo. Dev Biol 1991; 147(1):187-206.
39. Gonzalez ML, Silver J. Axon-glia interactions regulate ECM patterning in the postnatal rat olfactory bulb. J Neurosci 1994; 14(10):6121-6131.
40. Gonzalez Mde L, Malemud CJ, Silver J. Role of astroglial extracellular matrix in the formation of rat olfactory bulb glomeruli. Exp Neurol 1993; 123(1):91-105.
41. Steindler DA. Glial boundaries in the developing nervous system. Annu Rev Neurosci 1993; 16:445-470.
42. Steindler DA, Settles D, Erickson HP et al. Tenascin knockout mice: Barrels, boundary molecules, and glial scars. J Neurosci 1995; 15(3 Pt 1):1971-1983.

43. Crandall JE, Misson JP, Butler D. The development of radial glia and radial dendrites during barrel formation in mouse somatosensory cortex. Brain Res Dev Brain Res 1990; 55(1):87-94.
44. Steindler DA, Cooper NG, Faissner A et al. Boundaries defined by adhesion molecules during development of the cerebral cortex: The J1/tenascin glycoprotein in the mouse somatosensory cortical barrel field. Dev Biol 1989; 131(1):243-260.
45. Fitch MT, Silver J. Glial cell extracellular matrix: boundaries for axon growth in development and regeneration. Cell Tissue Res 1997; 290(2):379-384.
46. Pindzola RR, Doller C, Silver J. Putative inhibitory extracellular matrix molecules at the dorsal root entry zone of the spinal cord during development and after root and sciatic nerve lesions. Dev Biol 1993; 156(1):34-48.
47. Levine JM. Increased expression of the NG2 chondroitin-sulfate proteoglycan after brain injury. J Neurosci 1994; 14(8):4716-4730.
48. McKeon RJ, Jurynec MJ, Buck CR. The chondroitin sulfate proteoglycans neurocan and phosphacan are expressed by reactive astrocytes in the chronic CNS glial scar. J Neurosci 1999; 19(24):10778-10788.
49. Haas CA, Rauch U, Thon N et al. Entorhinal cortex lesion in adult rats induces the expression of the neuronal chondroitin sulfate proteoglycan neurocan in reactive astrocytes. J Neurosci 1999; 19(22):9953-9963.
50. Asher RA, Morgenstern DA, Fidler PS et al. Neurocan is upregulated in injured brain and in cytokine-treated astrocytes. J Neurosci 2000; 20(7):2427-2438.
51. Stichel CC, Kappler J, Junghans U et al. Differential expression of the small chondroitin/dermatan sulfate proteoglycans decorin and biglycan after injury of the adult rat brain. Brain Res 1995; 704(2):263-274.
52. Asher RA, Morgenstern DA, Shearer MC et al. Versican is upregulated in CNS injury and is a product of oligodendrocyte lineage cells. J Neurosci 2002; 22(6):2225-2236.
53. Jaworski DM, Kelly GM, Hockfield S. Intracranial injury acutely induces the expression of the secreted isoform of the CNS-specific hyaluronan-binding protein BEHAB/brevican. Exp Neurol 1999; 157(2):327-337.
54. Thon N, Haas CA, Rauch U et al. The chondroitin sulphate proteoglycan brevican is upregulated by astrocytes after entorhinal cortex lesions in adult rats. Eur J Neurosci 2000; 12(7):2547-2558.
55. Bahr M, Bonhoeffer F. Perspectives on axonal regeneration in the mammalian CNS. Trends Neurosci 1994; 17(11):473-479.
56. Fawcett JW. Astrocytic and neuronal factors affecting axon regeneration in the damaged central nervous system. Cell Tissue Res 1997; 290(2):371-377.
57. Fawcett JW, Asher RA. The glial scar and central nervous system repair. Brain Res Bull 1999; 49(6):377-391.
58. Fawcett JW. Spinal cord repair: from experimental models to human application. Spinal Cord 1998; 36(12):811-817.
59. Brodkey JA, Laywell ED, O'Brien TF et al. Focal brain injury and upregulation of a developmentally regulated extracellular matrix protein. J Neurosurg 1995; 82(1):106-112.
60. Reier P. Gliosis following CNS injury: The anatomy of glial scars and their influences on axonal elongation. In: Federoff S, Vernadakis A, eds. Astrocytes. New York: Academic Press; 1986:263-324.
61. Landis DM. The early reactions of non-neuronal cells to brain injury. Annu Rev Neurosci 1994; 17:133-151.
62. Rudge JS, Silver J. Inhibition of neurite outgrowth on astroglial scars in vitro. J Neurosci 1990; 10(11):3594-3603.
63. Bähr M, Przyrembel C, Bastmeyer M. Astrocytes from adult rat optic nerves are nonpermissive for regenerating retinal ganglion cell axons. Exp Neurol 1995; 131:211-220.
64. Liuzzi FJ, Lasek RJ. Astrocytes block axonal regeneration in mammals by activating the physiological stop pathway. Science 1987; 237(4815):642-645.
65. Davies SJ, Field PM, Raisman G. Regeneration of cut adult axons fails even in the presence of continuous aligned glial pathways. Exp Neurol 1996; 142(2):203-216.
66. Gallo V, Bertolotto A, Levi G. The proteoglycan chondroitin sulfate is present in a subpopulation of cultured astrocytes and in their precursors. Dev Biol 1987; 123(1):282-285.

67. McKeon RJ, Schreiber RC, Rudge JS et al. Reduction of neurite outgrowth in a model of glial scarring following CNS injury is correlated with the expression of inhibitory molecules on reactive astrocytes. J Neurosci 1991; 11(11):3398-3411.
68. McKeon RJ, Hoke A, Silver J. Injury-induced proteoglycans inhibit the potential for laminin-mediated axon growth on astrocytic scars. Exp Neurol 1995; 136(1):32-43.
69. Bovolenta P, Wandosell F, Nieto-Sampedro M. Neurite outgrowth inhibitors associated with glial cells and glial cell lines. Neuroreport 1993; 5(3):345-348.
70. Bovolenta P, Wandosell F, Nieto-Sampedro M. Characterization of a neurite outgrowth inhibitor expressed after CNS injury. Eur J Neurosci 1993;5(5):454-465.
71. Snow DM, Lemmon V, Carrino DA et al. Sulfated proteoglycans in astroglial barriers inhibit neurite outgrowth in vitro. Exp Neurol 1990; 109(1):111-130.
72. Davies SJ, Fitch MT, Memberg SP et al. Regeneration of adult axons in white matter tracts of the central nervous system. Nature 1997; 390(6661):680-683.
73. Davies SJ, Goucher DR, Doller C et al. Robust regeneration of adult sensory axons in degenerating white matter of the adult rat spinal cord. J Neurosci 1999; 19(14):5810-5822.
74. Zuo J, Ferguson TA, Hernandez YJ et al. Neuronal matrix metalloproteinase-2 degrades and inactivates a neurite- inhibiting chondroitin sulfate proteoglycan. J Neurosci 1998; 18(14):5203-5211.
75. Zuo J, Neubauer D, Dyess K et al. Degradation of chondroitin sulfate proteoglycan enhances the neurite- promoting potential of spinal cord tissue. Exp Neurol 1998; 154(2):654-662.
76. Yick LW, Wu W, So KF et al. Chondroitinase ABC promotes axonal regeneration of Clarke's neurons after spinal cord injury. Neuroreport 2000; 11(5):1063-1067.
77. Moon LD, Brecknell JE, Franklin RJ et al. Robust regeneration of CNS axons through a track depleted of CNS glia. Exp Neurol 2000; 161(1):49-66.
78. Groves AK, Entwistle A, Jat PS et al. The characterization of astrocyte cell lines that display properties of glial scar tissue. Dev Biol 1993; 159(1):87-104.
79. Smith-Thomas LC, Fok-Seang J, Stevens J et al. An inhibitor of neurite outgrowth produced by astrocytes. J Cell Sci 1994; 107(Pt 6):1687-1695.
80. Smith-Thomas LC, Stevens J, Fok-Seang J et al. Increased axon regeneration in astrocytes grown in the presence of proteoglycan synthesis inhibitors. J Cell Sci 1995; 108 (Pt 3):1307-1315.
81. Powell EM, Meiners S, DiProspero NA et al. Mechanisms of astrocyte-directed neurite guidance. Cell Tissue Res 1997; 290(2):385-393.
82. Meiners S, Powell EM, Geller HM. A distinct subset of tenascin/CS-6-PG-rich astrocytes restricts neuronal growth in vitro. J Neurosci 1995; 15(12):8096-8108.
83. Fidler PS, Schuette K, Asher RA et al. Comparing astrocytic cell lines that are inhibitory or permissive for axon growth: The major axon-inhibitory proteoglycan is NG2. J Neurosci 1999; 19(20):8778-8788.
84. Levine JM, Stallcup WB. Plasticity of developing cerebellar cells in vitro studied with antibodies against the NG2 antigen. J Neurosci 1987; 7(9):2721-2731.
85. Levine JM, Reynolds R, Fawcett JW. The oligodendrocyte precursor cell in health and disease. Trends Neurosci 2001; 24(1):39-47.
86. Dou CL, Levine JM. Differential effects of glycosaminoglycans on neurite growth on laminin and L1 substrates. J Neurosci 1995; 15(12):8053-8066.
87. Stallcup WB, Beasley L. Bipotential glial precursor cells of the optic nerve express the NG2 proteoglycan. J Neurosci 1987; 7(9):2737-2744.
88. Lander AD. Proteoglycans in the nervous system. Curr Opin Neurobiol 1993; 3(5):716-723.
89. Bandtlow CE, Zimmermann DR. Proteoglycans in the developing brain: new conceptual insights for old proteins. Physiol Rev 2000; 80(4):1267-1290.
90. Garwood J, Schnadelbach O, Clement A et al. DSD-1-proteoglycan is the mouse homolog of phosphacan and displays opposing effects on neurite outgrowth dependent on neuronal lineage. J Neurosci 1999; 19(10):3888-3899.
91. Maeda N, Noda M. Involvement of receptor-like protein tyrosine phosphatase zeta/RPTPbeta and its ligand pleiotrophin/heparin-binding growth-associated molecule (HB-GAM) in neuronal migration. J Cell Biol 1998; 142(1):203-216.
92. Harroch S, Palmeri M, Rosenbluth J et al. No obvious abnormality in mice deficient in receptor protein tyrosine phosphatase beta. Mol Cell Biol 2000; 20(20):7706-7715.

93. Faissner A, Clement A, Lochter A et al. Isolation of a neural chondroitin sulfate proteoglycan with neurite outgrowth promoting properties. J Cell Biol 1994; 126(3):783-799.
94. Hoffman S, Edelman GM. A proteoglycan with HNK-1 antigenic determinants is a neuron-associated ligand for cytotactin. Proc Natl Acad Sci USA 1987; 84(8):2523-2527.
95. Hoffman S, Crossin KL, Edelman GM. Molecular forms, binding functions, and developmental expression patterns of cytotactin and cytotactin-binding proteoglycan, an interactive pair of extracellular matrix molecules. J Cell Biol 1988; 106(2):519-532.
96. Clement AM, Nadanaka S, Masayama K et al. The DSD-1 carbohydrate epitope depends on sulfation, correlates with chondroitin sulfate D motifs, and is sufficient to promote neurite outgrowth. J Biol Chem 1998; 273(43):28444-28453.
97. Maurel P, Rauch U, Flad M et al. Phosphacan, a chondroitin sulfate proteoglycan of brain that interacts with neurons and neural cell-adhesion molecules, is an extracellular variant of a receptor-type protein tyrosine phosphatase. Proc Natl Acad Sci USA 1994; 91(7):2512-2516.
98. Krueger NX, Saito H. A human transmembrane protein-tyrosine-phosphatase, PTP zeta, is expressed in brain and has an N-terminal receptor domain homologous to carbonic anhydrases. Proc Natl Acad Sci USA 1992; 89(16):7417-7421.
99. Barnea G, Grumet M, Milev P et al. Receptor tyrosine phosphatase beta is expressed in the form of proteoglycan and binds to the extracellular matrix protein tenascin. J Biol Chem 1994; 269(20):14349-14352.
100. Canoll PD, Barnea G, Levy JB et al. The expression of a novel receptor-type tyrosine phosphatase suggests a role in morphogenesis and plasticity of the nervous system. Brain Res Dev Brain Res 1993; 75(2):293-298.
101. Canoll PD, Petanceska S, Schlessinger J et al. Three forms of RPTP-beta are differentially expressed during gliogenesis in the developing rat brain and during glial cell differentiation in culture. J Neurosci Res 1996; 44(3):199-215.
102. Meyer-Puttlitz B, Milev P, Junker E et al. Chondroitin sulfate and chondroitin/keratan sulfate proteoglycans of nervous tissue: Developmental changes of neurocan and phosphacan. J Neurochem 1995; 65(5):2327-2337.
103. Nagata S, Saito R, Yamada Y et al. Multiple variants of receptor-type protein tyrosine phosphatase beta are expressed in the central nervous system of Xenopus. Gene 2001; 262(1-2):81-88.
104. Maeda N, Noda M. 6B4 proteoglycan/phosphacan is a repulsive substratum but promotes morphological differentiation of cortical neurons. Development 1996; 122(2):647-658.
105. Meyer-Puttlitz B, Junker E, Margolis RU et al. Chondroitin sulfate proteoglycans in the developing central nervous system. II. Immunocytochemical localization of neurocan and phosphacan. J Comp Neurol 1996; 366(1):44-54.
106. Garwood J, Rigato F, Heck N et al. Tenascin glycoproteins and the complementary ligand DSD-1-PG/phosphacan – structuring the neural extracellular matrix during development and repair. Restor Neurol Neurosci 2001; 19(1,2):51-64.
107. Faissner A. Monoclonal antibody identifies a proteoglycan expressed by a subclass of glial cells. Soc Neurosci Abstr 1988; 14:920.
108. Schnadelbach O, Mandl C, Faissner A. Expression of DSD-1-PG in primary neural and glial-derived cell line cultures, upregulation by TGF-beta, and implications for cell-substrate interactions of the glial cell line Oli-neu. Glia 1998; 23(2):99-119.
109. Sakurai T, Friedlander DR, Grumet M. Expression of polypeptide variants of receptor-type protein tyrosine phosphatase beta: The secreted form, phosphacan, increases dramatically during embryonic development and modulates glial cell behavior in vitro. J Neurosci Res 1996; 43(6):694-706.
110. Engel M, Maurel P, Margolis RU et al. Chondroitin sulfate proteoglycans in the developing central nervous system. I. cellular sites of synthesis of neurocan and phosphacan. J Comp Neurol 1996; 366(1):34-43.
111. Rauch U, Gao P, Janetzko A et al. Isolation and characterization of developmentally regulated chondroitin sulfate and chondroitin/keratan sulfate proteoglycans of brain identified with monoclonal antibodies. J Biol Chem 1991; 266(22):14785-14801.
112. Heyman I, Faissner A, Lumsden A. Cell and matrix specialisations of rhombomere boundaries. Dev Dyn 1995; 204(3):301-315.

113. Wintergerst ES, Faissner A, Celio MR. The proteoglycan DSD-1-PG occurs in perineuronal nets around parvalbumin-immunoreactive interneurons of the rat cerebral cortex. Int J Dev Neurosci 1996; 14(3):249-255.
114. Maleski M, Hockfield S. Glial cells assemble hyaluronan-based pericellular matrices in vitro. Glia 1997; 20(3):193-202.
115. Rauch U. Modeling an extracellular environment for axonal pathfinding and fasciculation in the central nervous system. Cell Tissue Res 1997; 290(2):349-356.
116. Celio MR, Blumcke I. Perineuronal nets—a specialized form of extracellular matrix in the adult nervous system. Brain Res Brain Res Rev 1994; 19(1):128-145.
117. Gates MA, Thomas LB, Howard EM et al. Cell and molecular analysis of the developing and adult mouse subventricular zone of the cerebral hemispheres. J Comp Neurol 1995; 361(2):249-266.
118. Goldman SA, Luskin MB. Strategies utilized by migrating neurons of the postnatal vertebrate forebrain. Trends Neurosci 1998; 21(3):107-114.
119. Thomas LB, Gates MA, Steindler DA. Young neurons from the adult subependymal zone proliferate and migrate along an astrocyte, extracellular matrix-rich pathway. Glia 1996; 17(1):1-14.
120. Shintani T, Watanabe E, Maeda N et al. Neurons as well as astrocytes express proteoglycan-type protein tyrosine phosphatase zeta/RPTPbeta: Analysis of mice in which the PTPzeta/RPTPbeta gene was replaced with the LacZ gene. Neurosci Lett 1998; 247(2-3):135-138.
121. Maeda N, Matsui F, Oohira A. A chondroitin sulfate proteoglycan that is developmentally regulated in the cerebellar mossy fiber system. Dev Biol 1992; 151(2):564-574.
122. Haunso A, Celio MR, Margolis RK et al. Phosphacan immunoreactivity is associated with perineuronal nets around parvalbumin-expressing neurones. Brain Res 1999; 834(1-2):219-222.
123. Snyder SE, Li J, Schauwecker PE et al. Comparison of RPTP zeta/beta, phosphacan, and trkB mRNA expression in the developing and adult rat nervous system and induction of RPTP zeta/beta and phosphacan mRNA following brain injury. Brain Res Mol Brain Res 1996; 40(1):79-96.
124. Lips K, Stichel CC, Muller HW. Restricted appearance of tenascin and chondroitin sulphate proteoglycans after transection and sprouting of adult rat postcommissural fornix. J Neurocytol 1995; 24(6):449-464.
125. Laywell ED, Dorries U, Bartsch U et al. Enhanced expression of the developmentally regulated extracellular matrix molecule tenascin following adult brain injury. Proc Natl Acad Sci USA 1992; 89(7):2634-2638.
126. Barker RA, Dunnett SB, Faissner A et al. The time course of loss of dopaminergic neurons and the gliotic reaction surrounding grafts of embryonic mesencephalon to the striatum. Exp Neurol 1996; 141(1):79-93.
127. Deller T, Haas CA, Frotscher M. Reorganization of the rat fascia dentata after a unilateral entorhinal cortex lesion. Role of the extracellular matrix. Ann NY Acad Sci 2000; 911:207-220.
128. Deller T, Haas CA, Naumann T et al. Up-regulation of astrocyte-derived tenascin-C correlates with neurite outgrowth in the rat dentate gyrus after unilateral entorhinal cortex lesion. Neuroscience 1997; 81(3):829-846.
129. Rauch U, Karthikeyan L, Maurel P et al. Cloning and primary structure of neurocan, a developmentally regulated, aggregating chondroitin sulfate proteoglycan of brain. J Biol Chem 1992; 267(27):19536-19547.
130. Oohira A, Matsui F, Katoh-Semba R. Inhibitory effects of brain chondroitin sulfate proteoglycans on neurite outgrowth from PC12D cells. J Neurosci 1991; 11(3):822-827.
131. Milev P, Friedlander DR, Sakurai T et al. Interactions of the chondroitin sulfate proteoglycan phosphacan, the extracellular domain of a receptor-type protein tyrosine phosphatase, with neurons, glia, and neural cell adhesion molecules. J Cell Biol 1994; 127(6 Pt 1):1703-1715.
132. Friedlander DR, Milev P, Karthikeyan L et al. The neuronal chondroitin sulfate proteoglycan neurocan binds to the neural cell adhesion molecules Ng-CAM/L1/NILE and N-CAM, and inhibits neuronal adhesion and neurite outgrowth. J Cell Biol 1994; 125(3):669-680.
133. Brittis PA, Canning DR, Silver J. Chondroitin sulfate as a regulator of neuronal patterning in the retina. Science 1992; 255(5045):733-736.
134. Fichard A, Verna JM, Olivares J et al. Involvement of a chondroitin sulfate proteoglycan in the avoidance of chick epidermis by dorsal root ganglia fibers: a study using beta-D-xyloside. Dev Biol 1991; 148(1):1-9.

135. Grumet M, Flaccus A, Margolis RU. Functional characterization of chondroitin sulfate proteoglycans of brain: interactions with neurons and neural cell adhesion molecules. J Cell Biol 1993; 120(3):815-824.
136. Miller B, Sheppard AM, Pearlman AL. Developmental expression of keratan sulfate-like immunoreactivity distinguishes thalamic nuclei and cortical domains [published erratum appears in J Comp Neurol 1997 Sep 1;385(3):490-1]. J Comp Neurol 1997; 380(4):533-552.
137. Cole GJ, McCabe CF. Identification of a developmentally regulated keratan sulfate proteoglycan that inhibits cell adhesion and neurite outgrowth. Neuron 1991; 7(6):1007-1018.
138. Carbonetto S, Gruver MM, Turner DC. Nerve fiber growth in culture on fibronectin, collagen, and glycosaminoglycan substrates. J Neurosci 1983; 3(11):2324-2335.
139. Verna JM. In vitro analysis of interactions between sensory neurons and skin: Evidence for selective innervation of dermis and epidermis. J Embryol Exp Morphol 1985; 86:53-70.
140. Muir D, Engvall E, Varon S et al. Schwannoma cell-derived inhibitor of the neurite-promoting activity of laminin. J Cell Biol 1989; 109(5):2353-2362.
141. O'Brien TF, Faissner A, Schachner M et al. Afferent-boundary interactions in the developing neostriatal mosaic. Brain Res Dev Brain Res 1992; 65(2):259-267.
142. Crossin KL, Hoffman S, Tan SS et al. Cytotactin and its proteoglycan ligand mark structural and functional boundaries in somatosensory cortex of the early postnatal mouse. Dev Biol 1989; 136(2):381-392.
143. Inatani M, Honjo M, Otori Y et al. Inhibitory effects of neurocan and phosphacan on neurite outgrowth from retinal ganglion cells in culture. Invest Ophthalmol Vis Sci 2001; 42(8):1930-1938.
144. Chung KY, Taylor JS, Shum DK et al. Axon routing at the optic chiasm after enzymatic removal of chondroitin sulfate in mouse embryos. Development 2000; 127(12):2673-2683.
145. Moon L, Asher R, Rhodes K et al. Regeneration of CNS axons back to their target following treatment of adult rat brain with chondroitinase ABC. Nat Neurosci 2001; 4:465-466.
146. Emerling DE, Lander AD. Inhibitors and promoters of thalamic neuron adhesion and outgrowth in embryonic neocortex: functional association with chondroitin sulfate. Neuron 1996; 17(6):1089-1100.
147. Sheppard AM, Hamilton SK, Pearlman AL. Changes in the distribution of extracellular matrix components accompany early morphogenetic events of mammalian cortical development. J Neurosci 1991; 11(12):3928-3942.
148. Bicknese AR, Sheppard AM, O'Leary DD et al. Thalamocortical axons extend along a chondroitin sulfate proteoglycan-enriched pathway coincident with the neocortical subplate and distinct from the efferent path. J Neurosci 1994; 14(6):3500-3510.
149. Fukuda T, Kawano H, Ohyama K et al. Immunohistochemical localization of neurocan and L1 in the formation of thalamocortical pathway of developing rats. J Comp Neurol 1997; 382(2):141-152.
150. Ring C, Lemmon V, Halfter W. Two chondroitin sulfate proteoglycans differentially expressed in the developing chick visual system. Dev Biol 1995; 168(1):11-27.
151. McAdams BD, McLoon SC. Expression of chondroitin sulfate and keratan sulfate proteoglycans in the path of growing retinal axons in the developing chick. J Comp Neurol 1995; 352(4):594-606.
152. Streit A, Nolte C, Rasony T et al. Interaction of astrochondrin with extracellular matrix components and its involvement in astrocyte process formation and cerebellar granule cell migration. J Cell Biol 1993; 120(3):799-814.
153. Iijima N, Oohira A, Mori T et al. Core protein of chondroitin sulfate proteoglycan promotes neurite outgrowth from cultured neocortical neurons. J Neurochem 1991; 56(2):706-708.
154. Oohira A, Matsui F, Matsuda M et al. Occurrence of three distinct molecular species of chondroitin sulfate proteoglycan in the developing rat brain. J Biol Chem 1988; 263(21):10240-10246.
155. Sakurai T, Lustig M, Nativ M et al. Induction of neurite outgrowth through contactin and Nr-CAM by extracellular regions of glial receptor tyrosine phosphatase beta. J Cell Biol 1997; 136(4):907-918.
156. Peles E, Nativ M, Campbell PL et al. The carbonic anhydrase domain of receptor tyrosine phosphatase beta is a functional ligand for the axonal cell recognition molecule contactin. Cell 1995; 82(2):251-260.
157. Peles E, Nativ M, Lustig M et al. Identification of a novel contactin-associated transmembrane receptor with multiple domains implicated in protein-protein interactions. Embo J 1997; 16(5):978-988.

158. Streit A, Yuen CT, Loveless RW et al. The Le(x) carbohydrate sequence is recognized by antibody to L5, a functional antigen in early neural development. J Neurochem 1996; 66(2):834-844.
159. Damon DH, D'Amore PA, Wagner JA. Sulfated glycosaminoglycans modify growth factor-induced neurite outgrowth in PC12 cells. J Cell Physiol 1988; 135(2):293-300.
160. Hantaz-Ambroise D, Vigny M, Koenig J. Heparan sulfate proteoglycan and laminin mediate two different types of neurite outgrowth. J Neurosci 1987; 7(8):2293-2304.
161. Fernaud-Espinosa I, Nieto-Sampedro M, Bovolenta P. Differential effects of glycosaminoglycans on neurite outgrowth from hippocampal and thalamic neurones. J Cell Sci 1994; 107(Pt 6):1437-1448.
162. Lafont F, Prochiantz A, Valenza C et al. Defined glycosaminoglycan motifs have opposite effects on neuronal polarity in vitro. Dev Biol 1994; 165(2):453-468.
163. Lafont F, Rouget M, Triller A et al. In vitro control of neuronal polarity by glycosaminoglycans. Development 1992; 114(1):17-29.
164. Nadanaka S, Clement A, Masayama K et al. Characteristic hexasaccharide sequences in octasaccharides derived from shark cartilage chondroitin sulfate D with a neurite outgrowth promoting activity. J Biol Chem 1998; 273(6):3296-3307.
165. Clement AM, Sugahara K, Faissner A. Chondroitin sulfate E promotes neurite outgrowth of rat embryonic day 18 hippocampal neurons. Neurosci Lett 1999; 269(3):125-128.
166. Faissner A, Kruse J. J1/tenascin is a repulsive substrate for central nervous system neurons. Neuron 1990; 5(5):627-637.
167. Gotz B, Scholze A, Clement A et al. Tenascin-C contains distinct adhesive, anti-adhesive, and neurite outgrowth promoting sites for neurons. J Cell Biol 1996; 132(4):681-699.
168. Lochter A, Vaughan L, Kaplony A et al. J1/tenascin in substrate-bound and soluble form displays contrary effects on neurite outgrowth. J Cell Biol 1991; 113(5):1159-1171.
169. Snow DM, Letourneau PC. Neurite outgrowth on a step gradient of chondroitin sulfate proteoglycan (CS-PG). J Neurobiol 1992; 23(3):322-336.
170. Walsh FS, Doherty P. Neural cell adhesion molecules of the immunoglobulin superfamily: role in axon growth and guidance. Annu Rev Cell Dev Biol 1997; 13:425-456.
171. Adamsky K, Schilling J, Garwood J et al. Glial tumor cell adhesion is mediated by binding of the FNIII domain of receptor protein tyrosine phosphatase beta (RPTPbeta) to tenascin C. Oncogene 2001; 20(5):609-618.
172. Kawachi H, Fujikawa A, Maeda N et al. Identification of GIT1/Cat-1 as a substrate molecule of protein tyrosine phosphatase zeta /beta by the yeast substrate-trapping system. Proc Natl Acad Sci USA 2001; 98(12):6593-6598.
173. Kawachi H, Tamura H, Watakabe I et al. Protein tyrosine phosphatase zeta/RPTPbeta interacts with PSD-95/SAP90 family. Brain Res Mol Brain Res 1999; 72(1):47-54. bin/cas/tree/store/bresm/cas_sub/browse/browse.cgi?year=1999&volume=1972&issue=1991&aid=72194.
174. Ranscht B. Sequence of contactin, a 130-kD glycoprotein concentrated in areas of interneuronal contact, defines a new member of the immunoglobulin supergene family in the nervous system. J Cell Biol 1988; 107(4):1561-1573.
175. Gennarini G, Cibelli G, Rougon G et al. The mouse neuronal cell surface protein F3: A phosphatidylinositol-anchored member of the immunoglobulin superfamily related to chicken contactin. J Cell Biol 1989; 109(2):775-788.
176. Gennarini G, Durbec P, Boned A et al. Transfected F3/F11 neuronal cell surface protein mediates intercellular adhesion and promotes neurite outgrowth. Neuron 1991; 6(4):595-606.
177. Milev P, Maurel P, Haring M et al. TAG-1/axonin-1 is a high-affinity ligand of neurocan, phosphacan/protein-tyrosine phosphatase-zeta/beta, and N-CAM. J Biol Chem 1996; 271(26):15716-15723.
178. Roberts C, Platt N, Streit A et al. The L5 epitope: An early marker for neural induction in the chick embryo and its involvement in inductive interactions. Development 1991; 112(4):959-970.
179. Streit A, Faissner A, Gehrig B et al. Isolation and biochemical characterization of a neural proteoglycan expressing the L5 carbohydrate epitope. J Neurochem 1990; 55(5):1494-1506.
180. Allendoerfer KL, Magnani JL, Patterson PH. FORSE-1, an antibody that labels regionally restricted subpopulations of progenitor cells in the embryonic central nervous system, recognizes the Le(x) carbohydrate on a proteoglycan and two glycolipid antigens. Mol Cell Neurosci 1995; 6(4):381-395.

181. Milev P, Meyer-Puttlitz B, Margolis RK et al. Complex-type asparagine-linked oligosaccharides on phosphacan and protein-tyrosine phosphatase-zeta/beta mediate their binding to neural cell adhesion molecules and tenascin. J Biol Chem 1995; 270(42):24650-24653.
182. Zeng L, D'Alessandri L, Kalousek MB et al. Protein tyrosine phosphatase alpha (PTPalpha) and contactin form a novel neuronal receptor complex linked to the intracellular tyrosine kinase fyn. J Cell Biol 1999; 147(4):707-714.
183. Rios JC, Melendez-Vasquez CV, Einheber S et al. Contactin-associated protein (Caspr) and contactin form a complex that is targeted to the paranodal junctions during myelination. J Neurosci 2000; 20(22):8354-8364.
184. Fujikawa A, Watanabe E, Sakaguchi G et al. Dopaminergic dysfunction in the mice lacking the receptor tyrosine phosphatase zeta/RPTP-beta gene. Soc Neurosci Abstr 2001; 31:539.514.
185. Ratcliffe CF, Qu Y, McCormick KA et al. A sodium channel signaling complex: modulation by associated receptor protein tyrosine phosphatase beta. Nat Neurosci 2000; 3(5):437-444.
186. Milev P, Chiba A, Haring M et al. High affinity binding and overlapping localization of neurocan and phosphacan/protein-tyrosine phosphatase-zeta/beta with tenascin-R, amphoterin, and the heparin-binding growth-associated molecule. J Biol Chem 1998; 273(12):6998-7005.
187. Joester A, Faissner A. Evidence for combinatorial variability of tenascin-C isoforms and developmental regulation in the mouse central nervous system. J Biol Chem 1999; 274(24):17144-17151.
188. Joester A, Faissner A. The structure and function of tenascins in the nervous system. Matrix Biol 2001; 20(1):13-22.
189. Milev P, Fischer D, Haring M et al. The fibrinogen-like globe of tenascin-C mediates its interactions with neurocan and phosphacan/protein-tyrosine phosphatase-zeta/beta. J Biol Chem 1997; 272(24):15501-15509.
190. Meng K, Rodriguez-Pena A, Dimitrov T et al. Pleiotrophin signals increased tyrosine phosphorylation of beta—catenin through inactivation of the intrinsic catalytic activity of the receptor-type protein tyrosine phosphatase beta /zeta [In Process Citation]. Proc Natl Acad Sci USA 2000; 97(6):2603-2608.
191. Lemons ML, Howland DR, Anderson DK. Chondroitin sulfate proteoglycan immunoreactivity increases following spinal cord injury and transplantation. Exp Neurol 1999; 160(1):51-65.
192. Li J, Tullai JW, Yu WH et al. Regulated expression during development and following sciatic nerve injury of mRNAs encoding the receptor tyrosine phosphatase HPTPzeta/RPTPbeta. Brain Res Mol Brain Res 1998; 60(1):77-88.
193. Wu YP, Siao CJ, Lu W et al. The tissue plasminogen activator (tPA)/plasmin extracellular proteolytic system regulates seizure-induced hippocampal mossy fiber outgrowth through a proteoglycan substrate. J Cell Biol 2000; 148(6):1295-1304.
194. Kurazono S, Okamoto M, Sakiyama J et al. Expression of brain specific chondroitin sulfate proteoglycans, neurocan and phosphacan, in the developing and adult hippocampus of Ihara's epileptic rats. Brain Res 2001; 898(1):36-48.
195. Laywell ED, Steindler DA. Boundaries and wounds, glia and glycoconjugates. Cellular and molecular analyses of developmental partitions and adult brain lesions. Ann NY Acad Sci 1991; 633:122-141.
196. Perry VH, Brown MC. Macrophages and nerve regeneration. Curr Opin Neurobiol 1992; 2(5):679-682.
197. Hu S, Martella A, Anderson WR et al. Role of cytokines in lipopolysaccharide-induced functional and structural abnormalities of astrocytes. Glia 1994; 10(3):227-234.
198. Campbell IL. Cytokine-mediated inflammation and signaling in the intact central nervous system. Prog Brain Res 2001; 132:481-498.
199. Merrill JE, Benveniste EN. Cytokines in inflammatory brain lesions: helpful and harmful. Trends Neurosci 1996; 19(8):331-338.
200. Schonherr E, Jarvelainen HT, Sandell LJ et al. Effects of platelet-derived growth factor and transforming growth factor-beta 1 on the synthesis of a large versican-like chondroitin sulfate proteoglycan by arterial smooth muscle cells. J Biol Chem 1991; 266(26):17640-17647.
201. Redini F, Daireaux M, Mauviel A et al. Characterization of proteoglycans synthesized by rabbit articular chondrocytes in response to transforming growth factor-beta (TGF-beta). Biochim Biophys Acta 1991; 1093(2-3):196-206.

202. Romaris M, Heredia A, Molist A et al. Differential effect of transforming growth factor beta on proteoglycan synthesis in human embryonic lung fibroblasts. Biochim Biophys Acta 1991; 1093(2-3):229-233.
203. Benton HP, Tyler JA. Inhibition of cartilage proteoglycan synthesis by interleukin I. Biochem Biophys Res Commun 1988; 154(1):421-428.
204. Jung M, Kramer E, Grzenkowski M et al. Lines of murine oligodendroglial precursor cells immortalized by an activated neu tyrosine kinase show distinct degrees of interaction with axons in vitro and in vivo. Eur J Neurosci 1995; 7(6):1245-1265.
205. Fok-Seang J, Mathews GA, ffrench-Constant C et al. Migration of oligodendrocyte precursors on astrocytes and meningeal cells. Dev Biol 1995; 171(1):1-15.
206. Flanders KC, Ren RF, Lippa CF. Transforming growth factor-betas in neurodegenerative disease. Prog Neurobiol 1998; 54(1):71-85.
207. Flanders KC, Ludecke G, Engels S et al. Localization and actions of transforming growth factor-beta s in the embryonic nervous system. Development 1991; 113(1):183-191.
208. Krieglstein K, Suter-Crazzolara C, Fischer WH et al. TGF-beta superfamily members promote survival of midbrain dopaminergic neurons and protect them against MPP+ toxicity. Embo J 1995; 14(4):736-742.
209. Krieglstein K, Rufer M, Suter-Crazzolara C et al. Neural functions of the transforming growth factors beta. Int J Dev Neurosci 1995; 13(3-4):301-315.
210. Unsicker K, Meier C, Krieglstein K et al. Expression, localization, and function of transforming growth factor-beta s in embryonic chick spinal cord, hindbrain, and dorsal root ganglia. J Neurobiol 1996; 29(2):262-276.
211. Rapraeger A. Transforming growth factor (type beta) promotes the addition of chondroitin sulfate chains to the cell surface proteoglycan (syndecan) of mouse mammary epithelia. J Cell Biol 1989; 109(5):2509-2518.
212. Lindholm D, Castren E, Kiefer R et al. Transforming growth factor-beta 1 in the rat brain: Increase after injury and inhibition of astrocyte proliferation. J Cell Biol 1992; 117(2):395-400.
213. Logan A, Berry M, Gonzalez AM et al. Effects of transforming growth factor beta 1 on scar production in the injured central nervous system of the rat. Eur J Neurosci 1994; 6(3):355-363.
214. Bronner-Fraser M. Neural crest cell formation and migration in the developing embryo. Faseb J 1994; 8(10):699-706.
215. Perris R, Johansson S. Inhibition of neural crest cell migration by aggregating chondroitin sulfate proteoglycans is mediated by their hyaluronan-binding region. Dev Biol 1990; 137(1):1-12.
216. Perris R, Perissinotto D, Pettway Z et al. Inhibitory effects of PG-H/aggrecan and PG-M/versican on avian neural crest cell migration. Faseb J 1996; 10(2):293-301.
217. Small RK, Riddle P, Noble M. Evidence for migration of oligodendrocyte—type-2 astrocyte progenitor cells into the developing rat optic nerve. Nature 1987; 328(6126):155-157.
218. Pringle NP, Richardson WD. A singularity of PDGF alpha-receptor expression in the dorsoventral axis of the neural tube may define the origin of the oligodendrocyte lineage. Development 1993; 117(2):525-533.
219. Wolswijk G, Riddle PN, Noble M. Coexistence of perinatal and adult forms of a glial progenitor cell during development of the rat optic nerve. Development 1990; 109(3):691-698.
220. Hardy RJ, Friedrich VL Jr. Oligodendrocyte progenitors are generated throughout the embryonic mouse brain, but differentiate in restricted foci. Development 1996; 122(7):2059-2069.
221. Milner R, Edwards G, Streuli C et al. A role in migration for the alpha V beta 1 integrin expressed on oligodendrocyte precursors. J Neurosci 1996; 16(22):7240-7252.
222. Wang C, Rougon G, Kiss JZ. Requirement of polysialic acid for the migration of the O-2A glial progenitor cell from neurohypophyseal explants. J Neurosci 1994; 14(7):4446-4457.
223. Wang C, Pralong WF, Schulz MF et al. Functional N-methyl-D-aspartate receptors in O-2A glial precursor cells: A critical role in regulating polysialic acid-neural cell adhesion molecule expression and cell migration. J Cell Biol 1996; 135(6 Pt 1):1565-1581.
224. Milner R, Anderson HJ, Rippon RF et al. Contrasting effects of mitogenic growth factors on oligodendrocyte precursor cell migration. Glia 1997; 19(1):85-90.
225. McKinnon RD, Piras G, Ida JA Jr et al. A role for TGF-beta in oligodendrocyte differentiation. J Cell Biol 1993; 121(6):1397-1407.

226. Frost E, Kiernan BW, Faissner A et al. Regulation of oligodendrocyte precursor migration by extracellular matrix: evidence for substrate-specific inhibition of migration by tenascin-C. Dev Neurosci 1996; 18(4):266-273.
227. Luckenbill-Edds L. Laminin and the mechanism of neuronal outgrowth. Brain Res Brain Res Rev 1997; 23(1-2):1-27.
228. Lee EC, Lotz MM, Steele GD Jr et al. The integrin alpha 6 beta 4 is a laminin receptor. J Cell Biol 1992; 117(3):671-678.
229. Delwel GO, Hogervorst F, Kuikman I et al. Expression and function of the cytoplasmic variants of the integrin alpha 6 subunit in transfected K562 cells. Activation-dependent adhesion and interaction with isoforms of laminin. J Biol Chem 1993; 268(34):25865-25875.
230. Muller U, Bossy B, Venstrom K et al. Integrin alpha 8 beta 1 promotes attachment, cell spreading, and neurite outgrowth on fibronectin. Mol Biol Cell 1995; 6(4):433-448.
231. Varnum-Finney B, Venstrom K, Muller U et al. The integrin receptor alpha 8 beta 1 mediates interactions of embryonic chick motor and sensory neurons with tenascin-C. Neuron 1995; 14(6):1213-1222.
232. Yokosaki Y, Palmer EL, Prieto AL et al. The integrin alpha 9 beta 1 mediates cell attachment to a non-RGD site in the third fibronectin type III repeat of tenascin. J Biol Chem 1994; 269(43):26691-26696.
233. Bauvois B, Van Weyenbergh J, Rouillard D et al. TGF-beta 1-stimulated adhesion of human mononuclear phagocytes to fibronectin and laminin is abolished by IFN-gamma: Dependence on alpha 5 beta 1 and beta 2 integrins. Exp Cell Res 1996; 222(1):209-217.
234. Kumar NM, Sigurdson SL, Sheppard D et al. Differential modulation of integrin receptors and extracellular matrix laminin by transforming growth factor-beta 1 in rat alveolar epithelial cells. Exp Cell Res 1995; 221(2):385-394.
235. Paulus W, Sage EH, Jellinger K et al. Type VIII collagen in the normal and diseased human brain. Acta Histochem Suppl 1992; 42:195-199.
236. Iida J, Skubitz AP, Furcht LT et al. Coordinate role for cell surface chondroitin sulfate proteoglycan and alpha 4 beta 1 integrin in mediating melanoma cell adhesion to fibronectin. J Cell Biol 1992; 118(2):431-444.
237. Milner R, Ffrench-Constant C. A developmental analysis of oligodendroglial integrins in primary cells: changes in alpha v-associated beta subunits during differentiation. Development 1994; 120(12):3497-3506.
238. Palecek SP, Loftus JC, Ginsberg MH et al. Integrin-ligand binding properties govern cell migration speed through cell-substratum adhesiveness. Nature 1997; 385(6616):537-540.
239. Revest JM, Faivre-Sarrailh C, Maeda N et al. The interaction between F3 immunoglobulin domains and protein tyrosine phosphatases zeta/beta triggers bidirectional signalling between neurons and glial cells. Eur J Neurosci 1999; 11(4):1134-1147.
240. Snow DM, Brown EM, Letourneau PC. Growth cone behavior in the presence of soluble chondroitin sulfate proteoglycan (CSPG), compared to behavior on CSPG bound to laminin or fibronectin. Int J Dev Neurosci 1996; 14(3):331-349.
241. Dobbertin A, Rhodes KE, Garwood J et al. Regulation of RPTPbeta/phosphacan expression and glycosaminoglycan epitopes in injured brain and cytokine-treated glia. Mol Cell Neurosci 2003; 24(4):951-971.

Chapter 4

Regeneration Failure in the CNS:

Cellular and Molecular Mechanisms

Anne D. Zurn and Christine E. Bandtlow

Introduction

Traumatic injuries to the brain and spinal cord cause severe and irreversible disabilities due to the inability of the CNS, in contrast to the peripheral nervous system, to regenerate injured fibers. CNS axons initially start to sprout, but they fail to regenerate over long distances and cannot contact appropriate target cells, resulting in a permanent loss of function. This lack of regeneration appears to be due to both the intrinsic inability of central fibers to grow, and an inhospitable environment. The present review focuses on the cellular and molecular aspects of intrinsic differences in growth potential and of axonal growth inhibition, in particular on the recent advances in the characterization of growth-inhibitory molecules and their mode of action. Additional details may be found in other recent reviews.[1-6]

Intrinsic Neuronal Properties

Neuronal Survival Following Injury

Injury or trauma of the nervous system can vary considerably in extent and severity from minor contusions or crushes to open injuries associated with partial or even complete disruption of the nerves. Any type of axonal injury induces an array of dramatic molecular and cellular responses at the level of the cell bodies and at the site of the injury that have a direct impact on the potential for successful regeneration. One of the most immediate changes is the destruction of tissue at the site of injury, which can vary greatly depending on the type of injury. Damage to the axons may consist in the complete transection of the nerve fibres, followed in some cases by the death of the neuronal cell body. However, the most important requirement for regeneration of the injured axons to occur is the survival of the nerve cell body and its capacity to induce gene expression leading to axonal regrowth and reinnervation of the original targets (for a review see ref. 7). The closer the axonal insult occurs to the cell body, the less likely the neuron can be maintained in a healthy state. Injuries that sever axons relatively distal to the cell body produce a severe reduction in neuronal size (atrophy), but the neurons survive the lesion.[8] They undergo morphological and molecular changes that involve major rearrangements of the endoplasmic reticulum and Golgi apparatus, a process referred to as *chromatolysis*. Chromatolysis is often accompanied by an increase in the expression of a set of genes associated with axon growth. However, even if the neuron ultimately survives the injury,

Brain Repair, edited by Mathias Bähr. ©2006 Eurekah.com and Kluwer Academic / Plenum Publishers.

its nerve terminals and the entire segment distal to the lesion will degenerate, a process called *Wallerian degeneration.*[9]

Developmental Loss of Regenerative Ability

A number of studies have demonstrated that the intrinsic growth ability of mammalian CNS neurons is developmentally regulated. While embryonic CNS axons can regenerate quite readily, they lose this capacity with age.[10-13] For instance, successful regeneration and sprouting of nerve terminals following lesion gives rise to functional recovery in the CNS of young, but not adult mammals.[14-16] In addition, defined spinal cord lesions in chicken embryos or newborn mammals lead to extensive regeneration of fibres, whereas no such growth is observed in the adult. Adult neurons seem to loose their ability to grow during postnatal development and their regenerative response becomes abortive. The loss in capacity to grow appears to coincide with the maturation of CNS glial cells, both astrocytes and oligodendrocytes, and with the onset of myelination. Delaying spinal cord myelination, for instance, delays the end of the permissive period in chick embryos.[17] However, myelin and associated molecules are not the only candidates that account for the developmental loss of regenerative capacity. Indeed, in an in vitro model system devoid of myelin, axotomized Purkinje cells retain their age-dependent decrease in ability to regenerate, despite the absence of myelin.[18] In the retina, amacrine cells appear to be responsible for the loss of intrinsic growth capacity of retinal ganglion neurons.[19] Neurons can thus signal other neurons to lose their axonal growth ability.

Developmental loss of regenerative ability is also correlated with an intrinsic decrease in the neuronal expression of growth-associated proteins such as GAP-43 and SCG10[20] and a decrease in endogenous cAMP levels.[21] Interestingly, there may also be a developmental increase in the capacity to respond to extrinsic inhibitory cues since increases in intracellular calcium concentrations in response to the myelin protein Nogo are small and transient in cultured embryonic chick retinal ganglion neurons compared to mature neurons.[22] Similarly, a difference in receptor levels or responsiveness to trophic or guidance signals could explain these differences. Exogenously elevating TrkB levels, however, fails to revert postnatal axonal growth rates to embryonic levels.[23] Down-regulation of the expression of the anti-apoptotic protein Bcl-2 was also proposed to be responsible for the decreased axonal growth rates of retinal ganglion cells.[24] However, Bcl-2 overexpression in these neurons in vitro and in vivo neither promotes axon growth nor enhances growth in response to neurotrophic factors.[23] In summary, both the intrinsic neuronal growth state (developmental expression of genes involved in axonal growth and/or in limiting axonal growth), as well as the neuronal environment (oligodendrocytes and myelin or astrocytes) appears to account for the age-related switch from a successful to an abortive regeneration.[7]

Expression of Growth-Associated Molecules

Upon injury, adult CNS axons undergo a spontaneous, but short-lived and ultimately abortive attempt at repair called regenerative sprouting. These sprouts extend for only a few μm before growth is aborted and the new sprouts are gradually resorbed. The reasons for this unsuccessful regenerative process remain largely unknown, but they indicate that injured CNS neurons have not necessarily lost their intrinsic property to grow, but that they fail to initiate and/or maintain a specific growth programme required for axonal elongation. Important regulators of axonal growth that are expressed during nervous system development (GAP-43, cytoskeletal proteins, transcription factors, neuropeptides, integrins, growth factors and neurotrophin receptors) are re-expressed upon nerve injury (for review see ref. 25) However, in contrast to the PNS, this event is often only transient or completely absent in the CNS. These observations suggest that prolonged expression of growth-associated genes may be required for successful regeneration.

Experimental sensory fiber lesions have brought additional evidence that expression of growth-associated molecules may be a prerequisite for axonal regeneration. Primary sensory neurons with their cell bodies in the dorsal root ganglia (DRG) possess two axonal branches, a peripheral axon that can regenerate, and a central axon, located in the spinal cord that cannot regenerate upon injury. Interestingly, lesion of the central ascending sensory axons, in contrast to lesion of the peripheral branches, does not lead to the up-regulation of GAP-43, c-Jun, and JAK in the DRG neurons.[26-29] Overexpression of GAP-43 alone, however, allows only sprouting and not regeneration of CNS neurons.[30-32] Only the combined expression of GAP-43 with another growth cone protein, CAP-23, permits regrowth of sensory axons from the spinal cord lesion site into a peripheral nerve graft.[33] In addition to a reactivation of growth associated genes expressed during development, peripheral axotomy leads to the up-regulation of regeneration-associated proteins (RAGs) including SPRR1A, a small proline-rich protein that associates with F-actin, and fibroblast growth factor-inducible-14.[34,35] These two proteins are not expressed during development, but promote axon growth when overexpressed in neurons and PC12 cells, respectively.[34,35] RAGs, in addition to growth-associated proteins, may therefore also be required to trigger successful nerve regeneration.

Up-regulation of regeneration-associated genes upon injury in the CNS is influenced by the distance between the cell body and the site of the lesion.[36,37] This is illustrated by the observation that corticospinal cell bodies respond to intracortical, i.e., proximal, axotomy by up-regulating a range of growth-associated molecules, while they fail to do so in response to spinal axotomy.[38] This may in part explain the lack of regeneration of corticospinal axons into peripheral nerve grafts placed in the spinal cord.[39]

Conditioning Lesion

As described in the previous chapter, peripheral, but not central processes of primary sensory neurons can regenerate after injury. This discrepancy is due both to the non-permissive character of the CNS environment, and the failure of central, in contrast to peripheral sensory processes, to induce the expression of growth-associated molecules upon CNS injury.[28,40,41] Interestingly, central sensory processes show improved regeneration through a peripheral nerve graft if their peripheral axons are previously cut (conditioning lesion), i.e., if their axonal growth program is activated. In addition, a conditioning peripheral nerve lesion prior to dorsal column injury results in some growth of ascending sensory axons into the spinal cord above the lesion.[42] It is interesting to note that regenerating ascending sensory axons appear to grow across areas with strong expression of tenascin-C and CSPG, but avoid areas containing scar-associated fibroblasts expressing Semaphorin3A, as well as areas with CNS myelin.[43]

Intrisic Differences in Growth Potential

Different subtypes of CNS neurons appear to have distinct growth potentials. For instance, peripheral nerve grafts implanted in the CNS promote regeneration of inferior olivary axons from deep cerebellar nuclei, but not from Purkinje cells.[44] In addition, distinct classes of neurons react differently to the same non-permissive CNS environment since biotin-dextran labeled ascending sensory axons, but not the subgroup labeled with CGRP, can regenerate after spinal cord crush in mice.[45]

The CNS Environment Is Non-Permissive for Axonal Growth

Cellular Response to Injury

Failure of regeneration in the CNS is also due to an environment unfavorable for axonal growth. For instance, peripheral neurons that grow well in peripheral tissues often show only limited growth when transplanted into the brain or spinal cord of adult vertebrates. On the other hand, some injured CNS neurons can grow through peripheral nerve grafts transplanted as bridges into the brain or spinal cord of adult rats, but they cease to grow when they re-enter the CNS tissue.[39,46,47] These experiments demonstrate that (1) adult CNS tissue is unfavorable for axonal growth and (2) some adult CNS neurons can extend long processes when provided with a permissive environment.

Traumatic spinal cord injury results in many cellular and molecular changes that extend over months and years after injury. During the acute phase of injury (first few days), there is disruption of the blood-brain barrier, ischemia, edema, leading to neuronal death, degeneration of axons, astrocyte proliferation, and infiltration of immune cells and meningeal fibroblasts. The subsequent secondary injury extends over a time course of several weeks and results in reactive gliosis, formation of a glial scar, strong inflammation, formation of a cystic cavity, and apoptotic cell death (for review see ref. 48).

The Glial Scar

CNS injury results in a rapid glial response around the injury site, leading to *gliosis* and formation of a *glial scar*. Although the functional role of glial scarring is not completely understood, it has been suggested that it protects neuronal function following injury.[49] Astrogliosis may for instance contribute to the clearance of glutamate, restoration and maintainance of ion concentrations, and provide anti-inflammatory cytokines as well as neurotrophic support.[50-52] However, glial scarring can also be detrimental since it may impede axonal regeneration and thus interfere with neuronal repair. Reactive gliosis involves astrocytes, oligodendrocyte precursor cells, microglia/macrophages, leptomeningeal fibroblasts, and eventually forms a scar that is a major impediment to axonal growth. Reactive astrocytes are characterized by a number of cellular changes, including hypertrophy, generation of long and tightly packed cytoplasmic processes, increased production of the intermediate filament proteins GFAP and vimentin, and up-regulation of a number of cell surface and extracellular matrix (ECM) molecules, in particular chondroitin sulfate proteoglycans (CSPGs) and tenascin (for review see refs. 6,53). The glial scar thus forms both a mechanical and molecular barrier to axon growth. Upon injury, microglial cells also rapidly divide, transform into phagocytic cells that remove neuronal and myelin debris, and release inflammatory cytokines, glutamate, and free radicals (reviewed in refs. 54,55). In addition, another type of glial cell is recruited to the site of injury and participates in the formation of the glial scar, the oligodendrocyte precursor cell (OPC) believed to be an immature cell of the oligodendrocyte lineage.[56] OPCs are present throughout the CNS and respond to injury by dividing, hypertrophy, and increase in chondroitin sulfate proteoglycan NG2 expression.[57] Furthermore, if the injury disrupts the meninges and blood vessels, meningeal fibroblasts invade the lesion area.[58,59] They contribute to the inhibitory nature of the glial scar by synthesizing semaphorin class III chemorepellent molecules, tenascin-C, and NG2 (reviewed in ref. 43).

Wallerian Degeneration: Oligodendrocytes and Myelin Debris

Following CNS lesion, the entire distal segment of the nerve undergoes Wallerian degeneration, with desintegration of the distal axonal segment, degradation of myelin sheaths and the axon cytoskeleton, and the subsequent apoptotic death of oligodendrocytes around the lesion site. This process occurs within the first few days after lesion and is probably the result of protease, such as calpain, activation.[60] While axonal injuries in the PNS trigger the rapid recruitment of macrophages from the peripheral blood to remove axonal and myelin debris,[61] the number of circulating macrophages that invade the CNS lesion site is considerably smaller.[62] Instead, resident CNS microglia are activated, migrate into the lesion site, and progressively acquire properties of fully competent macrophages to finally phagocytose axon and myelin debris. CNS injury therefore results in the long-term persistence of myelin debris at the lesion site.[63] Cell culture experiments have shown that both CNS myelin and myelin producing oligodendrocytes can inhibit axonal growth indicating that myelin debris may participate in the failure of nerve fibre regrowth after lesion.[64,65] Consistent with this view are the observations that CNS injuries in very young mammals, i.e., before the onset of fibre myelination, allow axonal regeneration and often lead to full functional restoration. Thus the onset of myelination and the presence of myelin-associated inhibitory proteins are considered to be involved in the developmental switch that marks the end of the growth permissive period.

Injury-Induced Inflammation

Traumatic spinal cord injury leads to hemorrhage, glial activation, ischemia, edema, and results in inflammation, cyst formation, and secondary loss of neurons and glia.[66] These secondary events may be mediated by microglia/macrophage activation and the subsequent recruitment of cells of the immune system, leading to the production of free radicals and the release of pro-inflammatory cytokines such as Il-1, Il-6, TNFα, and TGFβ.[67] Inflammatory mediators do not appear to be sufficient to cause oligodendrocyte loss in white matter tracts that are spared after contusion lesion of the spinal cord, although many macrophages and neutrophils are present.[68] However, oligodendrocyte death has been observed primarily associated with degenerating fiber tracts, i.e., where axons are also damaged. This death is apoptotic and depends on the activation of p75 and on the expression of Bax.[69,70] Astrocytes may also be severely damaged and lost following CNS injury prior to the development of reactive gliosis.[71] The most common early response to injury is astrocyte swelling.[72] Astrocytic edema develops as early as 3h following trauma, and persists for as long as 3 days.[73] Cellular brain edema results from an entry of water into the intracellular compartment due to a dysregulation of ion homeostasis, leading to impaired neurotransmitter uptake and release of excitotoxic amino acids such as glutamate by reactive astrocytes.[72] This contributes to neuronal and oligodendroglial death as well as to a transient loss of astrocytes.[71] However, inflammation, if properly controlled, may also have beneficial effects on secondary neuronal and glial cell loss, axonal regrowth, and remyelination, thus promoting CNS repair.[74,75]

Several inflammatory cytokines have been suggested to be involved in mediating glial scar formation and thus impede regeneration. For instance, interleukin-1, TGFα, TGFβ, TNFα, and IFN-γ all have been shown to be increased in scar tissue after brain lesion, to promote astrocyte proliferation in vitro, and/or to augment gliosis in vivo.[76-80] Moreover, a poor or slow appearance of inflammatory mediators in the CNS compared to the PNS may also contribute to the inefficient removal of myelin debris and thus add to the failure of regeneration in the brain.[61] Interestingly, the expression of tenascin and of neurocan, a CNS-specific chondroitin sulfate proteoglycan with growth-inhibitory properties, is increased in highly purified rat astrocytes treated with TGFβ.[81,82] However, the potential regulation of these and other myelin-associated growth-inhibitors by inflammatory mediators remains to be evaluated further.

Species-Specific Responses to CNS Injury

Among other mammals examined to date, mice exhibit a unique pathological response to spinal cord injury. This is probably due to a species-specific neuroinflammatory response that contributes to the formation of distinct tissue environments at the lesion site.[83] For instance, in contrast to rats and humans in which spinal cord lesion leads to tissue loss and the formation of large cystic cavities, the lesion site in mice is filled with a dense connective tissue matrix made up of macrophages and fibroblasts, leading to revascularization of the tissue over time. In addition, the genetic background also appears to be important. For example, the lesion area and cavitation following spinal cord injury are larger in FVB/N strains that are sensitive to kainate toxicity than in the kainate-resistant strains C57Bl/6 and Balb/c.[84,85] Interestingly, C57Bl/6 mice are deficient in peripheral axonal regeneration after sciatic nerve lesion compared to Balb/c and other strains of mice.

Molecular Components of Inhibition

Myelin-Associated Inhibitors

Growth inhibitors present in myelin seem to be involved particularly during the early phase that follows CNS injury, before the glial scar forms.[86] Several myelin-associated proteins have been identified that are thought to contribute to a hostile environment for regenerating nerve fibers. These proteins include Nogo/RTN-4, myelin-associated glycoprotein (MAG), oligodendrocyte-myelin glycoprotein (OMgp), tenascin-R, and chondroitin-sulfate proteoglycans (CSPGs). The relative contribution of these proteins to the inhibitory activity of myelin has been a subject of intense study. The main in vivo evidence for an inhibitory effect of myelin on CNS regeneration comes from experiments using either antibodies that neutralize myelin inhibitory molecules,[87-90] or a peptide that specifically acts as a NGR antagonist,[91] Both approaches have improved the regeneration of corticospinal tract axons and have shown some functional recovery after dorsal hemisection of the adult spinal cord (for recent reviews see refs. 3,5,92).

MAG

MAG belongs to the family of immunoglobulin-type lectins (I-type lectins) called siglecs, a sub-family of the immunoglobulin superfamily of cell adhesion molecules.[93] While all other siglecs play important roles in the immune system, MAG/siglec-4 is expressed by oligodendrocytes and Schwann cells of the central and peripheral nervous system, respectively. In the CNS, MAG is found exclusively in the periaxonal myelin membrane, while in the PNS, it is also found in the outermost membrane of the myelin sheath.[94] MAG was shown in a variety of in vitro studies to act as a bifunctional protein that either promotes or inhibits neurite outgrowth, depending on the age and type of neuron. For example, neurite outgrowth in response to MAG is enhanced in early postnatal dorsal root ganglion (DRG),[95] and embryonic spinal cord neurons.[96] In contrast, growth is inhibited in older DRG neurons and in neurons such as cerebellar granule cells, retinal, spinal and hippocampal neurons, i.e., cells that are myelinated in vivo.[97] Changes in endogenous cAMP levels are thought to account for the developmental switch in the neuronal response to MAG in vitro.[98,99] To what extent MAG contributes to growth failure in vivo is less clear. The presence of MAG in the PNS, where regeneration can occur, appears to question the function of MAG as an important inhibitor in vivo. Several studies, however, have demonstrated that peripheral nerve myelin is cleared rapidly and efficiently by macrophages after injury to the PNS (reviewed in ref. 100). PNS regeneration is restored in Wlds mice that have a slow myelin clearance and regenerate poorly, when they are crossed with MAG-deficient mice.[101] These studies support a role for MAG as an in vivo inhibitor. However, experiments in MAGnull mice suggest that its inhibitory contribution in the CNS may be minor, as the extent of axonal regrowth in lesioned optic nerve and corticospinal tract is not, or only moderately enhanced in MAG-deficient mice.[102,103]

Nogo

Perhaps the most widely studied myelin-associated inhibitor is Nogo. Nogo was independently cloned in different laboratories[104-106] following the publication of a partial protein sequence of a high molecular weight inhibitory myelin protein, bovine NI-220.[107] Three distinct isoforms of Nogo, Nogo A, B, and C, are generated by alternative splicing (Nogo-A/B) and promoter usage (Nogo-C). They share a C-terminus of 188 amino acids (aa), called reticulon-homology domain (RHD). This sequence characterizes Nogo as a member of the reticulon (RTN) family of proteins (Nogo is RTN4). Reticulons are very old in evolutionary terms and occur in all eukaryotes including plants and fungi.[108] With the exception of Nogo/RTN 4, little is known about the possible function of the other three known mammalian paralogues (RTN1, -2, -3). A hallmark is their association with the endoplasmic reticulum (ER) via an ER retention motif suggesting a role in membrane trafficking and/or sorting.[109] Indeed, the majority of Nogo-A is found as an intracellular pool associated with the ER and the Golgi complex.[110] While the physiological relevance of this intracellular distribution is not clear yet, cell membrane localized Nogo-A is best known for inhibition of neurite growth and induction of growth cone collapse. There are at least two domains of Nogo-A which exert inhibitory activity on neurons.[110,111] First, there is a loop-domain of 66 aa, which is spanned between two hydrophobic stretches and termed Nogo-66. It is located within the RHD and is thus common to all three Nogo isoforms. Second, there is a large stretch which is specific for Nogo-A (aa 174-979 for rat Nogo-A) and termed NiG or central inhibitory domain of Nogo-A. Recent work indicates that Nogo-A can adopt at least two different cell membrane topologies in oligodendrocytes with Nogo-66 always being exposed to the cell surface and with NiG facing once the cell surface, and once the cytoplasm.[110]

Oligodendrocytes of the adult CNS, specifically the innermost adaxonal and outermost myelin membranes, are the predominant, but not the only sites of expression of Nogo-A.[112] Interestingly, Nogo-A is also found in neurons, in particular during development.[112-115] The localization of Nogo-A to mature oligodendrocytes fits well with its role as a myelin associated inhibitor of nerve fibre growth, whereas expression of Nogo-A in other tissues, mostly neurons, suggests further functions in addition to nerve growth inhibition.

Several lines of evidence demonstrate the relevance of Nogo to axon regeneration in vivo. Antibodies directed against Nogo-A can promote axon growth and plasticity in the adult brain (for a detailed review see ref. 116). In addition, a Nogo-66 peptide antagonist increases axonal sprouting and functional recovery after spinal cord injury.[91,117] Furthermore, overexpression of Nogo-A in PNS Schwann cells delays regeneration following sciatic nerve lesion.[118] Similar results were obtained with transgenic animals overexpressing Nogo-C, the smallest Nogo isoform, supporting a general inhibitory role for Nogo.[119] Analysis of Nogo knockout mice, however, has complicated the picture. Three different laboratories independently generated Nogo deletion mutants using different strategies. Insertion of a gene trap vector created a Nogo A/B mutant line that disrupted Nogo-A and -B expression, but Nogo-C expression remained unaffected.[120] These mice show substantial corticospinal sprouting proximal to a spinal cord dorsal hemisection, and regenerating corticospinal axons below the lesion. The regenerative effects seem to be restricted to young mice. A second Nogo-A mutant line using conventional gene targeting disrupting the large Nogo-A-specific exon 3.[121] Although these mice lack Nogo-A, a compensatory increase in Nogo-B was detected in this strain, which may explain the significant, but more modest regenerative effects in these mice. Finally, two Nogo knockout strains were produced by Tessier-Lavigne and collaborators: one disrupted in Nogo-A/B without affecting Nogo C, and the other one disrupted in all three Nogo isoforms by deletion of their common C-terminal region.[122] Surprisingly, neither of these strains showed improved axonal regeneration after dorsal spinal cord hemisection. To date, there is no clear explanation for the disparate findings from the various strains of Nogo knockout mice. One may speculate that differences in genetic background or compensatory up-or down-regulation of unidentified inhibitory influences may be responsible. Backcrossing *nogo* deletion mutants to pure genetic background might therefore help to resolve this issue. In fact, a recent report on a *nogo-a* knock-out line in a pure (greater than 99%) genetic background[123] confirmed results on enhanced axonal regeneration as described previously.[121] Clearly, more in-depth analysis is required to resolve these discrepancies.

Oligodendrocyte-Myelin Glycoprotein (OMgp)

The most recently identified myelin-associated protein that inhibits nerve fiber growth in vitro is the oligodendrocyte-myelin glycoprotein (OMgp).[124,125] This GPI-linked protein has been previously described as a relatively minor component of CNS myelin, being expressed by myelinating oligodendrocytes and localized to the glial-axonal interface of myelinated axons.[126,127] The OMgp gene is placed within an intron of the neurofibromatosis type I gene on the opposite strand, but until recently, its precise function has remained unknown. Mouse and human OMgp are structurally very similar. The protein has a series of tandem leucine-rich repeats, like those in a variety of adhesion molecules and receptors, including the Nogo receptor NgR. Like Nogo, OMgp in the mouse brain is not only expressed by white matter oligodendrocytes, but also in diverse groups of neurons[125] such as the pyramidal cells of the hippocampus, the Purkinje cells of the cerebellum, motoneurons in the brainstem and spinal cord. Whether and to what extent neuronally expressed OMgp contributes to axon growth inhibition remains to be determined.

Glial Scar-Associated Inhibitors

Astrocyte-associated inhibitory molecules are possible candidates responsible for growth inhibition in the glial scar. This is supported by the demonstration that astrocytes, depending on culture condition, source, and age, have growth-inhibiting properties in vitro.[128] Similarly, various astrocytic cell lines have been obtained that are either permissive or inhibitory to axonal growth in vitro.[129] ECM components of reactive astrocytes in the glial scar include laminin, fibronectin, matrix metalloproteases and their inhibitors, chondroitin sulfate proteoglycans (CSPGs) and tenascins. Oligodendrocyte precursor cells, oligodendrocytes, and meningeal cells that also make up the glial scar synthesize inhibitory molecules as well, among which CSPGs, the chemorepulsive axon guidance molecules semaphorins, and collagen.[6,130-135] Many of these molecules have been shown to restrain neurite outgrowth of a variety of neurons in vitro and are therefore strong candidates as inhibitors of regenerative responses in the CNS in vivo.

CSPGs

Chondroitin sulfate proteoglycans (CSPGs) appear to be the main class of inhibitory molecules produced by activated astrocytes. Indeed, CSPGs are up-regulated in the glial scar upon injury and axon growth on astrocytic scars is inhibited in the presence of injury-induced proteoglycans.[136,137] In addition, axon growth can be increased on inhibitory astrocytic cell lines treated with chondroitinase ABC[138] or grown in the presence of proteoglycan synthesis inhibitors.[139]

CSPGs are extracellular matrix glycoproteins carrying varying degrees of covalently bound sulfated glycosaminoglycan (GAG) chains (for further details see reviews by refs. 140,141). The CSPGs brevican, neurocan, and versican are a subgroup of proteoglycans called lecticans/hyalectans which share similar globular N-terminal hyaluronan-binding domains and C-terminal selectin-like domains, with a central region which is not conserved among lectican family members.[142] While the N-terminal domain of hyalectans interacts with hyaluronan, the C-terminal globular domains interact with tenascin C and tenascin R, surface molecules such as L1 and N-CAM, integrins, and the ECM proteins fibulins.[141-143] The brain-specific link proteins BralI and BralII probably serve to stabilize hyaluronan/versican and hyaluronan/brevican aggregates, respectively, by binding to both molecules and forming hyaluronan/Bral/hyalectan complexes.[144,145]

Brevican is the most abundant lectican in the adult CNS. It is present in different isoforms, a secreted form primarily produced by neurons and deposited in perineuronal nets where it interacts with hyaluronan and tenascinR,[146] and a GPI-linked form mainly produced by oligodendrocytes.[147,148] Brevican may also be expressed by adult astrocytes[149-151] and has been isolated from brain myelin.[152] Neurocan is primarily expressed by CNS neurons.[153] It undergoes proteolytic processing and is deposited in perineuronal nets where it binds to tenascin C and tenascin R. Versican exists in 4 isoforms resulting from differential splicing, V0, V1, V2, and V3 (reviewed in ref. 141) Versican V2 is exclusively present in the CNS where it appears to be expressed both by oligodendrocyte precursor cells and mature oligodendrocytes.[152,154] Versican V2 has also been isolated from bovine CNS myelin.[152,154] Versican V2 appears to be colocalized with the brain-specific link protein BralI to the CNS white matter, especially at the nodes of Ranvier.[145] While Brevican, Neurocan, and Versican V2 are all CNS-specific, Versican V1 has a wider tissue distribution, being expressed along neural crest pathways and in the adult aorta. The NG2 proteoglycan is a unique transmembrane CSPG that can also undergo proteolytic cleavage to produce a secreted form. NG2 is expressed mainly by oligodendrocyte precursor cells during brain development as well as throughout the adult brain.[155-159] Its possible expression on microglia and subpopulations of activated astrocytes remains to be confirmed. Like versican V1, NG2 is not CNS-specific since it has recently been described to be present on Schwann cells and/or perineurial fibroblasts in peripheral nerve and to be increased at the site of sciatic nerve lesion.[160]

Brevican, neurocan, versican V2, and NG2 have all been described to be potent growth inhibitors in vitro. Furthermore, neurocan, brevican, and NG2 are up-regulated in the glial scar after various types of CNS lesions. Whether Versican V2 expression is modulated upon CNS injury is still controversial since a knife lesion in the cerebral cortex appears to increase versican V2 expression,[161] whereas unilateral transection of the spinal cord leads to a decrease in versican V2 production.[162] CSPGs may also be important growth-inhibitors for sensory neurons since grafts of adult sensory neurons can regenerate axons in degenerating white matter tracts of the spinal cord, but not through a glial scar containing CSPGs.[163] Evidence that CSPGs are key players in preventing CNS regeneration has also recently been brought by the demonstration that extensive regeneration occurs after nigrostriatal axotomy or spinal cord lesion in rats treated with chondroitinase ABC, the enzyme which cleaves the GAG side chains of CSPGs.[164-166] CSPGs may therefore, together with other inhibitory molecules of CNS myelin, contribute to growth inhibition in the brain and spinal cord.

Tenascins

Tenascins are a family of large extracellular matrix glycoproteins including tenascin-C and tenascin-R that are expressed during development, down-regulated upon maturation, and reexpressed after injury (for more details see refs. 167-170) Tenascin-C is present both in the PNS and CNS, while tenascin-R is predominantly expressed in the CNS. Both tenascin-C and R, and the recently described new family member, tenascin-N, are repulsive for neurite growth in vitro.[171,172] Enhanced expression of tenascin C by activated astrocytes after CNS lesion may contribute to failed regeneration through the glial scar. Tenascin-R is expressed by oligodendrocytes and by subpopulations of neurons where it is deposited in perineuronal nets. Upon injury, it accumulates around the lesion site but is down-regulated in motoneurons of the spinal cord.[173]

Semaphorins

Semaphorins are one of the largest families of axon guidance molecules.[174-176] Sema3A, a chemorepellent during nervous system development, is up-regulated in meningeal fibroblasts in the glial scar and is therefore likely to contribute to the inhibitory properties of this tissue.[177,178] Sema3A, but not Sema3C, has recently been demonstrated to inhibit neurite outgrowth of embryonic DRGs in vitro.[179] Sema4D, a membrane-bound semaphorin, has been shown to induce the collapse of growth cones of CNS axons and to be up-regulated upon CNS injury.[180] In contrast to Sem3A, Sema4D is expressed by oligodenrocytes and myelin and is therefore a newly identified myelin-associated growth inhibitor. More recently, an additional member of the semaphorin family, Sema5A, has been found to be expressed by oligodendrocyte lineage cells and to impede axon growth from adult optic nerve explants.[181]

Transmission of Inhibitory Signals

Receptors for Myelin-Associated Inhibitors

A key step in understanding how axons respond to inhibitory influences is the identification of axonal receptors that bind myelin-associated inhibitory molecules. So far, NgR is the only known high affinity neuronal receptor originally identified as a binding molecule for Nogo-66.[182] NgR is a GPI-linked neuronal protein that consists of eight consecutive leucine-rich repeat (LRR) domains followed by a carboxy-terminal LRR. NgR is not only necessary to render neurons sensitive to Nogo-66, but its axonal expression pattern is consistent with NgR acting as a functional Nogo receptor.[115] Unexpectedly, recent studies have identified NgR as a component required also in the signalling of the inhibitory effects of MAG and OMgp.[125,183,184] Thus, the same molecule acts as a high affinity binding receptor for three different myelin-associated ligands. This finding is somewhat surprising since MAG, Nogo- 66, and OMgp are structurally unrelated proteins. Based on the crystal structure of the NgR ligand-binding domain, it was proposed that the highly symmetrical arrangement of conserved aromatic residues on the concave face of the curved ectodomain could provide generic binding sites for multiple ligands.[185,186] Future studies employing site-directed mutagenesis of NgR will be necessary to gain further insight into the promiscuous nature of the NgR binding protein. Independently, however, the NGR homolog NgR2 was presented as another receptor for MAG.[187] Like all NgR family members it is a GPI-linked membrane protein, but unlike NGR it is selective for MAG by binding MAG but not Nogo-66.

Although NgR is essential for binding of the myelin-associated inhibitors, it was predicted early on that it signals through the action of coreceptors, one of which was identified as the transmembrane neurotrophin receptor $p75^{NTR}$.[188,189] Another constituent was revealed to be LINGO-1, a nervous system specific leucine rich repeat (LRR) and Ig-domain containing transmembrane protein with a very short intracellular moiety (38 aa; (190). Thus, while NgR is the interacting subunit of this receptor complex, LINGO-1 and especially $p75^{NTR}$ have been shown to be essential for relaying the inhibitory signal across the membrane. Importantly, the NgR/$p75^{NTR}$ complex interacts with and mediates the effect of Nogo-66, MAG, and OMGp, but not of amino-Nogo or CSPGs.[188,189,191,192] The knowledge about the molecular interaction of the three main inhibitors of myelin with the same NgR/$p75^{NTR}$ receptor complex implies that there is functional redundancy between these inhibitors. This redundancy is likely to be even greater if the three inhibitors have the same or overlapping binding sites on NgR. This may explain previous findings in MAG or Nogo deficient mice, where no or only a small amount of spontaneous long distance axonal regeneration was observed. Possibly the presence of other, unblocked ligands for NgR/$p75^{NTR}$, or of other unidentified receptors, can still prevent most axons from regenerating. Most likely, the picture of myelin inhibitors and their signalling machinery is still far from complete. Recently, TROY/TAJ has been identified as another constituent of a myelin inhibitory receptor complex.[193,194] TROY/TAJ is an orphan TNF receptor family member and $p75^{NTR}$ homologue which apparently can replace the neurotrophin receptor in the NgR/$p75^{NTR}$/LINGO-1 complex and mediate the inhibitory effects of myelin inhibitors. Other than $p75^{NTR}$, TROY/TAJ is broadly expressed in the post-natal and adult CNS which might explain the responsiveness of neurons to myelin inhibitors that do not express $p75^{NTR}$.

Moreover, MAG has also been described to bind to neurons via a sialic acid linkage involving the gangliosides GT1b and GD1a, a subclass of glycosphingolipids containing one or more sialic acid residues. These gangliosides have therefore been proposed to act as functional MAG receptors involved in inhibition.[195-198] Further evidence that gangliosides may be involved in mediating growth inhibition is supported by the finding that clustering of ganglioside GT1b with antibodies in the absence of ligand is sufficient to inhibit neurite outgrowth.[196-198] However, sialic-acid-dependent binding of MAG to neurons is neither necessary nor sufficient for MAG-mediated inhibition of axonal regeneration since binding of MAG to NgR-expressing cells is unaffected by removal of sialic acid residues, and since the interaction between MAG and NgR does not require sialic acid.[183] Nevertheless, MAG binding to gangliosides may potentiate its inhibitory effects by clustering signaling molecules at the intracellular membrane face.

Signalling Pathways of Myelin-Associated Inhibitors

Rho-GTPases and Rho Kinase

Consistent with the redundancy of myelin-associated growth inhibitors is the finding that the effects of all three inhibitors can be overcome by altering certain intracellular signalling molecules. One such molecule is the small GTPase, Rho. Rho family GTPases transduce extracellular signals to the actin cytoskeleton to modulate growth cone motility. The most intensely studied members are Rho, Rac, and Cdc42. Rho GTPases cycle between an active GTP-bound state and an inactive GDP-bound state. This transition is positively or negatively controlled by several GTPase activating proteins (GAPs), by guanine nucleotide exchange factors (GEFs), and by guanine nucleotide dissociation inhibitor (GDI). These modulators maintain Rho GTPases in their inactive GDP-bound state.[199] In general, activation of Cdc42 and Rac is associated with growth cone attraction via promotion of F-actin polymerization in filopodia and lamellipodia. In contrast, activation of RhoA causes growth cone repulsion and collapse, possibly by enhancing retrograde F-actin flow.

Involvement of RhoA in myelin-dependent outgrowth inhibition and growth cone collapse is supported by various findings. For instance, inactivation of RhoA by the clostridial toxin C3 transferase protects CNS neurons from myelin-induced growth cone collapse and outgrowth inhibition in vitro.[200,201] A problem with C3 transferase however, is that it is poorly membrane permeable. C3 could penetrate injured axons soon after spinal cord injury, but this may be limited to a short period after the trauma. Application of C3 in vivo has been described to promote regeneration after optic nerve lesion or spinal cord injury.[200,202] However, the marked improvement in function was seen already two days post-injury, i.e., too early to be accounted for by axonal regeneration. Whether these early effects are the consequence of an early effect of C3 on neuronal survival remains to be confirmed.

Biochemical studies have shown that amino-Nogo-A, Nogo-66, MAG, Versican V2 and presumably other CSPGs directly activate RhoA[191,201,203,204] and concomitantly inactivate Rac1. Although it is reasonable to assume that a specific GEF is recruited by activated RhoA during neurite inhibition, the identity of such a guanine nucleotide exchange factor remains speculative. However, the guanine nucleotide dissociation inhibitor Rho-GDI seems to directly interact with $p75^{NTR}$, and is enhanced by binding of myelin-associated inhibitors to NgR.[205] This suggests a model whereby Rho-GDI is sequestered by $p75^{NTR}$, leading to Rho activation. This model is consistent with previous results demonstrating that $p75^{NTR}$ activation stimulates Rho activity.[204]

RhoA acts on several downstream effectors to regulate the underlying cytoskeleton. One of these effectors, termed Rho kinase, or ROCK, has been implicated in neurite outgrowth inhibition by CNS myelin components.[201] Treatment with Y-27632, an ATP competitive antagonist that blocks ROCK activation, promotes neuronal outgrowth on myelin-associated inhibitory substrates in vitro as well as in in vivo models of spinal cord injury.[201,202,206,207] Moreover, glial-scar derived inhibitors such as CSPGs, can also be blocked by inhibitors of ROCK.[206] How ROCK induces neurite retraction is a complex issue because it has been shown to phosphorylate and signal through multiple downstream effectors.[208] In summary, regardless of its mechanism of activation, RhoA seems to be the point of convergence for all presently identified myelin and glial scar derived inhibitors and might therefore be an attractive molecular target for small molecules that interfere with its activation. One caveat however is that small GTPases and their effector proteins are universally expressed and involved in regulating many distinct cellular functions such as cell migration, proliferation, and adhesion. Thus, therapeutically manipulating RhoA and/or ROCK to improve axonal regeneration, may negatively influence other physiological events.

cAMP/PKA

Neuronal cAMP levels decrease spontaneously and sharply with development at a time that correlates with their loss of regenerative ability and the appearance of an inhibitory response to MAG and myelin. Furthermore, elevation of intracellular cAMP levels by intraganglionic administration of cAMP overcomes inhibition by MAG and myelin and results in extensive regeneration of injured dorsal column axons.[98,99] This process initially involves phosphorylation of PKA, but eventually becomes PKA-independent.[209] The precise molecular mechanisms are not yet fully understood, but are presumably transcription dependent which would also explain the time period that is required for cAMP elevation to come into effect.[210,211] Importantly, other than with RhoA, a variety of low-molecular-weight compounds are available which modulate the activity status of the cAMP-PKA signalling module. One of them, rolipram, a phosphodiesterase inhibitor which allows accumulation of cAMP by blocking its degradation, was reported to promote axonal regeneration and functional recovery upon spinal cord injury, even in a post-lesion manner.[212]

Experimental Strategies to Promote Nerve Regeneration Following CNS Injury

Several experimental strategies have been developed to overcome inhibition after CNS injury. They involve the use of antibodies to neutralize growth inhibitors, Nogo peptides binding to NGR and blocking its activity, and blockers of the intracellular pathways mediating growth inhibition (for recent reviews see refs. 3,5,213,214) In addition, rather than blocking inhibitors, receptors, or their signaling pathways, growth of dorsal column ascending sensory axons has been improved by intraganglionic administration of cAMP[98,209] or by application of rolipram.[212] To decrease its inhibitory properties, the biochemical composition of the ECM has also been altered by treatment with chondroitinase ABC, the enzyme which cleaves the GAG side chains of CSPGs.[164-166] Furthermore, attempts to reduce glial scarring have been made using pharmacological agents to decrease collagen deposition in the ECM.[228] Matrix and ADAMs metalloproteinases, enzymes that can degrade ECM proteins and CSPGs, are also potential candidates for remodeling the glial scar. Other strategies to promote CNS repair including application of neurotrophic factors, implantation of olfactory bulb ensheathing cells, or grafting of peripheral nerve, have recently been reviewed elsewhere.[215-227,229,230]

The discovery that the NgR/p75NTR complex binds and mediates the effects of the three most prominent myelin-derived ligands has made this receptor complex a very attractive target for therapeutic drug development. Therapeutic candidates for targeting NgR include protein antagonists and small-molecule antagonists, but an antagonist of NgR that blocks binding of all three myelin proteins has not yet been found. One likely explanation is the lack of compelling evidence for the existence of a single molecular binding site on NgR for all myelin inhibitors. Recently, inhibition of NgR was achieved utilising a peptide corresponding to residues 1-40 of Nogo-66 (NEP1-40) that acts as a competitive antagonist of the receptor in vitro and, more importantly, promotes axonal regeneration and functional recovery following CNS injury in vivo. A soluble version of a recombinant NgR could also be a possible solution to effectively inhibit the action of all three myelin proteins interacting with NgR. Several independent reports have shown that the soluble full length NgR ectodomain lacking the GPI anchor, and an C-terminally truncated NgR ectodomain comprising the ligand binding domain alone (NgR-310), are potent antagonists of neurite outgrowth inhibition mediated by Nogo-66, MAG and whole myelin in vitro[183,231] and in vivo.[232] Similarly, preventing the interaction of NgR with its coreceptor $p75^{NTR}$ could be an alternative site for intervention to promote axonal regeneration. The recent finding that the interaction between $p75^{NTR}$ and Rho- GDI requires the fifth alpha helix of $p75^{NTR}$, has opened additional strategies. A peptide ligand, Pep5, for instance, has been shown to bind this helix and has been used to inhibit the interaction between Rho-GDI and $p75^{NTR}$.[205] This peptide blocks both Rho activation and the growth inhibitory effects of MAG and Nogo on cerebellar granule cells. Pep5 therefore has potential as a therapeutic agent to block inhibitory cues that impede regeneration in the CNS.

Increased deposition of a number of ECM molecules such as collagen IV and CSPGs during glial scar formation is considered to contribute to the local impediment of axonal growth. Two main in vivo approaches have been taken to reduce the inhibitory properties of the ECM (1) reduction of collagen deposition, and (2) enzymatic digestion of CSPGs. Injection into a fimbria fornix lesion of the iron chelator 2,2 -dipyridine (DPY), an inhibitor of collagen synthesis and triple helix formation, significantly reduced lesion-induced collagen deposition. This effect was correlated with a pronounced regeneration of the lesioned fimbria fornix axons.[233] However, in a spinal cord lesion model in which basement membrane deposition is more extensive, injections of DPY did not result in sufficient reduction in collagen deposition to allow regrowth of corticospinal tract axons.[234,235] Furthermore, because basal membrane ECM proteins such as collagen IV are structural elements of blood vessels, it is uncertain whether the vascular blood supply is compromised by the application of DPY to the injured area. To remove additional inhibitory components of the glial scar, the bacterial enzyme Chondroitinase ABC (ChABC) was used to digest CSPGs. ChABC hydrolyzes glycosaminoglycan side-chains of CSPGs and has been shown to block the inhibitory activity of CSPGs in vitro. Application of ChABC via a gelfoam to the contused dorsal thoracic cord resulted in decreased CSPG immunoreactivity, but no axonal regrowth was reported.[236] Some growth of spinal cord neurons into a peripheral nerve graft was achieved when gelfoam soaked with ChABC was placed on the interface between the thoracic hemisection and a peripheral nerve graft.[165] More recently, Bradbury and collaborators have reported that intrathecal administration of ChABC after dorsal column lesion in rats, could successfully promote regeneration of corticospinal tract axons as well as functional recovery of locomotor and proprioceptive function.[164]

Conclusion

In the last few years, enormous advances have been made to revise the dogma that axons in the adult mammalian CNS cannot regenerate after injury. In various experimental CNS, and in particular spinal cord lesion models, administration of neurotrophic factors, implantation of olfactory ensheathing cells, blockade of myelin-associated growth inhibitors and/or their receptors, or removal of inhibitory extracellular matrix molecules by enzymatic treatment has had very encouraging success, leading to some functional recovery. Several of the experimental interventions that promote axonal regeneration have also enhanced compensatory sprouting of nonlesioned nerve fibers, resulting in innervation of denervated targets (see review ref. 237). This type of plasticity may therefore contribute to functional recovery after injury. Aberrant sprouting of sensory fibers, for instance, may however lead to debilitating pain. Hence, a delicate balance between beneficial versus detrimental sprouting/regeneration needs to be achieved for optimal outcomes. Although the final aim for the treatment of human spinal injuries is to promote long fiber regeneration, restore myelination, and to achieve appropriate connectivity, a more limited repair may already improve the quality of life of paraplegic patients. Thus, the many promising results obtained in the last decade offer new strategies for developing clinically effective therapies, either alone or in combination, for many patients.

References

1. Bareyre FM, Schwab ME. Inflammation, degeneration and regeneration in the injured spinal cord: insights from DNA microarrays. Trends Neurosci 2003; 26:555-63.
2. De Winter F, Holtmaat AJ, Verhaagen J. Neuropilin and class 3 semaphorins in nervous system regeneration. Adv Exp Med Biol 2002; 515:115-39.
3. Filbin MT. Myelin-associated inhibitors of axonal regeneration in the adult mammalian CNS. Nat Rev Neurosci 2003; 4:703-13.
4. Huber AB, Schwab ME. Nogo-A, a potent inhibitor of neurite outgrowth and regeneration. Biol Chem 2000; 381:407-19.
5. Hunt D, Coffin RS, Anderson PN. The Nogo receptor, its ligands and axonal regeneration in the spinal cord; a review. J Neurocytol 2002; 31:93-120.
6. Morgenstern DA., Asher RA, Fawcett JW. Chondroitin sulphate proteoglycans in the CNS injury response. Prog Brain Res 2002; 137:313-32.
7. Goldberg JL, Barres BA. The relationship between neuronal survival and regeneration. Annu Rev Neurosci 2000; 23:579-612.
8. Sofroniew MV, Isacson O. Distribution of degeneration of cholinergic neurons in the septum following axotomy in different portions of the fimbria-fornix: a correlation between degree of cell loss and proximity of neuronal somata to the lesion. J Chem Neuroanat 1988; 1:327-37.
9. Stoll G, Jander S, Myers RR. Degeneration and regeneration of the peripheral nervous system: from Augustus Waller's observations to neuroinflammation. J Peripher Nerv Syst 2002; 7:13-27.
10. Chen DF, Jhaveri S, Schneider GE. Intrinsic changes in developing retinal neurons result in regenerative failure of their axons. Proc Natl Acad Sci USA 1995; 92:7287-91.
11. Dusart I, Airaksinen MS, Sotelo C. Purkinje cell survival and axonal regeneration are age dependent: an in vitro study. J Neurosci 1997; 17:3710-26.
12. Gianola S, Rossi F. Evolution of the Purkinje cell response to injury and regenerative potential during postnatal development of the rat cerebellum. J Comp Neurol 2001; 430:101-17.
13. Steeves JD, Keirstead HS, Ethell DW et al. Permissive and restrictive periods for brainstem-spinal regeneration in the chick. Prog Brain Res 1994; 103:243-62.
14. Bates CA, Stelzner DJ. Extension and regeneration of corticospinal axons after early spinal injury and the maintenance of corticospinal topography. Exp Neurol 1993; 123:106-17.
15. Firkins SS, Bates CA, Stelzner DJ. Corticospinal tract plasticity and astroglial reactivity after cervical spinal injury in the postnatal rat. Exp Neurol 1993; 120:1-15.
16. Hasan SJ, Keirstead HS, Muir GD et al. Axonal regeneration contributes to repair of injured brainstem-spinal neurons in embryonic chick. J Neurosci 1993; 13:492-507.
17. Keirstead HS, Hasan SJ, Muir GD et al. Suppression of the onset of myelination extends the permissive period for the functional repair of embryonic spinal cord. Proc Natl Acad Sci USA 1992; 89:11664-8.
18. Bouslama-Oueghlani L, Wehrle R, Sotelo C et al. The developmental loss of the ability of Purkinje cells to regenerate their axons occurs in the absence of myelin: an in vitro model to prevent myelination. J Neurosci 2003; 23:8318-29.

19. Goldberg JL, Klassen MP, Hua Y et al. Amacrine-signaled loss of intrinsic axon growth ability by retinal ganglion cells. Science 2002; 296:1860-4.
20. Mori N, Morii H. SCG10-related neuronal growth-associated proteins in neural development, plasticity, degeneration, and aging. J Neurosci Res 2002; 70:264-73.
21. Cai D, Qiu J, Cao Z et al. Neuronal cyclic AMP controls the developmental loss in ability of axons to regenerate. J Neurosci 2001; 21:4731-9.
22. Bandtlow CE, Loschinger J. Developmental changes in neuronal responsiveness to the CNS myelin- associated neurite growth inhibitor NI-35/250. Eur J Neurosci 1997; 9:2743-52.
23. Goldberg JL, Espinosa JS, Xu Y et al. Retinal ganglion cells do not extend axons by default: promotion by neurotrophic signaling and electrical activity. Neuron 2002; 33:689-702.
24. Chen DF, Schneider GE, Martinou JC et al. Bcl-2 promotes regeneration of severed axons in mammalian CNS. Nature 1997; 385:434-9.
25. Fernandes K. Gene expression in axotomized neurons: identifying the intrinsic determinants of axonal growth. In: Ingoglia NA, ed. Axonal regeneration in the central nervous system. New York: Marcel Dekker Inc., 2000:219-266.
26. Broude E, McAtee M, Kelley MS et al. c-Jun expression in adult rat dorsal root ganglion neurons: differential response after central or peripheral axotomy. Exp Neurol 1997; 148:367-77.
27. Chong MS, Reynolds ML, Irwin N et al. GAP-43 expression in primary sensory neurons following central axotomy. J Neurosci 1994; 14:4375-84.
28. Schreyer DJ, Skene JH. Injury-associated induction of GAP-43 expression displays axon branch specificity in rat dorsal root ganglion neurons. J Neurobiol 1993; 24:959-70.
29. Schwaiger FW, Hager G, Schmitt AB et al. Peripheral but not central axotomy induces changes in Janus kinases (JAK) and signal transducers and activators of transcription (STAT). Eur J Neurosci 2000; 12:1165-76.
30. Aigner L, Arber S, Kapfhammer JP et al. Overexpression of the neural growth-associated protein GAP-43 induces nerve sprouting in the adult nervous system of transgenic mice. Cell 1995; 83:269-78.
31. Buffo A, Holtmaat AJ, Savio T et al. Targeted overexpression of the neurite growth-associated protein B-50/GAP-43 in cerebellar Purkinje cells induces sprouting after axotomy but not axon regeneration into growth-permissive transplants. J Neurosci 1997; 17:8778-91.
32. Mason MR, Campbell G, Caroni P et al. Overexpression of GAP-43 in thalamic projection neurons of transgenic mice does not enable them to regenerate axons through peripheral nerve grafts. Exp Neurol 2000; 165:143-52.
33. Bomze HM, Bulsara KR, Iskandar BJ et al. Spinal axon regeneration evoked by replacing two growth cone proteins in adult neurons. Nat Neurosci 2001; 4:38-43.
34. Bonilla IE, Tanabe K, Strittmatter SM. Small proline-rich repeat protein 1A is expressed by axotomized neurons and promotes axonal outgrowth. J Neurosci 2002; 22:1303-15.
35. Tanabe K, Bonilla I, Winkles JA et al. Fibroblast growth factor-inducible-14 is induced in axotomized neurons and promotes neurite outgrowth. J Neurosci 2003; 23:9675-86.
36. Hull M, Bahr M. Differential regulation of c-JUN expression in rat retinal ganglion cells after proximal and distal optic nerve transection. Neurosci Lett 1994; 178:39-42.
37. Fernandes KJ, Fan DP, Tsui BJ et al. Influence of the axotomy to cell body distance in rat rubrospinal and spinal motoneurons: differential regulation of GAP-43, tubulins, and neurofilament-M. J Comp Neurol 1999; 414:495-510.
38. Mason MR, Lieberman AR, Anderson PN. Corticospinal neurons up-regulate a range of growth-associated genes following intracortical, but not spinal, axotomy. Eur J Neurosci 2003; 18:789-802.
39. Richardson PM, McGuinness UM, Aguayo AJ. Axons from CNS neurons regenerate into PNS grafts. Nature 1980; 284:264-5.
40. Chong MS, Reynolds ML, Irwin N et al. GAP-43 expression in primary sensory neurons following central axotomy. J Neurosci 1994; 14:4375-84.
41. Mason MR, Lieberman AR, Grenningloh G et al. Transcriptional upregulation of SCG10 and CAP-23 is correlated with regeneration of the axons of peripheral and central neurons in vivo. Mol Cell Neurosci 2002; 20:595-615.
42. Neumann S, Woolf CJ. Regeneration of dorsal column fibers into and beyond the lesion site following adult spinal cord injury. Neuron 1999; 23:83-91.

43. Pasterkamp RJ, Verhaagen J. Emerging roles for semaphorins in neural regeneration. Brain Res Brain Res Rev 2001; 35:36-54.
44. Rossi F, Strata P. Reciprocal trophic interactions in the adult climbing fibre-Purkinje cell system. Prog Neurobiol 1995; 47:341-69.
45. Inman DM, Steward O. Ascending sensory, but not other long-tract axons, regenerate into the connective tissue matrix that forms at the site of a spinal cord injury in mice. J Comp Neurol 2003; 462:431-49.
46. David S, Aguayo AJ. Axonal elongation into peripheral nervous system "bridges" after central nervous system injury in adult rats. Science 1981; 214:931-3.
47. Pettigrew DB, Crutcher KA. White matter of the CNS supports or inhibits neurite outgrowth in vitro depending on geometry. J Neurosci 1999; 19:8358-66.
48. Schwab ME, Bartholdi D. Degeneration and regeneration of axons in the lesioned spinal cord. Physiol Rev 1996; 76:319-70.
49. Faulkner JR, Herrmann JE, Woo MJ et al. Reactive astrocytes protect tissue and preserve function after spinal cord injury. J Neurosci 2004; 24:2143-55.
50. Cui W, Allen ND, Skynner M et al. Inducible ablation of astrocytes shows that these cells are required for neuronal survival in the adult brain. Glia 2001; 34:272-82.
51. Constam DB, Philipp J, Malipiero UV et al. Differential expression of transforming growth factor-beta 1, -beta 2, and -beta 3 by glioblastoma cells, astrocytes, and microglia. J Immunol 1992; 148:1404-10.
52. Muller HW, Junghans U, Kappler J. Astroglial neurotrophic and neurite-promoting factors. Pharmacol Ther 1995; 65:1-18.
53. Hoke A, Silver J. Proteoglycans and other repulsive molecules in glial boundaries during development and regeneration of the nervous system. Prog Brain Res 1996; 108:149-63.
54. Kreutzberg GW. Microglia: a sensor for pathological events in the CNS. Trends Neurosci 1996; 19:312-8.
55. Raivich G, Jones LL, Werner A et al. Molecular signals for glial activation: pro- and anti-inflammatory cytokines in the injured brain. Acta Neurochir Suppl 1999; 73:21-30.
56. McTigue DM, Wei P, Stokes BT. Proliferation of NG2-positive cells and altered oligodendrocyte numbers in the contused rat spinal cord. J Neurosci 2001; 21:3392-400.
57. Levine JM, Reynolds R, Fawcett JW. The oligodendrocyte precursor cell in health and disease. Trends Neurosci 2001; 24:39-47.
58. Carbonell AL, Boya J. Ultrastructural study on meningeal regeneration and meningo-glial relationships after cerebral stab wound in the adult rat. Brain Res 1988; 439:337-44.
59. Shearer MC, Fawcett JW. The astrocyte/meningeal cell interface—a barrier to successful nerve regeneration? Cell Tissue Res 2001; 305:267-73.
60. Coleman MP, Perry VH. Axon pathology in neurological disease: a neglected therapeutic target. Trends Neurosci 2002; 25:532-7.
61. Shamash S, Reichert F, Rotshenker S. The cytokine network of Wallerian degeneration: tumor necrosis factor-alpha, interleukin-1alpha, and interleukin-1beta. J Neurosci 2002; 22:3052-60.
62. Stoll G, Jander S. The role of microglia and macrophages in the pathophysiology of the CNS. Prog Neurobiol 1999; 58:233-47.
63. George R, Griffin JW. Delayed macrophage responses and myelin clearance during Wallerian degeneration in the central nervous system: the dorsal radiculotomy model. Exp Neurol 1994; 129:225-36.
64. Bandtlow C, Zachleder T, Schwab ME. Oligodendrocytes arrest neurite growth by contact inhibition. J Neurosci 1990; 10:3837-48.
65. Caroni P, Savio T, Schwab ME. Central nervous system regeneration: oligodendrocytes and myelin as non-permissive substrates for neurite growth. Prog Brain Res 1988; 78:363-70.
66. Beattie MS, Hermann GE, Rogers RC et al. Cell death in models of spinal cord injury. Prog Brain Res 2002; 137:37-47.
67. Allan SM, Rothwell NJ. Cytokines and acute neurodegeneration. Nat Rev Neurosci 2001; 2:734-44.
68. Frei E, Klusman I, Schnell L et al. Reactions of oligodendrocytes to spinal cord injury: cell survival and myelin repair. Exp Neurol 2000; 163:373-80.
69. Dong H, Fazzaro A, Xiang C et al. Enhanced oligodendrocyte survival after spinal cord injury in Bax-deficient mice and mice with delayed Wallerian degeneration. J Neurosci 2003; 23:8682-91.

70. Casha S, Yu WR, Fehlings MG. Oligodendroglial apoptosis occurs along degenerating axons and is associated with FAS and p75 expression following spinal cord injury in the rat. Neuroscience 2001; 103:203-18.
71. Grossman SD, Rosenberg LJ, Wrathall JR. Temporal-spatial pattern of acute neuronal and glial loss after spinal cord contusion. Exp Neurol 2001; 168:273-82.
72. Kimelberg HK. Astrocytic edema in CNS trauma. J Neurotrauma 1992; 9 Suppl 1, S71-81.
73. Bullock R, Maxwell WL, Graham DI et al. Glial swelling following human cerebral contusion: an ultrastructural study. J Neurol Neurosurg Psychiatry 1991; 54:427-34.
74. Wyss-Coray T, Mucke L. Inflammation in neurodegenerative disease—a double-edged sword. Neuron 2002; 35:419-32.
75. Schwartz M, Lazarov-Spiegler O, Rapalino O et al. Potential repair of rat spinal cord injuries using stimulated homologous macrophages. Neurosurgery 1999; 44:1041-5; discussion 1045-6.
76. Rostworowski M, Balasingam V, Chabot S et al. Astrogliosis in the neonatal and adult murine brain post-trauma: elevation of inflammatory cytokines and the lack of requirement for endogenous interferon-gamma. J Neurosci 1997; 17:3664-74.
77. Yong VW, Moumdjian R, Yong FP et al. Gamma-interferon promotes proliferation of adult human astrocytes in vitro and reactive gliosis in the adult mouse brain in vivo. Proc Natl Acad Sci USA 1991; 88:7016-20.
78. Logan A, Berry M, Gonzalez AM et al. Effects of transforming growth factor beta 1 on scar production in the injured central nervous system of the rat. Eur J Neurosci 1994; 6:355-63.
79. Lagord C, Berry M, Logan A. Expression of TGFbeta2 but not TGFbetal correlates with the deposition of scar tissue in the lesioned spinal cord. Mol Cell Neurosci 2002; 20:69-92.
80. Giulian D, Lachman LB. Interleukin-1 stimulation of astroglial proliferation after brain injury. Science 1985; 228:497-9.
81. Asher RA, Morgenstern DA, Fidler PS et al. Neurocan is upregulated in injured brain and in cytokine-treated astrocytes. J Neurosci 2000; 20:2427-38.
82. Smith GM, Hale JH. Macrophage/Microglia regulation of astrocytic tenascin: synergistic action of transforming growth factor-beta and basic fibroblast growth factor. J Neurosci 1997; 17:9624-33.
83. Sroga JM, Jones TB, Kigerl KA et al. Rats and mice exhibit distinct inflammatory reactions after spinal cord injury. J Comp Neurol 2003; 462:223-40.
84. Inman D, Guth L, Steward O. Genetic influences on secondary degeneration and wound healing following spinal cord injury in various strains of mice. J Comp Neurol 2002; 451:225-35.
85. Steward O, Schauwecker PE, Guth L et al. Genetic approaches to neurotrauma research: opportunities and potential pitfalls of murine models. Exp Neurol 1999; 157:19-42.
86. Caroni P, Schwab ME. Two membrane protein fractions from rat central myelin with inhibitory properties for neurite growth and fibroblast spreading. J Cell Biol 1988; 106:1281-8.
87. Schnell L, Schwab ME. Axonal regeneration in the rat spinal cord produced by an antibody against myelin-associated neurite growth inhibitors. Nature 1990; 343:269-72.
88. Sicotte M, Tsatas O, Jeong SY et al. Immunization with myelin or recombinant Nogo-66/MAG in alum promotes axon regeneration and sprouting after corticospinal tract lesions in the spinal cord. Mol Cell Neurosci 2003; 23:251-63.
89. Huang DW, McKerracher L, Braun PE et al. A therapeutic vaccine approach to stimulate axon regeneration in the adult mammalian spinal cord. Neuron 1999; 24:639-47.
90. Bregman BS, Kunkel-Bagden E, Schnell L et al. Recovery from spinal cord injury mediated by antibodies to neurite growth inhibitors. Nature 1995; 378:498-501.
91. GrandPre T, Li S, Strittmatter SM. Nogo-66 receptor antagonist peptide promotes axonal regeneration. Nature 2002; 417:547-51.
92. Grados-Munro EM, Fournier AE. Myelin-associated inhibitors of axon regeneration. J Neurosci Res 2003; 74:479-85.
93. Salzer JL, Holmes WP, Colman DR. The amino acid sequences of the myelin-associated glycoproteins: homology to the immunoglobulin gene superfamily. J Cell Biol 1987; 104:957-65.
94. Willison HJ, Ilyas AI, O'Shannessy DJ et al. Myelin-associated glycoprotein and related glycoconjugates in developing cat peripheral nerve: a correlative biochemical and morphometric study. J Neurochem 1987; 49:1853-62.

95. Johnson PW, Abramow-Newerly W, Seilheimer B et al. Recombinant myelin-associated glycoprotein confers neural adhesion and neurite outgrowth function. Neuron 1989; 3:377-85.
96. Turnley AM, Bartlett PF. MAG and MOG enhance neurite outgrowth of embryonic mouse spinal cord neurons. Neuroreport 1998; 9:1987-90.
97. Mukhopadhyay G, Doherty P, Walsh FS et al. A novel role for myelin-associated glycoprotein as an inhibitor of axonal regeneration. Neuron 1994; 13:757-67.
98. Neumann S, Bradke F, Tessier-Lavigne M et al. Regeneration of sensory axons within the injured spinal cord induced by intraganglionic cAMP elevation. Neuron 2002; 34:885-93.
99. Qiu J, Cai D, Filbin MT. A role for cAMP in regeneration during development and after injury. Prog Brain Res 2002; 137:381-7.
100. Hirata K, Kawabuchi M. Myelin phagocytosis by macrophages and nonmacrophages during Wallerian degeneration. Microsc Res Tech 2002; 57:541-7.
101. Schafer M, Fruttiger M, Montag D et al. Disruption of the gene for the myelin-associated glycoprotein improves axonal regrowth along myelin in C57BL/Wlds mice. Neuron 1996; 16:1107-13.
102. Li M, Shibata A, Li C et al. Myelin-associated glycoprotein inhibits neurite/axon growth and causes growth cone collapse. J Neurosci Res 1996; 46:404-14.
103. Bartsch U, Bandtlow CE, Schnell L et al. Lack of evidence that myelin-associated glycoprotein is a major inhibitor of axonal regeneration in the CNS. Neuron 1995; 15:1375-81.
104. Prinjha R, Moore SE, Vinson M et al. Inhibitor of neurite outgrowth in humans. Nature 2000; 403:383-4.
105. GrandPre T, Nakamura F, Vartanian T et al. Identification of the Nogo inhibitor of axon regeneration as a Reticulon protein. Nature 2000; 403:439-44.
106. Chen MS, Huber AB, van der Haar ME et al. Nogo-A is a myelin-associated neurite outgrowth inhibitor and an antigen for monoclonal antibody IN-1. Nature 2000; 403:434-9.
107. Spillmann AA, Bandtlow CE, Lottspeich F et al. Identification and characterization of a bovine neurite growth inhibitor (bNI-220). J Biol Chem 1998; 273:19283-93.
108. Oertle T, Schwab ME. Nogo and its paRTNers. Trends Cell Biol 2003; 13:187-94.
109. Oertle T, Klinger M, Stuermer CA et al. A reticular rhapsody: phylogenic evolution and nomenclature of the RTN/Nogo gene family. Faseb J 2003; 17:1238-47.
110. Oertle T, van der Haar ME, Bandtlow CE et al. Nogo-A inhibits neurite outgrowth and cell spreading with three discrete regions. J Neurosci 2003; 23:5393-406.
111. GrandPré T, Nakamura F, Vartanian T et al. Identification of the Nogo inhibitor of axon regeneration as a Reticulon protein. Nature 2000; 403:439-44.
112. Huber AB, Weinmann O, Brosamle C et al. Patterns of Nogo mRNA and protein expression in the developing and adult rat and after CNS lesions. J Neurosci 2002; 22:3553-67.
113. Hunt D, Coffin RS, Prinjha RK et al. Nogo-A expression in the intact and injured nervous system. Mol Cell Neurosci 2003; 24:1083-102.
114. Josephson A, Widenfalk J, Widmer HW et al. NOGO mRNA expression in adult and fetal human and rat nervous tissue and in weight drop injury. Exp Neurol 2001; 169:319-28.
115. Wang X, Chun SJ, Treloar H et al. Localization of Nogo-A and Nogo-66 receptor proteins at sites of axon- myelin and synaptic contact. J Neurosci 2002; 22:5505-15.
116. Schwab ME. Increasing plasticity and functional recovery of the lesioned spinal cord. Prog Brain Res 2002; 137:351-9.
117. Li S, Strittmatter SM. Delayed systemic Nogo-66 receptor antagonist promotes recovery from spinal cord injury. J Neurosci 2003; 23:4219-27.
118. Pot C, Simonen M, Weinmann O et al. Nogo-A expressed in Schwann cells impairs axonal regeneration after peripheral nerve injury. J Cell Biol 2002; 159:29-35.
119. Kim JE, Bonilla IE, Qiu D et al. Nogo-C is sufficient to delay nerve regeneration. Mol Cell Neurosci 2003; 23:451-9.
120. Kim JE, Li S, GrandPre T et al. Axon regeneration in young adult mice lacking Nogo-A/B. Neuron 2003; 38:187-99.
121. Simonen M, Pedersen V, Weinmann O et al. Systemic deletion of the myelin-associated outgrowth inhibitor Nogo-A improves regenerative and plastic responses after spinal cord injury. Neuron 2003; 38:201-11.

122. Zheng B, Ho C, Li S et al. Lack of enhanced spinal regeneration in Nogo-deficient mice. Neuron 2003; 38:213-24.
123. Schnell L, Liebscher T, Weinmann OR et al. Increased regeneration and functional recovery in rats treated with antibodies to NOGO-A and IN NOGO-A knockout mice. In: Program No. 43.11. 2004 Abstract Viewer/Itinerary Planner. Washington, D.S.f.N., 2004. Online, ed.
124. Kottis V, Thibault P, Mikol D et al. Oligodendrocyte-myelin glycoprotein (OMgp) is an inhibitor of neurite outgrowth. J Neurochem 2002; 82:1566-9.
125. Wang KC, Koprivica V, Kim JA et al. Oligodendrocyte-myelin glycoprotein is a Nogo receptor ligand that inhibits neurite outgrowth. Nature 2002; 417:941-4.
126. Mikol DD, Gulcher JR, Stefansson K. The oligodendrocyte-myelin glycoprotein belongs to a distinct family of proteins and contains the HNK-1 carbohydrate. J Cell Biol 1990; 110:471-9.
127. Mikol DD, Stefansson K. A phosphatidylinositol-linked peanut agglutinin-binding glycoprotein in central nervous system myelin and on oligodendrocytes. J Cell Biol 1988; 106:1273-9.
128. Fawcett JW, Asher RA. The glial scar and central nervous system repair. Brain Res Bull 1999; 49:377-91.
129. Fidler PS, Schuette K, Asher RA et al. Comparing astrocytic cell lines that are inhibitory or permissive for axon growth: the major axon-inhibitory proteoglycan is NG2. J Neurosci 1999; 19:8778-88.
130. Logan A, Berry M. Cellular and molecular determinants of glial scar formation. Adv Exp Med Biol 2002; 513:115-58.
131. Muir EM, Adcock KH, Morgenstern DA et al. Matrix metalloproteases and their inhibitors are produced by overlapping populations of activated astrocytes. Brain Res Mol Brain Res 2002; 100:103-17.
132. Properzi F, Asher RA, Fawcett JW. Chondroitin sulphate proteoglycans in the central nervous system: changes and synthesis after injury. Biochem Soc Trans 2003; 31:335-6.
133. Rudge JS, Smith GM, Silver J. An in vitro model of wound healing in the CNS: analysis of cell reaction and interaction at different ages. Exp Neurol 1989; 103:1-16.
134. Silver J, Miller JH. Regeneration beyond the glial scar. Nat Rev Neurosci 2004; 5:146-56.
135. Stichel CC, Niermann H, D'Urso D et al. Basal membrane-depleted scar in lesioned CNS: characteristics and relationships with regenerating axons. Neuroscience 1999; 93:321-33.
136. McKeon RJ, Hoke A, Silver J. Injury-induced proteoglycans inhibit the potential for laminin-mediated axon growth on astrocytic scars. Exp Neurol 1995; 136:32-43.
137. McKeon RJ, Schreiber RC, Rudge JS et al. Reduction of neurite outgrowth in a model of glial scarring following CNS injury is correlated with the expression of inhibitory molecules on reactive astrocytes. J Neurosci 1991; 11:3398-411.
138. Smith-Thomas LC, Fok-Seang J, Stevens J et al. An inhibitor of neurite outgrowth produced by astrocytes. J Cell Sci 1994; 107(Pt 6):1687-95.
139. Smith-Thomas LC, Stevens J, Fok-Seang J et al. Increased axon regeneration in astrocytes grown in the presence of proteoglycan synthesis inhibitors. J Cell Sci 1995; 108(Pt 3):1307-15.
140. Hartmann U, Maurer P. Proteoglycans in the nervous system—the quest for functional roles in vivo. Matrix Biol 2001; 20:23-35.
141. Bandtlow CE, Zimmermann DR. Proteoglycans in the developing brain: new conceptual insights for old proteins. Physiol Rev 2000; 80:1267-90.
142. Yamaguchi Y. Lecticans: organizers of the brain extracellular matrix. Cell Mol Life Sci 2000; 57:276-89.
143. Timpl R, Sasaki T, Kostka G et al. Fibulins: a versatile family of extracellular matrix proteins. Nat Rev Mol Cell Biol 2003; 4:479-89.
144. Bekku Y, Su WD, Hirakawa S et al. Molecular cloning of Bral2, a novel brain-specific link protein, and immunohistochemical colocalization with brevican in perineuronal nets. Mol Cell Neurosci 2003; 24:148-59.
145. Oohashi T, Hirakawa S, Bekku Y et al. Bral1, a brain-specific link protein, colocalizing with the versican V2 isoform at the nodes of Ranvier in developing and adult mouse central nervous systems. Mol Cell Neurosci 2002; 19:43-57.

146. Hagihara K, Miura R, Kosaki R et al. Immunohistochemical evidence for the brevican-tenascin-R interaction: colocalization in perineuronal nets suggests a physiological role for the interaction in the adult rat brain. J Comp Neurol 1999; 410:256-64.
147. Seidenbecher CI, Richter K, Rauch U et al. Brevican, a chondroitin sulfate proteoglycan of rat brain, occurs as secreted and cell surface glycosylphosphatidylinositol-anchored isoforms. J Biol Chem 1995; 270:27206-12.
148. Seidenbecher CI, Smalla KH, Fischer N et al. Brevican isoforms associate with neural membranes. J Neurochem 2002; 83:738-46.
149. Yamada H, Fredette B, Shitara K et al. The brain chondroitin sulfate proteoglycan brevican associates with astrocytes ensheathing cerebellar glomeruli and inhibits neurite outgrowth from granule neurons. J Neurosci 1997; 17:7784-95.
150. Schwartz NB, Domowicz M. Proteoglycans in brain development. Glycoconj J 2004; 21:329-41.
151. Ogawa T, Hagihara K, Suzuki M et al. Brevican in the developing hippocampal fimbria: differential expression in myelinating oligodendrocytes and adult astrocytes suggests a dual role for brevican in central nervous system fiber tract development. J Comp Neurol 2001; 432:285-95.
152. Niederost BP, Zimmermann DR, Schwab ME et al. Bovine CNS myelin contains neurite growth-inhibitory activity associated with chondroitin sulfate proteoglycans. J Neurosci 1999; 19:8979-89.
153. Rauch U, Feng K, Zhou XH. Neurocan: a brain chondroitin sulfate proteoglycan. Cell Mol Life Sci 2001; 58:1842-56.
154. Schmalfeldt M, Bandtlow CE, Dours-Zimmermann MT et al. Brain derived versican V2 is a potent inhibitor of axonal growth. J Cell Sci 2000; 113:807-16.
155. Chang A, Nishiyama A, Peterson J et al. NG2-positive oligodendrocyte progenitor cells in adult human brain and multiple sclerosis lesions. J Neurosci 2000; 20:6404-12.
156. Chen ZJ, Negra M, Levine A et al. Oligodendrocyte precursor cells: reactive cells that inhibit axon growth and regeneration. J Neurocytol 2002; 31:481-95.
157. Dawson MR, Levine JM, Reynolds R. NG2-expressing cells in the central nervous system: are they oligodendroglial progenitors? J Neurosci Res 2000; 61:471-9.
158. Greenwood K, Butt AM. Evidence that perinatal and adult NG2-glia are not conventional oligodendrocyte progenitors and do not depend on axons for their survival. Mol Cell Neurosci 2003; 23:544-58.
159. Mallon BS, Shick HE, Kidd GJ et al. Proteolipid promoter activity distinguishes two populations of NG2-positive cells throughout neonatal cortical development. J Neurosci 2002; 22:876-85.
160. Jones LL, Yamaguchi Y, Stallcup WB et al. NG2 is a major chondroitin sulfate proteoglycan produced after spinal cord injury and is expressed by macrophages and oligodendrocyte progenitors. J Neurosci 2002; 22:2792-803.
161. Asher RA, Morgenstern DA, Shearer MC et al. Versican is upregulated in CNS injury and is a product of oligodendrocyte lineage cells. J Neurosci 2002; 22:2225-36.
162. Tang X, Davies JE, Davies SJ. Changes in distribution, cell associations, and protein expression levels of NG2, neurocan, phosphacan, brevican, versican V2, and tenascin-C during acute to chronic maturation of spinal cord scar tissue. J Neurosci Res 2003; 71:427-44.
163. Davies SJ, Goucher DR, Doller C et al. Robust regeneration of adult sensory axons in degenerating white matter of the adult rat spinal cord. J Neurosci 1999; 19:5810-22.
164. Bradbury EJ, Moon LD, Popat RJ et al. Chondroitinase ABC promotes functional recovery after spinal cord injury. Nature 2002; 416:636-40.
165. Yick LW, Wu W et al. Chondroitinase ABC promotes axonal regeneration of Clarke's neurons after spinal cord injury. Neuroreport 2000; 11:1063-67.
166. Moon LD, Asher RA et al. Regeneration of CNS axons back to their target following treatment of adult rat brain with chondroitinase ABC. Nat Neurosci 2001; 4:465-466.
167. Ajemian A, Ness R, David S. Tenascin in the injured rat optic nerve and in non-neuronal cells in vitro: potential role in neural repair. J Comp Neurol 1994; 340:233-42.
168. Chiquet-Ehrismann R. Tenascins. Int J Biochem Cell Biol 2004; 36:986-90.
169. Joester A, Faissner A. The structure and function of tenascins in the nervous system. Matrix Biol 2001; 20:13-22.

170. Pesheva P, Gloor S, Probstmeier R. Tenascin-R as a regulator of CNS glial cell function. Prog Brain Res 2001; 132:103-14.
171. Lochter A, Vaughan L, Kaplony A et al. J1/tenascin in substrate-bound and soluble form displays contrary effects on neurite outgrowth. J Cell Biol 1991; 113:1159-71.
172. Neidhardt J, Fehr S, Kutsche M, Lohler J et al. Tenascin-N: characterization of a novel member of the tenascin family that mediates neurite repulsion from hippocampal explants. Mol Cell Neurosci 2003; 23:193-209.
173. Probstmeier R, Braunewell K, Pesheva P. Involvement of chondroitin sulfates on brain-derived tenascin-R in carbohydrate-dependent interactions with fibronectin and tenascin-C. Brain Res 2000; 863:42-51.
174. de Wit J, Verhaagen J. Role of semaphorins in the adult nervous system. Prog Neurobiol 2003; 71:249-67.
175. Pasterkamp RJ, Verhaagen J. Emerging roles for semaphorins in neural regeneration. Brain Res Brain Res Rev 2001; 35:36-54.
176. Pasterkamp RJ, Kolodkin AL. Semaphorin junction: making tracks toward neural connectivity. Curr Opin Neurobiol 2003; 13:79-89.
177. De Winter F, Oudega M, Lankhorst AJ et al. Injury-induced class 3 semaphorin expression in the rat spinal cord. Exp Neurol 2002; 175:61-75.
178. Pasterkamp RJ, Giger RJ, Ruitenberg MJ et al. Expression of the gene encoding the chemorepellent semaphorin III is induced in the fibroblast component of neural scar tissue formed following injuries of adult but not neonatal CNS. Mol Cell Neurosci 1999; 13:143-66.
179. Niclou SP, Franssen EH, Ehlert EM et al. Meningeal cell-derived semaphorin 3A inhibits neurite outgrowth. Mol Cell Neurosci 2003; 24:902-12.
180. Moreau-Fauvarque C, Kumanogoh A, Camand E et al. The transmembrane semaphorin Sema4D/CD100, an inhibitor of axonal growth, is expressed on oligodendrocytes and upregulated after CNS lesion. J Neurosci 2003; 23:9229-39.
181. Goldberg JL, Vargas ME, Wang JT et al. An oligodendrocyte lineage-specific semaphorin, Sema5A, inhibits axon growth by retinal ganglion cells. J Neurosci 2004; 24:4989-99.
182. Fournier AE, GrandPre T, Strittmatter SM. Identification of a receptor mediating Nogo-66 inhibition of axonal regeneration. Nature 2001; 409:341-6.
183. Domeniconi M, Cao Z, Spencer T et al. Myelin-associated glycoprotein interacts with the Nogo66 receptor to inhibit neurite outgrowth. Neuron 2002; 35:283-90.
184. Liu BP, Fournier A, GrandPre T et al. Myelin-Associated Glycoprotein as a Functional Ligand for the Nogo-66 Receptor. Science 2002; 1073031.
185. Barton WA, Liu BP, Tzvetkova D et al. Structure and axon outgrowth inhibitor binding of the Nogo-66 receptor and related proteins. Embo J 2003; 22:3291-302.
186. He XL, Bazan JF, McDermott G et al. Structure of the Nogo receptor ectodomain: a recognition module implicated in myelin inhibition. Neuron 2003; 38:177-85.
187. Venkatesh K, Chivatakarn O, Lee H et al. The Nogo-66 receptor homolog NgR2 is a sialic acid-dependent receptor selective for myelin-associated glycoprotein. J Neurosci 2005; 25:808-22.
188. Wang KC, Kim JA, Sivasankaran R et al. P75 interacts with the Nogo receptor as a co-receptor for Nogo, MAG and OMgp. Nature 2002; 420:74-8.
189. Wong ST, Henley JR, Kanning KC et al. A p75(NTR) and Nogo receptor complex mediates repulsive signaling by myelin-associated glycoprotein. Nat Neurosci 2002; 5:1302-8.
190. Mi S, Lee X, Shao Z et al. LINGO-1 is a component of the Nogo-66 receptor/p75 signaling complex. Nat Neurosci 2004; 7:221-8.
191. Schweigreiter R, Walmsley AR, Niederost B et al. Versican V2 and the central inhibitory domain of Nogo-A inhibit neurite growth via p75NTR/NgR-independent pathways that converge at RhoA. Mol Cell Neurosci 2004; 27:163-74.
192. Yamashita T, Higuchi H, Tohyama M. The p75 receptor transduces the signal from myelin-associated glycoprotein to Rho. J Cell Biol 2002; 157:565-70.
193. Park JB, Yiu G, Kaneko S et al. A TNF receptor family member, TROY, is a coreceptor with Nogo receptor in mediating the inhibitory activity of myelin inhibitors. Neuron 2005; 45:345-51.
194. Shao Z, Browning JL, Lee X et al. TAJ/TROY, an orphan TNF receptor family member, binds Nogo-66 receptor 1 and regulates axonal regeneration. Neuron 2005; 45:353-9.

195. McKerracher L. Ganglioside rafts as MAG receptors that mediate blockade of axon growth. Proc Natl Acad Sci USA 2002; 99:7811-3.
196. Vinson M, Strijbos PJ, Rowles A et al. Myelin-associated glycoprotein interacts with ganglioside GT1b. A mechanism for neurite outgrowth inhibition. J Biol Chem 2001; 276:20280-5.
197. Vyas AA, Patel HV, Fromholt SE et al. Gangliosides are functional nerve cell ligands for myelin-associated glycoprotein (MAG), an inhibitor of nerve regeneration. Proc Natl Acad Sci USA 2002; 99:8412-7.
198. Yang LJ, Zeller CB, Shaper NL et al. Gangliosides are neuronal ligands for myelin-associated glycoprotein. Proc Natl Acad Sci USA 1996; 93:814-8.
199. Etienne-Manneville S, Hall A. Integrin-mediated activation of Cdc42 controls cell polarity in migrating astrocytes through PKCzeta. Cell 2001; 106:489-98.
200. Lehmann M, Fournier A, Selles-Navarro I et al. Inactivation of Rho signaling pathway promotes CNS axon regeneration. J Neurosci 1999; 19:7537-47.
201. Niederost B, Oertle T, Fritsche J et al. Nogo-A and myelin-associated glycoprotein mediate neurite growth inhibition by antagonistic regulation of RhoA and Rac1. J Neurosci 2002; 22:10368-76.
202. Dergham P, Ellezam B, Essagian C et al. Rho signaling pathway targeted to promote spinal cord repair. J Neurosci 2002; 22:6570-7.
203. Monnier PP, Sierra A, Schwab JM et al. The Rho/ROCK pathway mediates neurite growth-inhibitory activity associated with the chondroitin sulfate proteoglycans of the CNS glial scar. Mol Cell Neurosci 2003; 22:319-30.
204. Yamashita T, Tucker KL, Barde YA. Neurotrophin binding to the p75 receptor modulates Rho activity and axonal outgrowth. Neuron 1999; 24:585-93.
205. Yamashita T, Tohyama M. The p75 receptor acts as a displacement factor that releases Rho from Rho-GDI. Nat Neurosci 2003; 6:461-7.
206. Borisoff JF, Chan CC, Hiebert GW et al. Suppression of Rho-kinase activity promotes axonal growth on inhibitory CNS substrates. Mol Cell Neurosci 2003; 22:405-16.
207. Fournier AE, Takizawa BT, Strittmatter SM. Rho kinase inhibition enhances axonal regeneration in the injured CNS. J Neurosci 2003; 23:1416-23.
208. Riento K, Ridley AJ. Rocks: multifunctional kinases in cell behaviour. Nat Rev Mol Cell Biol 2003; 4:446-56.
209. Qiu J, Cai D, Dai H et al. Spinal axon regeneration induced by elevation of cyclic AMP. Neuron 2002; 34:895-903.
210. Gao Y, Deng K, Hou J et al. Activated CREB is sufficient to overcome inhibitors in myelin and promote spinal axon regeneration in vivo. Neuron 2004; 44:609-21.
211. Spencer T, Filbin MT. A role for cAMP in regeneration of the adult mammalian CNS. J Anat 2004; 204:49-55.
212. Nikulina E, Tidwell JL, Dai HN et al. The phosphodiesterase inhibitor rolipram delivered after a spinal cord lesion promotes axonal regeneration and functional recovery. Proc Natl Acad Sci USA 2004; 101:8786-90.
213. Ellezam B, Dubreuil C, Winton M et al. Inactivation of intracellular Rho to stimulate axon growth and regeneration. Prog Brain Res 2002; 137:371-80.
214. Strittmatter S.M. Modulation of axonal regeneration in neurodegenerative disease: focus on Nogo. J Mol Neurosci 2002; 19:117-21.
215. Boyd JG, Doucette R, Kawaja MD. Defining the role of olfactory ensheathing cells in facilitating axon remyelination following damage to the spinal cord. Faseb J 2005; 19:694-703.
216. Reier PJ. Cellular Transplantation Strategies for Spinal Cord Injury and Translational Neurobiology. Neurorx 2004; 1:424-451.
217. Barnett SC. Olfactory ensheathing cells: unique glial cell types? J Neurotrauma 2004; 21:375-82.
218. Bunge MB, Pearse DD. Transplantation strategies to promote repair of the injured spinal cord. J Rehabil Res Dev 2003; 40:55-62.
219. Jones DG, Anderson ER, Galvin KA. Spinal cord regeneration: moving tentatively towards new perspectives. NeuroRehabilitation 2003; 18:339-51.
220. Hendriks WT, Ruitenberg MJ, Blits B et al. Viral vector-mediated gene transfer of neurotrophins to promote regeneration of the injured spinal cord. Prog Brain Res 2004; 146:451-76.

221. Barnett SC, Chang L. Olfactory ensheathing cells and CNS repair: going solo or in need of a friend? Trends Neurosci 2004; 27:54-60.
222. Barnett SC, Riddell JS. Olfactory ensheathing cells (OECs) and the treatment of CNS injury: advantages and possible caveats. J Anat 2004; 204:57-67.
223. Nieto-Sampedro M. Central nervous system lesions that can and those that cannot be repaired with the help of olfactory bulb ensheathing cell transplants. Neurochem Res 2003; 28:1659-76.
224. Santos-Benito FF, Ramon-Cueto A. Olfactory ensheathing glia transplantation: a therapy to promote repair in the mammalian central nervous system. Anat Rec B New Anat 2003; 271:77-85.
225. Blits B, Boer GJ, Verhaagen J. Pharmacological, cell, and gene therapy strategies to promote spinal cord regeneration. Cell Transplant 2002; 11:593-613.
226. Raisman G. Olfactory ensheathing cells - another miracle cure for spinal cord injury? Nat Rev Neurosci 2001; 2:369-75.
227. Fry EJ. Central nervous system regeneration: mission impossible? Clin Exp Pharmacol Physiol 2001; 28:253-8.
228. Stichel CC, Hermanns S, Luhmann HJ et al. Inhibition of collagen IV deposition promotes regeneration of injured CNS axons. Eur J Neurosci 1999; 11:632-646.
229. Blesch A, Lu P, Tuszynski MH. Neurotrophic factors, gene therapy, and neural stem cells for spinal cord repair. Brain Res Bull 2002; 57:833-8.
230. David S, Lacroix S. Molecular approaches to spinal cord repair. Annu Rev Neurosci 2003; 26:411-40.
231. Fournier AE, Gould GC, Liu BP et al. Truncated soluble Nogo receptor binds Nogo-66 and blocks inhibition of axon growth by myelin. J Neurosci 2002; 22:8876-83.
232. Li S, Liu BP, Budel S et al. Blockade of Nogo-66, myelin-associated glycoprotein, and oligodendrocyte myelin glycoprotein by soluble Nogo-66 receptor promotes axonal sprouting and recovery after spinal injury. J Neurosci 2004 24:10511-20.
233. Stichel CC, Lausberg F, Hermanns S et al. Scar modulation in subacute and chronic CNS lesions: Effects on axonal regeneration. Restor Neurol Neurosci 1999; 15:1-15.
234. Hermanns S, Reiprich P, Muller HW. A reliable method to reduce collagen scar formation in the lesioned rat spinal cord. J Neurosci Methods 2001; 110:141-6.
235. Weidner N, Grill RJ, Tuszynski MH. Elimination of basal lamina and the collagen "scar" after spinal cord injury fails to augment corticospinal tract regeneration. Exp Neurol 1999; 160:40-50.
236. Lemons ML, Howland DR, Anderson DK. Chondroitin sulfate proteoglycan immunoreactivity increases following spinal cord injury and transplantation. Exp Neurol 1999; 160:51-65.
237. Raineteau O, Schwab ME. Plasticity of motor systems after incomplete spinal cord injury. Nat Rev Neurosci 2001; 2:263-73.

CHAPTER 5

The Role of Ionotropic Purinergic Receptors (P2X) in Mediating Plasticity Responses in the Central Nervous System

Fulvio Florenzano, Maria Teresa Viscomi, Fabio Cavaliere, Cinza Volonté and Marco Molinari

The past few years have witnessed increasing interest in the field of purinergic signalling and have recognised ATP as an extracellular messenger eliciting a wide array of physiological effects in several different tissues. These effects range from simple biological events such as neurotransmission, secretion and vasodilatation, to complex biological phenomena such as development, cell death and regeneration. Purinergic actions are mediated by specific receptors, either for purine nucleotides (P2 receptors for ATP), or for purine nucleosides (P1 receptors for adenosine). The present review will discuss the involvement of purinoceptors (particularly P2X subtypes) in the CNS reaction to epigenetic insults. After briefly reviewing in vitro data, we will describe recently obtained results demonstrating modulation of the P2X receptor subunits following CNS damage in in vivo models. In addition, the role of P2X receptor in neural plasticity and in neuronal responses to damage will be discussed.

Neurobiology of P2 Receptors

Some properties of the adenosine 5'-triphosphate (ATP) molecule make it an ideal transmitter in cell-cell signalling: it is small and quickly diffusing, highly unstable and not abundant in the extracellular space. The ATP signalling function and the vesicular release of ATP during excitatory transmission have been definitively demonstrated in PNS neurons[1,2] and in CNS neurons.[3,4] The diffusion and the actions of the nucleotide ATP in the extracellular space is terminated by ectoATPases.[5] This family of enzymes sequentially hydrolyses phosphate groups until the nucleoside adenosine is liberated.[5,6]

In the nervous system, P2 receptors are widely expressed and two subclasses have been identified: P2X, fast ionotropic receptors responsible for fast excitatory neurotransmission, and slow P2Y, G-protein coupled metabotropic receptors.[7,8] Both receptor classes have been reported in the PNS and CNS.[6,9] Changes in the expression of purinergic receptors are frequently observed, not only as a function of neuronal maturation and differentiation,[10,11] but also in different pathological conditions, hence justifying the exploitation of purinergic agonists and antagonists as potential therapeutic agents.[12-18] P2X receptor structure and function have been mainly investigated in in vitro models. They have been reported to mediate fast synaptic transmission, neuromodulation,[6,8] cell death, differentiation and to act as trophic

Brain Repair, edited by Mathias Bähr. ©2006 Eurekah.com and Kluwer Academic / Plenum Publishers.

agents.[9,19,20] Recently, in vivo studies demonstrated P2X involvement in many different physiological functions, ranging from controlling normal male reproduction[21] to central chemosensory signalling.[22,23]

Functional Properties of the ATP Gated P2 Receptor Channels

P2X receptors are cation-selective channels mediating fast excitatory synaptic transmission (within milliseconds) and presenting low affinity, in the μm range, for ATP. These receptors present almost equal permeability to Na+ and K+ and substantial Ca2+ permeability.[2,6,8] Seven distinct P2X subunits ($P2X_{1-7}$) have been cloned from mammalian species and all can form either homomultimer or heteromultimer combinations, of which the minimum stoichiometric ratio is a trimer.[24,25] Different receptor subunit combinations yield different receptor phenotypes allowing diversity in transmission signalling. P2X receptor phenotypes differ in agonist and antagonist selectivity, desensitisation and other channel properties.[4,6,9] Therefore, it has been suggested that different P2X receptor combinations may be linked to different physiological functions.[20,25] This hypothesis was confirmed in recent studies on the P2X phenotypes expressed by peripheral neurons.[22,26,27] Dorsal root ganglion (DRG) neurons present differences in P2X receptor combinations associated with differences in cell sensitivity to capsaicin.[26,28] In particular, homomeric $P2X_3$ receptors are mainly expressed in nociceptive capsaicin-sensitive small DRG neurons, while the heteromeric $P2X_{2/3}$ receptors are mainly expressed in non-nociceptive capsaicin-insensitive medium DRG neurons.[28] Immunohistochemical and in situ hybridisation studies have demonstrated the co-expression of different P2X receptor subtypes in sensory trigeminal, dorsal root and nodose ganglion neurons,[4,26,29] as well in CNS neurons.[2,30,31] Peripheral neurons present considerable heterogeneity in their physiological responses to ATP.[4] It has been proposed that this heterogeneity depends on varying proportions of homomeric and heteromeric receptors in the different cell populations.[2,4,31]

Cellular expression of different P2X receptor phenotypes seems to be a common feature in the nervous system. However, also in those cells that express several P2X receptor subunits, the overall receptor functional properties correspond to a single phenotype.[6,24] Therefore, it has been suggested that, in the case of multiple expression of different subunits, a particular one dominates cell functioning.[2] Consistently, it has been argued that the receptor subunit profile may regulate the receptor function through modifications of the receptor molecular properties.

P2X Receptors: Interactions with other Transmitter Systems and Regulation of Ca^{2+} Flow

Despite the lack of structural homologies to the aminoacidic sequences of other ligand gated channels,[6,8] similarities in activation kinetics and permeability of P2X receptors to nicotinic and glutamate receptor superfamilies have been evidenced.[32] The above similarities, the wide expression in the nervous system and the presence of many well known P2X receptor pharmacological agonists and antagonists[33] led to the proposal that P2X receptors are a third class of ligand gated channels in the nervous system. In particular, the interest of investigators has been attracted not only by the P2X receptor similarities to other families of ligand gated channels, but also by the P2X receptor capacity to influence the functions of the other receptor families. Indeed, interactions between ATP and ACh receptors on the postsynaptic side have been suggested since the postsynaptic response, during co-transmission of ATP and Ach, is not the linear sum of the two individual synaptic potentials.[34,35] These studies indicated that the concept of receptor functional independence might not apply to P2X receptors. In a recent review Khakh et al[6] proposed the existence of structural and functional interdependence of the P2X receptors with other families of receptors starting from the demonstration of the interaction/co-assembly of P2X receptors with other families of ligand gated channels.

The co-release of ATP and different neurotransmitters on the presynaptic side,[2,36] and the contribution of P2X currents to glutamate currents[3] on the postsynaptic side have raised some interesting questions about ATP involvement in the generation of action potentials. It has been reported that the currents carried out by the P2X receptors are physiologically important for neurotransmission at least in the medial habenula and in sympathetic neurons.[37,38] Nevertheless, when compared to the currents elicited from glutamatergic or cholinergic inputs, P2X currents are relatively small; this suggests a minor contribution of P2X receptors to the generation of action potentials.[6] Taken together, these data suggest that P2X receptors do not directly affect the generation of action potentials, but more likely provide a channel for the modulation of cell excitability and neurotransmission.[2,6]

One of the most interesting functions of P2X receptors is their influence on Ca2+ influx and on the release of neurotransmitters. In cultured DRG neurons, Ca2+ influx through P2X receptors was shown to induce glutamate and substance P release.[3,39] ATP induced presynaptic currents in cholinergic presynaptic nerve terminals and co-release of the two transmitters were demonstrated in the chicken ciliary ganglion.[40] In cultured cerebellar granule neurons, extracellular ATP induces aspartate release.[41] Of particular relevance is the P2X receptor mediated Ca2+ entry observed after co-release of a fast excitatory and a fast inhibitory transmitter such as ATP and GABA. This experiment was the first demonstration of a co-release of excitatory and inhibitory transmitters from the same presynaptic cell[36,42] and marshalled some very interesting aspects regarding the modulation of the Ca2+ concentration in a hyperpolarised cell.[2] In cultured spinal dorsal horn neurons, ATP and GABA co-release was shown to lead to a membrane hyperpolarisation because of the greater conductance of the GABA receptor activation.[36] Neither voltage gated nor ligand gated channels, such as AMPA or NMDA receptors, permit Ca2+ influx under hyperpolarised conditions. On the other hand, due to the driving force for Ca2+ influx through the open P2X voltage independent channel, Ca2+ influx through the P2X receptors is maximised. In the same conditions, depolarisation owing to Ca2+ entrance through P2X receptors is minimal. Nevertheless, it is possible to achieve a fast local rise in Ca2+ under completely different conditions from those typical of other known Ca2+ signalling pathways. All the above considerations led Robertson et al[2] to suggest a possible synaptic physiological role of P2X receptor mediated currents mainly in modulating Ca2+ influx, rather than in cell depolarisation.

There is general agreement about the importance of Ca2+ concentrations in inducing plasticity responses. Ca2+ spatiotemporal patterns at presynaptic and postsynaptic sites are critical for different downstream signalling functions.[43-45] Different forms of short-term synaptic plasticity phenomena, such as facilitation, augmentation, and post-tetanic potentiation, depend on pre-synaptic Ca2+ levels.[46] Similarly, long-term synaptic functional changes, such as LTP and LTD, are related to postsynaptic Ca2+ content.[43] Presynaptic or postsynaptic elevation of Ca2+ can be elicited by Ca2+ release from intracellular sources or/and by Ca2+ entry from extracellular space through ligand gated channels.[45] The best characterised sources for Ca2+ entry are glutamate gated receptor channels and voltage gated calcium channels.[47] In addition, as indicated previously, P2X receptor channels can play a crucial role by mediating Ca2+ entry in conditions completely different from those of the other Ca2+ sources. In general, P2X receptors present substantial permeability to Ca2+ ions; this permeability is significantly modulated by their subunit composition, and P2X receptors can act on both presynaptic and postsynaptic sides.[2,31] However, as indicated in the previous paragraph, P2X receptors elicit interactions with the other families of ligand gated channels which can modulate intracellular Ca2+ concentration[6] and, consequently, cell plasticity responses.

P2X and Neuronal Morphological Plasticity

The influence of activity on sculpting specific patterns of dendritic arbors and on the development of circuits in the nervous system is among the most interesting and puzzling scientific topics. Spine plasticity and dendritic remodelling are considered the morphological manifestations of activity-dependent changes in neuronal behaviour.[44,48-52] The importance of P2X in mediating activity-dependent morphological adaptations is suggested by recent data demonstrating inducible redistribution of the neuronal localisation of P2X receptors.[53-55] ATP application has been shown to induce a redistribution of $P2X_1$ receptor clusters in cultured superior cervical ganglia neurons,[53] in Xenopus oocytes and in human embryonic kidney 293 cells.[54] In these studies, $P2X_1$-GFP adenoviral infection was used to follow the time course of $P2X_1$ receptor clustering in cytoplasmatic membranes. After a few seconds of ATP exposure, $P2X_1$-GFP receptor chimera spontaneously form synaptic-size clusters in the plasma membrane that are dispersed and, subsequently, internalised.[53]

Further, changes in $P2X_2$ receptor subcellular localisation affecting cellular morphology were reported in ex-vivo preparations.[55] In untreated dissociated cultures of embryonic hippocampal neurons, $P2X_2$ receptor was distributed mainly in the cytosolic domain. Following ATP administration, $P2X_2$ receptor aggregates formed varicosities which migrated over a few micrometers into neuronal filopodia, affecting dendritic morphology.[55] The observed redistribution was reversible and not directly linked to cellular depolarisation. The above data demonstrate that ATP availability influences P2X receptor cellular distribution and that changes in P2X receptor localisation are able to modify the morphological cellular aspect in in vitro systems. Confirmation of these data in in vivo models would be very interesting. Activity-dependent changes in dendritic morphology are of paramount importance for understanding the mechanisms that control adaptation in the mature CNS,[49] as well as during development[56] or in pathological conditions.[57]

P2X Distribution and Localisation

Inferences about the physiological functions of P2X receptors can be drawn from their CNS distribution features, from the cellular phenotypes expressing P2X receptors and from P2X receptors sub-cellular localisation. Early in situ hybridisation studies showed widespread expression of P2X receptor subunits in the CNS. With the availability of P2X subunit specific antibodies, the presence of P2X receptors in the CNS was confirmed and characterised using immunocytochemical methods. In one extensive immunocytochemical study on $P2X_2$ receptor subunit expression in the CNS, different structures were demonstrated to express this subunit.[58] The pattern of the $P2X_2$ receptor CNS expression is extremely widespread and includes broad sectors of the cerebral cortex, different rhinencephalic structures, many thalamic, hypothalamic and basal ganglia nuclei as well as sensory, motor and integration nuclei of the brainstem and the cerebellar cortex. This wide distribution and the evidence that different systems rely on $P2X_2$ receptor signalling strongly suggest its involvement in basic cellular mechanisms.

The topographical distribution of $P2X_{1\text{-}6}$ receptor subunits in the rat and common marmoset hindbrain was investigated by using antibodies directed against an extracellular portion of the different receptors.[59] This study showed much wider distribution of all six P2X receptor subunits than previously reported.[60,61] Within this general pattern, distinctions were found between brain nuclei and P2X receptor subunits expression. The medial vestibular nucleus and the nucleus of the solitary tract tended to present the highest level of positivity for all antisera in both species. These findings was recently confirmed in a study specifically devoted to the characterisation of $P2X_{1\text{-}5}$ receptor subunits expression in the nucleus of the solitary tract.[62] Of particular interest is the evidence that different subunits are differently expressed within the various subdivisions of the nucleus suggesting a possible correlation between P2X subunits expression and neuronal phenotype within the nucleus.

Furthermore, in agreement with the evidence that exogenously applied ATP causes hemodynamic changes in vivo,[63,64] immunopositive neurons for different P2X receptor subunits were reported in many brainstem structures involved in the central regulation of the cardiovascular mechanisms.[59] In addition, P2X receptor subunit immunopositivity was reported in different sensory (medial vestibular nucleus, spinal trigeminal nucleus) and motor (hypoglossal and motor trigeminal nuclei) brainstem structures. These latter findings are consistent with the hypothesised P2X receptors role in sensory[6,65] and motor[59,66] functions.

The possible involvement of purinergic receptors in hormone release and in membrane recycling was proposed in a recent study demonstrating the expression of the $P2X_6$ receptor subunit in neurons and axons of the rat hypothalamo-neurohypophysical system.[67]

Subcellular localisation is another important aspect which can help assess the P2X receptor actions on cellular functions. P2X receptor subunits were demonstrated by immunocytochemistry in neuronal somata, dendrites, axons and terminals.[29,58,59,61,68] Interestingly, immunohistochemical and ultrastructural studies showed $P2X_{2,4,6}$ receptor subunits targeted to subpopulations of the dendritic spines of Purkinje and CA1 pyramidal neurons.[69] In particular, in the Purkinje cells this arrangement of purinergic receptor subunits was opposed to parallel fibre terminals and was not observed in the spines opposed to climbing or basket terminals. This specificity in the subunit distribution among populations of spines within the same neuron was also observed in CA1 pyramidal neurons. In the latter case, only spines postsynaptic to Shaffer collateral terminals contained $P2X_{2,4,6}$ receptor subunits.[69] It is worth stressing that both systems are known to be capable of a high level of plasticity and thus the above mentioned data strongly suggest the involvement of P2X receptors in controlling adaptation of glutamatergic transmission. Further support for the importance of P2X receptors control over synaptic functions derives from immunogold electron microscopic data showing both presynaptic and postsynaptic distribution of the $P2X_{2,4,6}$ receptor subunits.[69] Therefore, P2X receptors can affect presynaptic modulation of transmitter release as well as postsynaptic excitatory fast currents.

P2X Signalling and Development

Activity-driven Ca2+ influx into the cell can lead to the induction of the new gene transcription[70] required to shape and wire the brain during development. Ca2+ spikes and waves directly control neuritogenesis in cultures of amphibian spinal neurons.[71,72] Intracellular Ca2+ spikes can modulate neuronal migration[73] or neurogenesis of neocortical precursor cells.[74] P2X receptors have been repeatedly indicated to play a role in development.[75,4] Therefore, Ca2+ influx through the P2X receptors appears to be a promising field of CNS development research. In the chick embryo P2X receptor subunits are expressed at early stages of skeletal muscle development; the expression disappears before the stage in which myoblasts fuse to form myotubes.[76] $P2X_1$ and $P2X_2$ receptor subunits were shown to play a key role in the development of afferent and efferent innervation and morphogenesis of the cochlea.[77,78] $P2X_1$ receptor subunit expression presents specific localisations in definite developmental time windows; from E16 to P6 in the spiral ganglion neurons, from E18 to P6 in the efferent and afferent nerve fibres to the spiral ganglion.[78] The $P2X_2$ receptor subunit is expressed both in the adult and in the developing rat cochlea. As early as E19, $P2X_2$ expression is detected in the spiral ganglion and in associated nerve fibres extending to the inner hair cells and underneath the outer hair cells. By P6, $P2X_2$ receptor subunit is seen in the synaptic regions of both types of hair cells. This expression becomes most prominent from the onset of cochlear functioning (P8-P12). By P21, the pattern of immunolabelling is similar to that found in the adult, i.e., there is a weak expression only in some spiral ganglion neurons.

Developmental changes of P2X receptors associated with changes in synaptic plasticity were reported in preparations of dissociated rat spinal cord neurons. Glycinergic presynaptic nerve terminals, possibly from dorsal horn interneurons, terminating on substantia gelatinosa neurons presented non-desensitising P2X receptors at early stages of postnatal development. During postnatal development, a shift was found toward a more desensitising organisation of P2X receptors.[79] These findings were interpreted as an indication that P2X receptor changes may contribute toward the regulation of neuronal excitability and in particular the regulation of pain signals.

$P2X_3$ receptor subunit expression has often been associated with sensory innervations and their transient expression during development has been reported in both CNS and PNS structures.[75,80] Surprisingly, until now the $P2X_3$ receptor subunit is the only subtype of purinergic receptor that has been reported in early CNS embryonic development. The $P2X_3$ receptor subunit is expressed as early as E11 in the neural tube and in the sensory ganglia. During successive embryonic development a strong association with neural crest derivatives is present and a major role of $P2X_3$ receptor subunit is proposed in sensory processing and possibly in somatic and autonomic functions during embryonic development.[75] In the early postnatal period, $P2X_3$ expression down regulates. Although all researchers agree that $P2X_3$ receptor subunit expression is less diffuse in the adult, conflicting results have been reported regarding its distribution mainly related to differences in the antibodies used (see ref. 75 for a discussion of this matter). Another point that needs to be mentioned is that co-expression of $P2X_2$ and $P2X_3$ receptor subunits was present in the sensory structures of E16 embryos.[75] This co-expression followed the first appearance of $P2X_3$ receptor subunit expression in the same structures after 5 days.

The transient expression and/or co-expression of different P2X receptor subunits during development, and in particular the presence of the $P2X_3$ receptor subunit in early embryonic stages, indicate that ATP signalling plays a significant role in CNS developmental mechanisms. The characterisation of purinergic receptors during development will certainly provide key elements for a better understanding of CNS development.

P2X, Glial Signalling and Plasticity Responses

Glial cells use ATP as an extracellular signal mediating communication in glial-glial and glial-neuronal interactions and ATP has been demonstrated to be the primary glial extracellular messenger.[14,19,81] The role of P2X receptors for ATP signalling between glial cells is suggested by the presence of different P2X receptor subunits on astrocytes[82] and on microglial cells.[83] The functioning of these receptors is shown by the block of calcium wave propagation in cultured astrocytes after the addition of purinergic antagonists.[84] Indeed, a recent study demonstrated that the spatial expansion of Ca2+ waves is mediated by ATP and subsequent activation of purinergic receptors.[85] Interestingly, Ca2+ entry through purinergic receptors and its diffusion through gap junctions were proposed as signalling mechanisms in glial cells.[14,86]

Purinergic signals can induce functional changes instrumental to development, differentiation and survival of both neurons and glial cells.[19,81] In this regard, it is interesting to stress that calcium waves, which are diffused through gap junctions and modulated by purinergic agonist/antagonists, can induce growth and differentiation in glial neuronal co-cultures.[85] Further, the direct influence of neuronal activity on glia can be shown by the capacity of DRG neurons to influence Schwann cell development and proliferation by releasing ATP. This release increases intracellular Ca2+ in Schwann cells through activation of P2 receptors.[87] Similarly, ATP analogues were shown to induce astrocyte differentiation in primary cultures of rat striatum.[33,88] In this experimental condition, the elongation of astrocyte processes was blocked by the administration of suramin, a potent P2X receptor antagonist.[88] These correlations between ATP release and glial responses prompted further investigations of the role of purinergic signalling in controlling astroglial cell differentiation during brain development and ischaemia- and trauma-associated hypergliosis.[89-93]

Microglia and astrocytes are the non-neuronal cells which respond to CNS injury[94,95] and ATP is a well known signal of cell damage. It was shown that P2X receptors are expressed in rat (ref. 96 and 97) and human microglial cells.[98] The capacity of ATP to influence microglial cells was shown in culture where ATP stimulation induced Ca2+ transients because of Ca2+ entry through P2X receptors[99] and induced microglia ramification.[100]

Some recent studies pointed out the selective involvement of the $P2X_7$ receptor subunit in the purinergic communication between astrocytes and microglial cells[83] and in microglial inflammatory cytokine release.[101,102] Astrocyte released ATP mediates paracrine activation of the microglial $P2X_7$ receptor which provokes a Ca2+ response in microglial cells.[83] Repeated stimulations of microglial cells by astrocyte-released ATP activate $P2X_7$ receptor on microglial cells and greatly increase membrane permeability, eventually leading to microglial apoptosis. IFN-gamma increases ATP release and potentiates the $P2X_7$ receptor mediated cytolytic effect. ADP and AMP stimulation can induce release of IL-1 from microglial cells and ATP stimulation of $P2X_7$ receptor strengthens this effect.[102]

Taken together these data indicate the strict links between purinergic signalling and glial mediated CNS inflammatory responses. The purinergic mediated release of inflammatory cytokines from microglial cells is of particular relevance taking into account the suggested role of cytokines in different CNS pathologies. Cytokine mediated inflammation mechanisms were suggested as pathogenic mechanisms in multiple sclerosis, Alzheimer's and Parkinson's disease.[103] The availability of specific purinergic agonists/antagonists and the definition of the P2X receptors role in controlling the inflammatory cascade is providing important data for planning new therapeutic strategies for neurodegenerative diseases.

P2X in in Vivo Plasticity Models

As previously indicated, in vitro and in vivo data suggest that P2X receptors may have a role in mediating neuronal responses to injuries. ATP release, shown in different pathological conditions, can exert both protective and neurotoxic effects; these effects depend at least partially on the activity of P2X receptors. Specific purinergic agonist/antagonists provide important experimental tools. A better understanding of the P2X receptors role in different pathological conditions may lead to the development of pharmacological strategies potentially useful in different neurological diseases.[15,18]

P2X Expression after Peripheral Nerve Injury

In recent years an increasing number of studies have reported changes in the expression of P2X receptors after PNS injuries, supporting the idea that purinergic receptors are involved in the modulation of the cellular response to damage. P2X receptor subunits are expressed physiologically in DRG neurons,[4,104] and they are involved in the processing of noxious information.[15,6] Transection of peripheral nerves induces a down-regulation of the $P2X_3$ subunit mRNA and protein content in DRG neurons. This down-regulation is blocked by intrathecal delivery of the glial derived neurotrophic factor (GDNF).[65,105] Two different mechanisms were proposed to explain the effect of GDNF administration on $P2X_3$ receptor subunit expression: an increase in protein synthesis or a protection against cell death.[106] Although at present it is not possible to discriminate between these two hypotheses, it is conceivable that $P2X_3$ receptors are involved in mediating cell responses to both injures and trophic factors.

Further support for the hypothesis of involvement of the $P2X_3$ receptor in neuronal postlesional plasticity derives from studies of uninjured DRG neurons. Administration of GDNF up-regulates the $P2X_3$ subunit in the DRG neurons that normally express this receptor. In contrast, administration of NGF induces de novo $P2X_3$ subunit expression.[107] These latter results agree with in vitro data demonstrating that under neurite regenerating conditions NGF administration to PC12 cells induces up-regulation of $P2X_{2,3,4}$ protein expression.[20,25] Furthermore, up-regulation of $P2X_3$ subunits in DRG neurons was shown following chronic constriction injury.[108] In this model, the increase in the $P2X_3$ subunit was interpreted as dependent on the protein transport block secondary to the nerve ligation. It was also shown that P2X receptors are involved in peripheral inflammation mechanisms. An injection of CFA into the hindpaw provoked the up-regulation of P2X receptors in DRG neurons.[109] It was suggested that the over expression of P2X receptors might mediate increased ATP responses which, in turn, might account for abnormal pain responses. Based on this hypothesis, P2X receptor blockers were suggested as useful tools for inflammatory and neuropathic pain therapy.[109]

It is well known that severed PNS neurons survive and regenerate their axons.[110] Understanding the signalling mechanisms that regulate PNS neuronal survival and regeneration may help in the development of approaches to support the regenerative attempts of central axons. Damaged cells release ATP and P2X receptors up-regulation may be a mechanism to increase or reduce cellular sensitivity to damage signals. An increase of P2X receptor expression augments the mediated currents and consequently the Ca2+ influx, which in turn promotes intracellular signalling and plasticity responses. The importance of this purinergic route is underlined by the capacity of trophic factors to modulate purinergic post-lesional responses, as shown by the dependence of the PNS neuronal P2X receptors expression on neurotrophin supply in cultured DRG neurons.[107] The role of neurotrophic factors in peripheral neuron survival and maintenance in normal and neuropathic conditions is well documented.[110] In fact, neurotrophin administration has been proposed for the treatment of peripheral neuropathies.[111-113] Up-regulation of P2X receptor expression after neurotrophin administration in normal and neuropathic conditions suggests a positive correlation for neuronal survival.[114] The interactions between trophic factors and P2X receptors open up interesting perspectives for the development of plasticity promoting drugs. Indeed, if neurotrophic factors act directly on P2X receptors expression, pharmacological interactions between trophic factors and P2X receptor agonist/antagonists could be effective in rescuing damaged neurons.[20,25]

P2X Expression after CNS Injury

Glial Reaction

Expression of P2X receptor subunits in glial cells is well documented (see paragraph P2X, glial signalling and plasticity responses) and in vitro manipulations can induce changes in the $P2X_7$ receptor subunit expression pattern[115] as well in purinergic receptor profiles.[99,116] Data from in vivo models are less common and conflicting. Intraperitoneal injection of kainic acid induces hippocampal gliosis proportional to the gravity of the induced seizures. Slice preparations can be obtained from this tissue that are apt for studying the electrical properties of reactive glial cells. In this model, Jabs and coworkers[117] reported a lack of P2 receptor involvement in the genesis of the reactive gliosis. Opposite findings were reported in a different in vivo model of induced astrogliosis. Injection of a mixture of artificial CSF and a P2 agonist into the nucleus accumbens of the rat was shown to produce a clear increment of astrogliosis compared to an injection of artificial CSF alone. This effect was clearly reduced if P2 selective antagonists were added to the injected mixture.[90] Findings supporting the idea of P2X involvement in controlling glial reactions were reported also in a stab wound model of reactive astrogliosis. After mechanical lesion of the nucleus accumbens, astrocytes presented an up-regulation of the immunofluorescence signal for the $P2X_{1-7}$ receptor subunit profile. This increment presented a characteristic distribution on astrocyte processes and cell bodies.[91,92]

The effects of axonal degeneration on the glial expression of purinergic receptors were recently studied by James and Butt.[118] Following neonatal enucleation, optic nerve axons degenerated provoking a parallel loss of oligodendrocytes. In this condition, the optic nerve cellular population is prevalently formed by astrocytes with a small proportion of microglial cells.[119] In this experimental model, it is possible to test the glial response to ATP by removing the optic nerve and setting a whole free nerve preparation. James and Butt[118] used the fura-2 ratiometric recording technique to assess intracellular Ca2+ levels after administering ATP to the optic nerve preparation. In this experimental setting, a rapid and large increase in intracellular Ca2+ concentration was recorded after ATP application. The potency of different purinergic receptor agonists in mimicking the ATP effect was used to assess the contribution of the different receptor subtypes in mediating the increased Ca2+ concentration. In the unlesioned optic nerve, ATP induced increment of Ca2+ concentration was mainly sustained by $P2Y_1$ receptors with only a minor contribution from $P2Y_{2,4}$, $P2X_{1,2/3}$ and P1 receptor subtypes. In the lesioned optic nerve, a clear reduction in the contribution of $P2Y_1$ receptors with a parallel increase in the contribution of $P2Y_{2,4}$ and $P2X_{1,2/3}$ receptors to the ATP induced Ca2+ influx was recorded.[118] Thus, reactive astrocytes present a reduction of the glial signalling associated P2Y receptors and an increment of P2X and $P2Y_{2,4}$ receptors. Further, on the basis of the duration of the ATP induced increase in the glial Ca2+ concentration the presence of the pore-forming $P2X_7$ purinoceptor in the reactive astrocytes was suggested.[118] In a nutshell, after a lesion astrocyte and microglial cells present a modulation of P2 receptor phenotypes with a shift in functions from glial signalling to control of Ca2+ intracellular concentration. In synthesis, in vivo data, although still scarce, indicate that P2 receptors are active in mediating astrocyte and microglial responses to different types of CNS injuries. The overall effect of the purinergic mediation is still largely unknown and, depending on the experimental conditions, both survival and degenerative actions have been described. The modulation of purinergic receptors in in vivo glial reaction paradigms suggests that these receptors mediate extracellular damage signals between glia and neurons and that both survival or degenerative effects can be mediated through this route.

P2X and Neuronal Reaction to Axotomy

As stated previously, the P2X receptor in vivo literature is still scarce and a few data are available to delineate their functions in central neurons. The presence of P2X receptors in central neurons has been ascertained but the complete picture of the distribution of the different subunits still needs to be clarified. Regarding P2X receptor mediated functions, Thomas and Spyer[120] demonstrated the involvement of $P2X_{1\text{-}7}$ receptor subunits in central chemoreception mechanisms in neurons of the Botzinger complex in the ventrolateral medulla. The location of $P2X_2$ in neuronal and glial cells induced Kanjhanand and coworkers[58] to propose a role for purinergic receptors in plasticity and CNS homeostasis.

The majority of adult CNS neurons die after disconnection from their target, but some are able to survive and can regenerate their axons.[110,121,122] Despite extensive descriptions of the morphological features of the neuronal reaction,[123-125] the cellular mechanisms that constitute the basis of neuronal sensitivity and fate after axotomy are still poorly understood. In recent years, several observations and experimental successes in regenerating CNS neurons stimulated interest in the field.[110,121,122,126-128] Among the different hypotheses advanced to explain the death of central neurons after loss of their axons, the most widely cited is the trophic deprivation hypothesis. It proposes that the death of axotomized neurons is caused by disconnection from a target which normally supplies trophic signals to the cell bodies.[57,110] In line with this hypothesis, completely axotomized neurons should die. However, if there are some spared axonal branches that can still deliver target-derived trophic signals, axotomized neurons can survive. On this basis, the exogenous application of trophic factors was proposed as a therapeutic approach to induce or enhance regenerative responses in axotomized CNS neurons.[110,129,130] In addition, it was observed that CNS and PNS neuronal survival may also depend on electrical activity and/or cAMP elevation.[131-133] This last observation implies that axotomized neurons receiving less synaptic input or becoming less active lose trophic sensitivity. Consequently, axotomized CNS neurons may fail not only to receive but also to respond to trophic stimuli. This can result either in the death of the axotomized neurons or in the failure of the axotomized axons to re-extend.[110]

At this time, two important research directions are being followed to promote CNS regeneration. The first approach uses nerve grafts or transplanted cells.[134,135] The second approach focuses on understanding the molecular mechanisms that act after neuronal injury. Indeed, injured adult CNS neurons up-regulate the expression of several markers associated with both degenerative and reparative molecular aspects.[110,128,136] Some of these proteins have been identified and their function is currently being investigated. For example, axotomized neurons can express c-Jun and other transcription factors, the enzyme NOS and the protein GAP-43 (see for review refs. 136,128). However, the functional meaning of these markers is not fully understood and they can be related to both degeneration and/or survival efforts depending on the experimental conditions.[137]

Cerebellar circuits constitute a well-known topographically organised system, particularly apt for investigating the functional meaning of injury markers through morphological methods. Either regressive or reactive neuronal modifications have been reported after different types of injuries (see for reviews refs. 138,139). Cellular atrophy and marked cell loss can be provoked in precerebellar neurons after cerebellar lesions.[128,138-141] However, at least a subset of precerebellar neurons resist such injury, do not undergo atrophy and express some injury markers. This neuronal population does not show cellular signs of degeneration and is capable of establishing compensatory phenomena in response to injury.[128,142,143]

As previously detailed, P2X receptors are (i) involved in the glial response to injury,[91,92,118] (ii) expressed in the cerebellar system,[58,59] (iii) can mediate trophic actions which induce neuritogenesis[10,20,55] and (iv) are up-regulated after axotomy in the PNS.[65,105] In particular, $P2X_{1,2}$ receptor subunits expression has been reported in precerebellar neurons[58,59] and in vitro data support their involvement in cell survival and neuritogenesis,[20] as well as in the mediation of pathological processes.[15,18]

The potential changes in $P2X_{1,2}$ receptor subunits expression in neurons of the precerebellar nuclei following unilateral ablation of their cerebellar targets was recently investigated.[141] $P2X_{1,2}$ immunoreactivity was assessed by immunocytochemical and quantitative analyses were performed by comparing immunolabelled neurons between experimental side and control side.

In normal rats no side differences were present in the $P2X_{1,2}$ immunoreactive brainstem. Starting from day 7 after the lesion, many precerebellar neurons started to express strong $P2X_1$ immunolabelling (Fig. 1). Accordingly, precerebellar populations of $P2X_2$ immunolabelled neurons showed clear changes, becoming more numerous and intensely stained. Interestingly, the neurons that expressed $P2X_1$ or $P2X_2$ immunoreactivity did not present morphological signs of degeneration. Quantitative evaluation demonstrated differences in time course of $P2X_1$ and $P2X_2$ induction (Fig. 2). From day 14 to day 28 the number of $P2X_1$ immunoreactive neurons increased progressively, reaching the peak between day 21 and day 28 followed by a progressive reduction. One month after the lesion, only a few sparse $P2X_1$ positive neurons were still detectable. For $P2X_2$ positive neurons, the peaks were on day 14 and the successive reduction was less evident.

Thus, damage of the cerebellum induces a transient and time dependent up-regulation of $P2X_{1,2}$ receptor subunits neuronal expression in the precerebellar nuclei. The two receptor subtypes considered in the cited study present differences in the time windows of their responses suggesting differences in their specific cellular functions and/or regulation. Importantly, the up-regulation follows the appearance of morphological cellular signs of degeneration. Thus, the healthy state of the $P2X_{1,2}$ immunoreactive neurons, the late onset and the relative persistence of the $P2X_{1,2}$ receptor subunit up-regulation at later time points is at odds with a possible involvement of the $P2X_{1,2}$ receptor subunits in precerebellar neuronal death mechanisms. In addition, the late onset and, particularly for the $P2X_2$ subunit, the prolonged activation are in favour of a role in mediating the survival/regenerative cellular responses to axotomy. This hypothesis, that up-regulation of $P2X_{1,2}$ receptor subunits mediates pro-survival and pro-regenerative cellular response to axotomy in precerebellar neurons, is also supported by several studies indicating that: (1) precerebellar neurons can survive and express regenerative

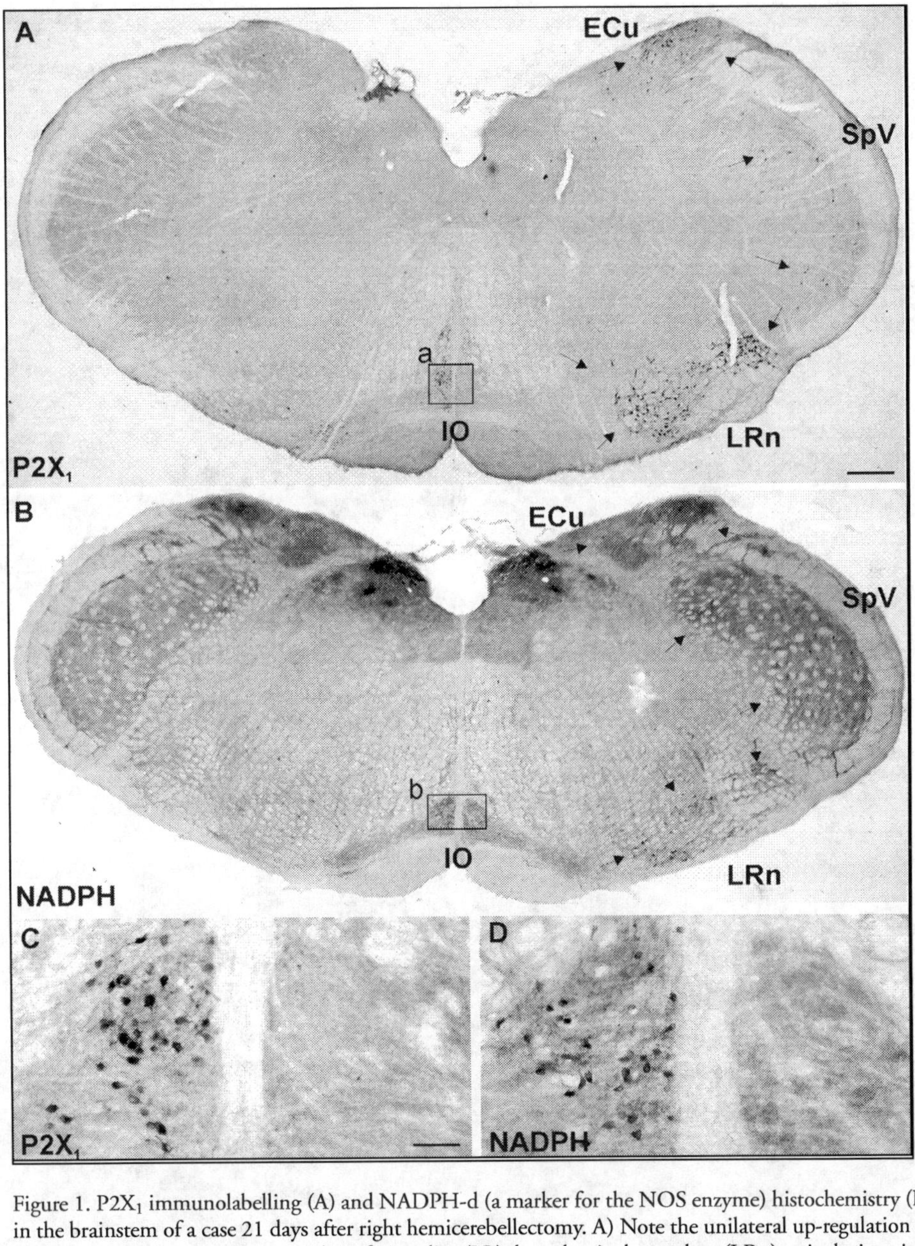

Figure 1. $P2X_1$ immunolabelling (A) and NADPH-d (a marker for the NOS enzyme) histochemistry (B) in the brainstem of a case 21 days after right hemicerebellectomy. A) Note the unilateral up-regulation of $P2X_1$ immunopositive neurons in the inferior olive (IO), lateral reticular nucleus (LRn), spinal trigeminal nucleus (SpV), external cuneate nucleus (ECu) (arrows). B) Note the unilateral up-regulation of NADPH-d histochemistry in IO, LRn, SpV and ECu (arrows). C) Inset a of A showing $P2X_1$ immunopositive cells in the IO of the experimental side (left side), and the absence of immunopositive cells in the IO of the control side (right side). D) Inset b of B showing NADPH-d positive cells in the IO of the experimental side (left side), and the absence of immunopositive cells in the IO of the control side (right side). Note that $P2X_1$ and NADPH-d positive neurons present a comparable distribution; compare A and B. Scale bars: A, B = 350 mm; C, D = 100 μm.

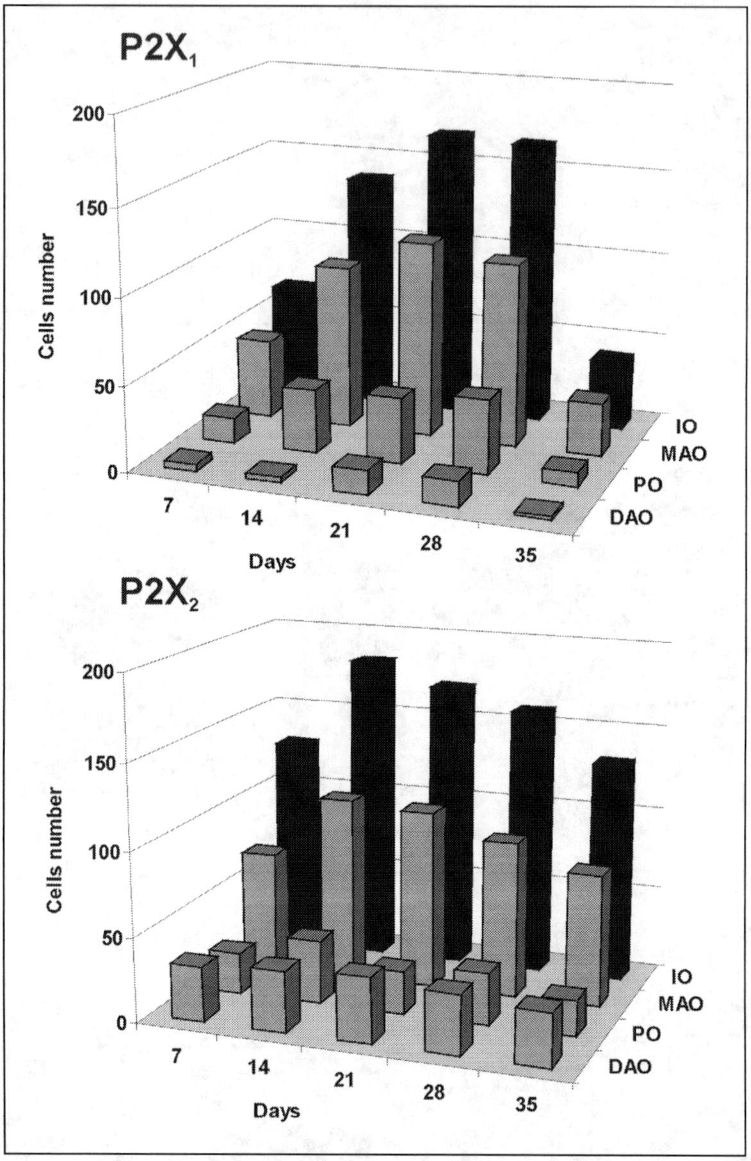

Figure 2. Relations between the total number of $P2X_1$ and $P2X_2$ immunopositive neurons in the inferior olive (IO) and the number of $P2X_1$ and $P2X_2$ immunopositive neurons in the different olivary subdivisions (MAO= medial accessory olive; PO= principal olive; DAO= dorsal accessory olive). Each bar represents the mean value of the difference between sides (experimental versus control) obtained from three cases per time point. Note the substantial contribution of the immunopositive neurons of the MAO subdivision to the total number of IO for $P2X_1$ and $P2X_2$ immunopositive neurons at all time points. Reprinted with permission from: Florenzano F, Viscomi MT, Cavaliere F et al. Neuroscience 2002;115(2):425-34. ©2002.

markers,[128,142,143] (2) a role has been established in regeneration and trophic actions for ATP and purinergic receptors,[15,18,19,81] (3) P2X receptors are up-regulated following injuries to sensory neurons.[65,105]

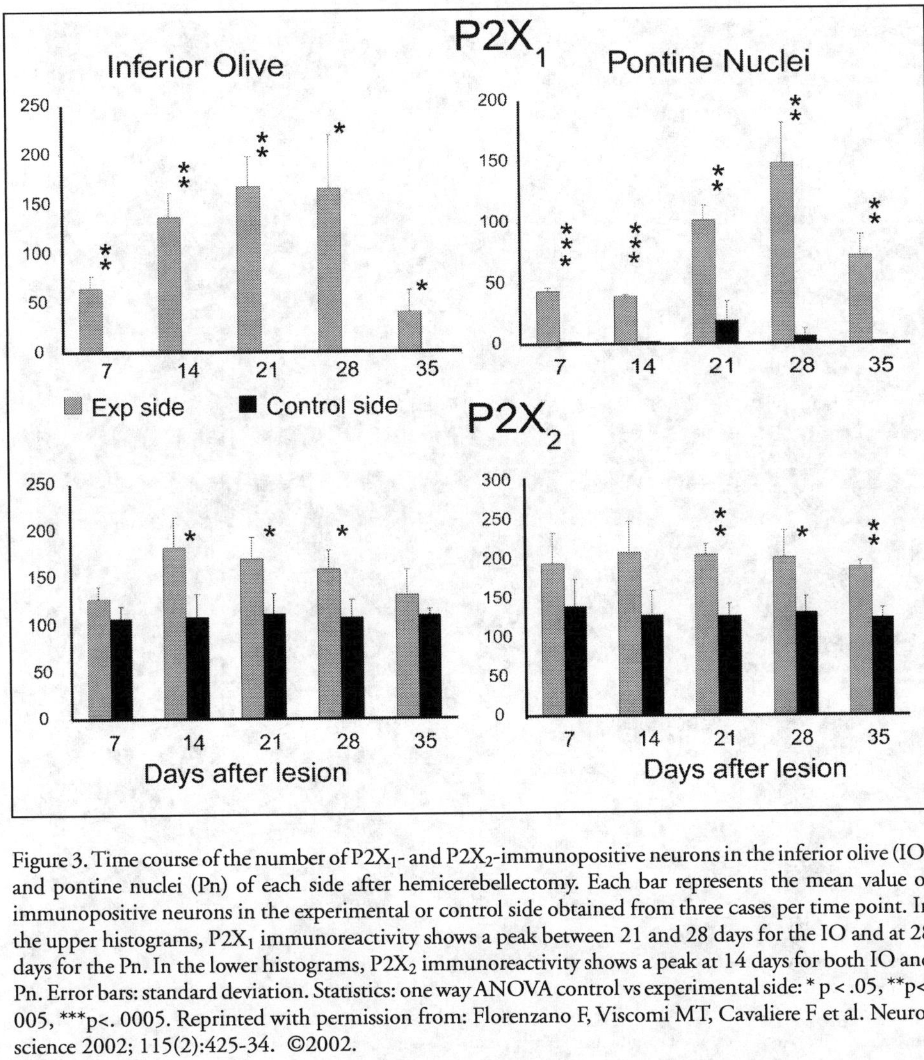

Figure 3. Time course of the number of $P2X_1$- and $P2X_2$-immunopositive neurons in the inferior olive (IO) and pontine nuclei (Pn) of each side after hemicerebellectomy. Each bar represents the mean value of immunopositive neurons in the experimental or control side obtained from three cases per time point. In the upper histograms, $P2X_1$ immunoreactivity shows a peak between 21 and 28 days for the IO and at 28 days for the Pn. In the lower histograms, $P2X_2$ immunoreactivity shows a peak at 14 days for both IO and Pn. Error bars: standard deviation. Statistics: one way ANOVA control vs experimental side: * $p < .05$, **p<. 005, ***p<. 0005. Reprinted with permission from: Florenzano F, Viscomi MT, Cavaliere F et al. Neuroscience 2002; 115(2):425-34. ©2002.

As stated previously, different data indicate that purinergic receptors are relevant in mediating glial reaction to injury in the CNS. The above mentioned study[141] is the first demonstration that in CNS the injury dependent P2X receptor subunit up-regulation is not limited to glial cells but also involves neurons. Thus, in agreement with the well known involvement of ATP in glia/glia and glia/neuron communication,[19] the marked expression of different $P2X_{1-7}$ receptor subunits in both neurons or glia after lesion suggests a primary role for ATP in CNS communication of cell damage among different cell types.

It has been shown that precerebellar neurons up-regulate the enzyme nNOS after cerebellar lesion.[143-145] It is of particular interest that $P2X_{1,2}$ receptor subunit up-regulation is localised in the same precerebellar structures that up-regulate the enzyme nNOS after cerebellar lesion (Fig. 1). Preliminary data from double labelling immunofluorencence experiments indicate that the same cells that present postlesional nNOS up-regulation also present up-regulation of the $P2X_{1,2}$ receptor (Florenzano et al, Soc. Neurosci. Abs. 2002) (Fig. 3).

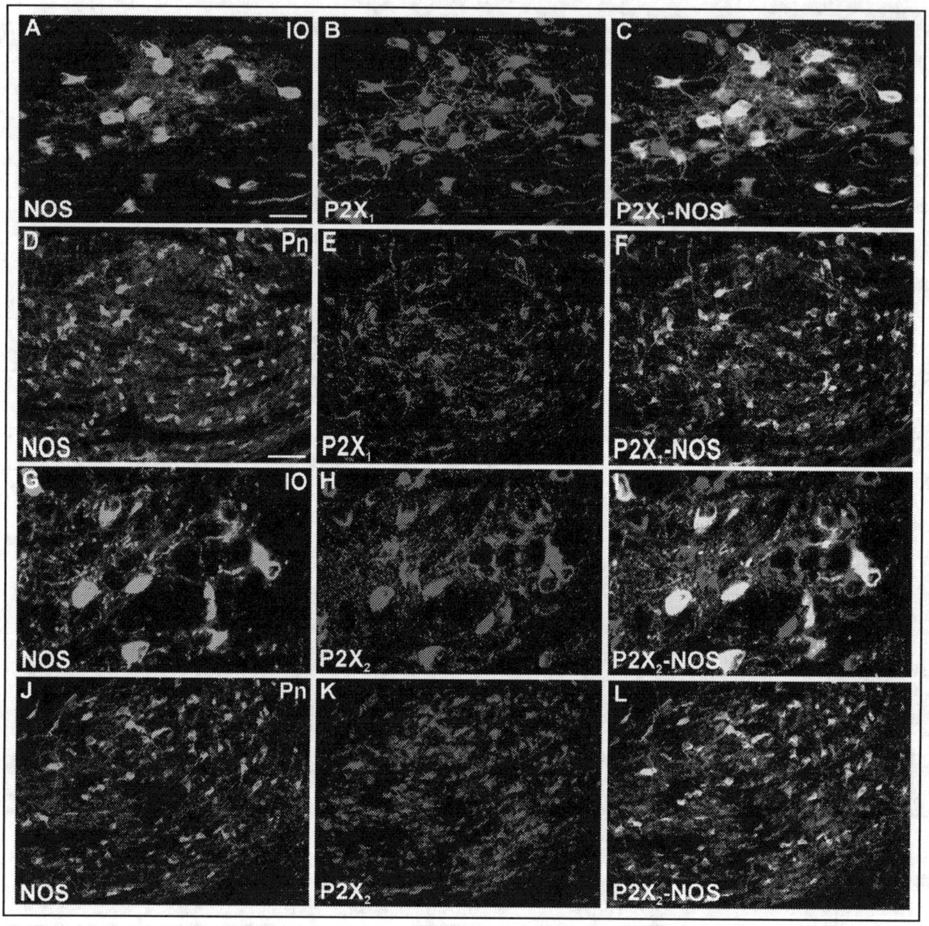

Figure 4. Confocal images of NOS (A,D,G,J) and $P2X_1$ (B,E) or $P2X_2$ (H,K) double immunofluorescence in the inferior olive (IO; A-C, G-I) and pontine nuclei (Pn; D-F, J-L) of a case 21 days after hemicerebellectomy. Note colocalizion of NOS and $P2X_1$ (C,F), or of NOS and $P2X_2$ (L,P) in the majority of neurons in the IO and Pn. Scale bars: A-C), G-I) = 40 µm; D-F), J-L) = 90 µm.

P2X and CNS Ischaemic Damage

Several studies have described a protective action for adenosine and A1 receptor agonists resulting from oxygen/glucose depletion secondary to an ischaemic insult.[146,147] Besides the involvement of adenosine, direct participation of extracellular ATP receptors in ischaemic stress has also been suggested.[16,17,148,149] In addition, in vitro data indicate that ATP stimulation exerts neurotoxic effects. In primary neuronal cultures high extracellular concentrations of ATP are toxic and induce both necrotic and apoptotic cell loss.[11] Further, primary neuronal cultures exposed to excessive glutamate[12] to serum/potassium deprivation,[13] to hypoglycaemia or chemical hypoxia[16,17] can be rescued by the addition of several P2 receptor antagonists.

Recently, the expression of $P2X_2$ and $P2X_4$ receptor subunits during cerebral ischaemia was investigated both in in vivo and in vitro models.[114] After hypoxia insult in gerbils, an intense induction of both $P2X_2$ and $P2X_4$ was observed (Fig. 4). In this condition $P2X_2$ receptors were up-regulated on fibres extending throughout the CA1-CA2 pyramidal cell layer and

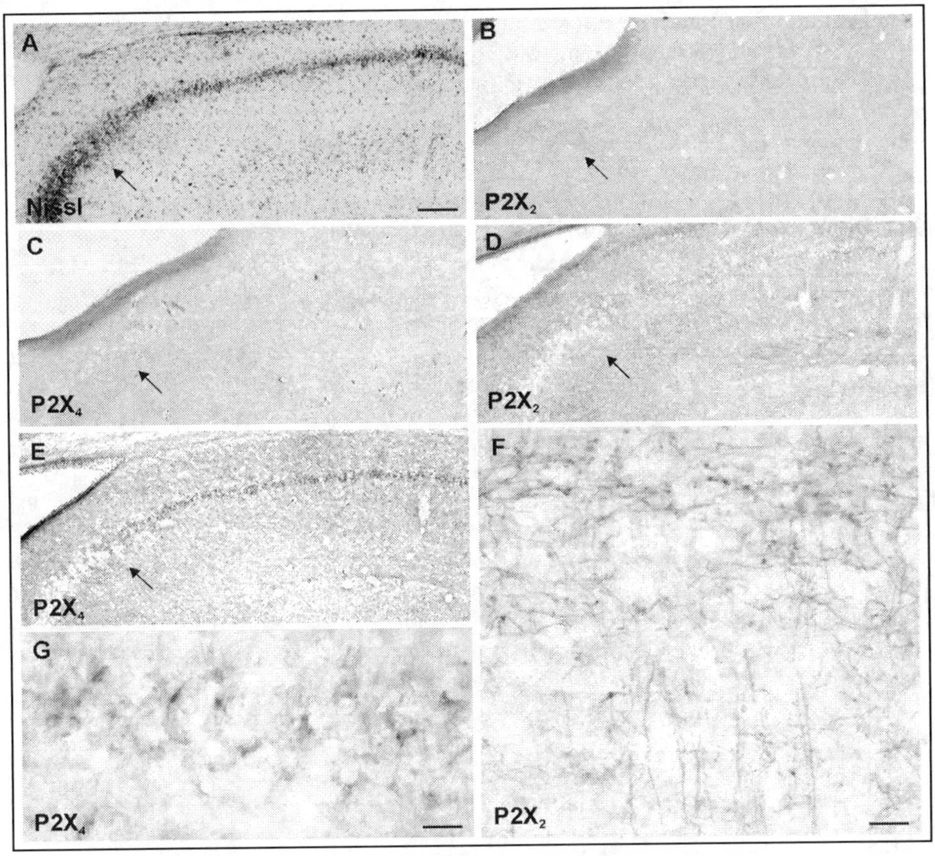

Figure 5. Transverse sections through the dorsal hippocampus of a gerbil six days after ischemia (A,D-G) or after sham operation (B-C). Arrows in A-E indicate CA1-CA2 transition zone. A) Nissl staining after ischemia. Note the massive neuronal loss in the CA1 pyramidal cell layer. B) $P2X_2$ immunoreactivity in a control animal, note the lack of immunostaining. C) P2X4 immunoreactivity in a control animal, note the lack of immunostaining. D) $P2X_2$ immunoreactivity in a ischemic animal, note the up-regulation with presence of immunopositive fibers in the CA1 region (compare to B). E) $P2X_4$ immunoreactivity in a ischemic animal, note the up-regulation with presence of immunopositive cellular processes in the CA1 region (compare to C). F) High magnification of $P2X_2$ immunoreactivity in the CA1 region of an ischemic animal, note the presence of a dense network of immunopositive fibers. G) High magnification of $P2X_4$ immunoreactivity in the CA1 region of an ischemic animal, note the presence of immunopositive structures resembling glial processes. Scale bars: A-E) = 250 μm; F) = 50 μm; G) 20 μm.

strata oriens and radiatum (Fig. 4). Conversely, $P2X_4$ receptor distribution after hypoxia was strictly confined to the pyramidal cell layer of the CA1 and to the transition zone of the CA2 sub fields (Fig. 4). In this area, the $P2X_4$ receptor was up-regulated in microglial cells that surrounded unlabelled cellular bodies (Figs. 4, 5). In the course of the same study, converging data were obtained in hippocampal organotypic cultures. In this model after 20 hours of oxygen/glucose deprivation a steep up-regulation, about 2.5 fold, of the expression of $P2X_2$ and $P2X_4$ receptors was reported.[114] Of particular relevance for potential therapeutic strategies is the evidence that the presence of P2 receptor antagonists, like suramin or in reactive blue 2, during the ischaemic insult not only abolishes P2X receptor up-regulation but also totally prevents the ischaemia induced cell death.[114]

Similar protective effects were also observed in slice cultures from different brain regions using various P2 antagonists.[114] In double organotypic cultures from cortex/striatum, oxygen-glucose deprivation induced massive cellular loss in the dorso-lateral region of the cortex and in the caudate putamen. Addition of basilen blue induced different degrees of neuroprotection to striatal and cortical cells. Indeed, while it totally prevented cell degeneration in the striatum, it only reduced cell death in the cortex (25%).[114] The relationship between neuronal cell types and degree of the neuroprotection exerted by P2X blockers is in agreement with results obtained in dissociated primary neuronal cultures.[11]

Thus, in vitro and in vivo data converge in focussing on P2X receptors as mediating key elements of CNS responses to different insults marshalling them as relevant elements in controlling cell fate. Interestingly, various neuronal populations present differences in the pattern of the different P2X subunits activated by ATP stimulation. In the hippocampus, toxic concentrations of extracellular ATP induced activation of $P2X_2$ and $P2X_4$ receptor subtypes.[11] In the cerebellum, in the same toxic conditions only $P2X_7$ and $P2Y_4$ presented up-regulation.[11] This multifarious framework indicates that P2 receptors are important in supporting cell specificity, particularly in explaining some of the differences observed among neuronal populations in the reaction to toxic stimuli.[11,16,17]

Of particular relevance is the evidence that P2 receptors are present and modulated by toxic insult not only in neurons[11,114] but also in glial cells.[114] Therefore, these different cell types share the same signalling and/or communicating mechanisms. This framework supports the concept of the interplay between neurons and glia in mediating ischaemia-evoked signals and proposes P2 receptor modulation as the common language between glia and neurons and ATP as the intercellular signalling molecule acting between neurons and glia.[14]

Hypothesis on P2X Role in CNS Toxic Insult Signalling

The intracellular energy pool is a key element in determining the ability of CNS cells to respond to stressful environmental conditions.[18,152] ATP is released under physiological or pathological conditions[148,153] and sources of extracellular ATP include both neurons and glial cells.[102] In turn, ATP concentration in the extracellular compartment can influence either physiological functions,[9,20,81] or have detrimental effects, thus acting as a toxic agent.[11,153] Furthermore, in the latter case purinergic signalling can induce either regressive or regenerative changes. In this respect the two experimental models reviewed here are prototypical.

The axonal reaction induced in the precerebellar neurons by lesion of their target areas is associated with an up-regulation of the $P2X_{1,2}$ receptor subunits. The two receptor subtypes present differences in the time course and in the features of their expression in different groups of precerebellar neurons. This evidence supports the hypothesis that P2X expression is an element in defining the specificity of the reaction to injury in different neuronal populations. Further, the dissociation between the time course of the regressive changes and the time course of $P2X_{1,2}$ receptors as well as the morphological features of $P2X_{1,2}$ receptor subunits expressing neurons indicates that, at least in this model, P2X receptors can mediate cellular regenerative efforts.

A completely different picture emerges from the analysis of the pattern of P2X receptor up-regulation observed under ischaemic conditions. First of all, in this case there is a temporal coincidence between cell death and P2X receptors over-expression. Second, the toxic nature of the information conveyed by P2X receptors is clearly indicated by the protective effect exerted by P2X blockers. In this view, it has to be underlined that the two P2X receptor subtypes that are up-regulated in the hippocampus after ischaemia, namely, $P2X_2$ and $P2X_4$, are expressed by neurons ($P2X_2$) and by microglial cells ($P2X_4$) (Fig. 6). Therefore, metabolic impairment shares signalling and/or communicating mechanisms in these different cell types. In this context, it is

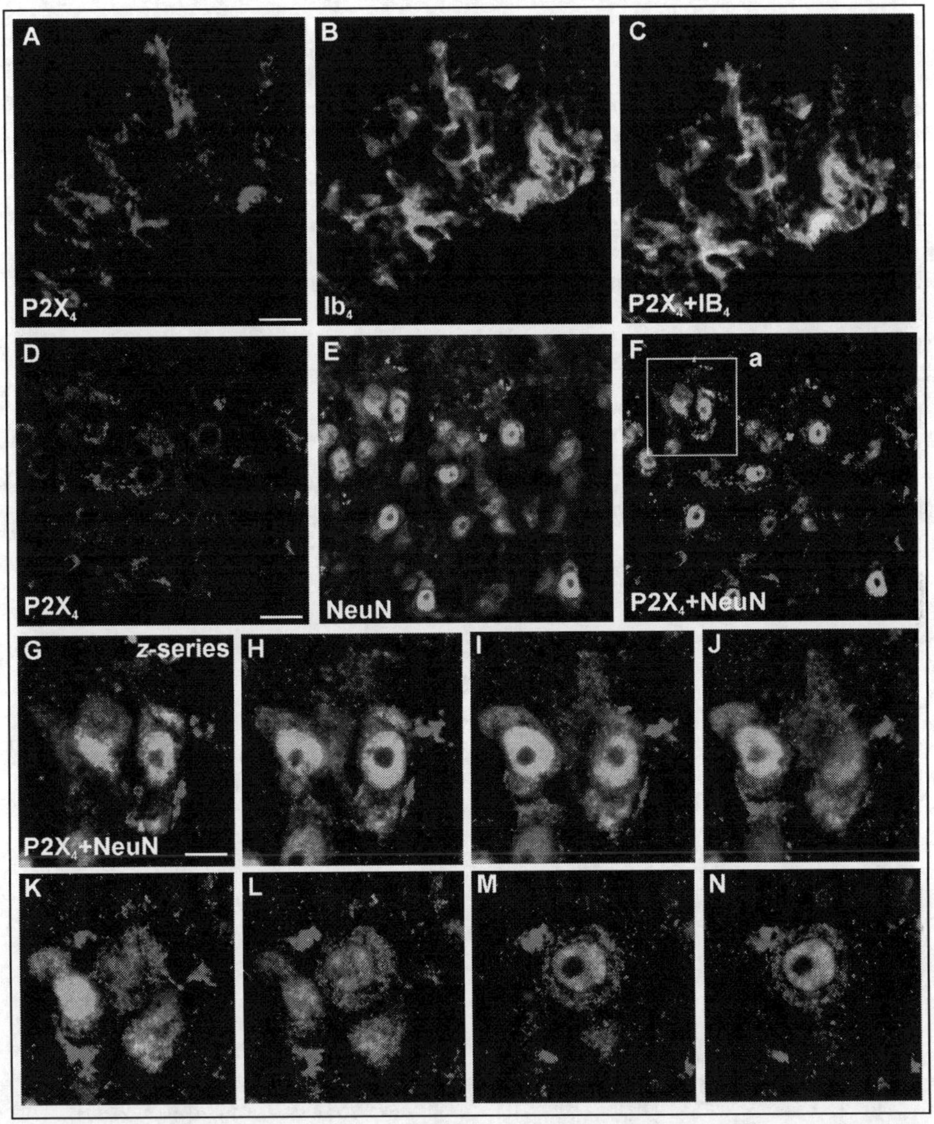

Figure 6. Confocal images of double immunofluorescence for $P2X_4$ and IB_4 (a marker for microglial cells) or NeuN (a marker for neuronal cells) in the CA1 pyramidal cell layer of a gerbil six days after ischemia. A),B),C), $P2X_4$ and IB_4 double labelling. Note the high degree of colocalization in processes of microglial cells. D),E),F), $P2X_4$ and NeuN double immunofluorescence. Note the almost complete segregation of the two markers, $P2X_4$ immunofluorescent processes surround with no overlap the NeuN immunofluorescent neuronal cell bodies. G),-N), z-series of optical sections 1 mm spaced of $P2X_4$ immunopositive processes from inset a of F. Note how activated migroglial process (red) tend to surround the CA1 neuronal cell body. Scale bars: A)-C) = 20 μm; D)-F) = 50 μm; G)-N) = 10 μm.

worth recalling that data from different models indicate neurons and glia as capable of releasing ATP in an activity-dependent manner.[102] Further, through P2 receptors ATP released from glial cells[14] can induce fast excitatory responses and Ca2+ signals in neurons. Therefore, ATP flowed out during ischaemia/cellular damage[154] may represent the intercellular signalling molecule between neurons and glia in different pathological conditions.[14]

In addition to neurons, all types of glial cells also possess membrane receptors for extracellular ATP.[91,92] In the in vivo models reviewed here different patterns of P2 receptor activation were linked to astrocytes or microglia reaction to insults. Different compositions of P2 receptor subunits seem to characterise astrogliosis when induced by axonal degeneration or by a stab wound. In the former case, the activated astroglial cells express the $P2X_{1-7}$ subunits; in the latter case, activated astrocytes present a shift from P2Y signalling subtypes to P2X, $P2Y_{2-4}$ subtypes. After anoxic insult, the P2X receptor pattern displayed by the activated glia is again different. In this case, glial up-regulation is mainly sustained by $P2X_4$ over-expression in microglia. This agrees with previous results showing that in microglial cells the $P2X_4$ receptor (together with the $P2X_7$)[99] is the major candidate for activation by toxic ATP release.

Evidence available from in vitro studies indicates that purinergic receptors are modulated and involved in responses to injury as well as regeneration.[93] This indication is also supported by recent in vivo evidence demonstrating up-regulation of different P2X receptor subtypes in glia and neurons following CNS lesions. This can correlate either with degeneration or with regenerating efforts, although neurons injured by stroke or trauma in the adult mammalian CNS normally fail to regenerate. After CNS damage, a glial reaction often occurs involving proliferation, expression of immunomolecules at the cellular surface, secretion of cytokines or growth factors and differentiation into brain phagocytes.[125,154] This glial reaction can in turn transmit signals through the common purinergic language to neurons favouring or opposing the neuronal reaction. Some of these events may mediate tissue damage, while others may play an important role in limiting the extent of damage and or promoting/inhibiting tissue repair. Although in some cases it is relatively easy to depict a potential role for P2X in triggering neuronal degeneration or conversely in mediating cell regenerative efforts, in others it is not so obvious whether P2X activation is beneficial or detrimental. Nevertheless, the use of selective P2 agonists and inhibitors in many instances clarifies the general effect of purinergic stimulation in a given context. In synthesis, in vivo data on purinergic involvement in the CNS reaction to injury are still scarce. However, purinergic research constitutes an extremely interesting and promising field for investigating the mechanisms that regulate CNS behaviour in different pathological conditions. Last but surely not least, purinergic research is a fruitful area for specific drugs capable of selective modulation of the CNS response to injury by inhibiting P2 mediated regressive changes or by enhancing P2 mediated regenerative actions.

Acknowledgements

We are grateful to Prof. G. Bernardi for his continuous support and encouragement. Thanks are due to Claire Montagna for style editing. This work was supported by Italian Ministry of Health grants (RA0085M – 2000 and Ricerca Corrente) to MM and by a grant from MURST Cofin 2001 on "Purinoceptors and Neuroprotection" to CV.

References

1. Edwards FA, Gibb AJ, Colquhoun D. ATP receptor-mediated synaptic currents in the central nervous system. Nature 1992; 359:144-147.
2. Robertson SJ, Ennion SJ, Evans RJ et al. Synaptic P2X receptors. Curr Opin Neurobiol 2001; 11:378-386.
3. Gu JG, MacDermott AB. Activation of ATP P2X receptors elicits glutamate release from sensory neuron synapses. Nature 1997; 389:749-753.
4. Dunn PM, Zhong Y, Burnstock G. P2X receptors in peripheral neurons. Prog Neurobiol 2001; 65:107-134.
5. Zimmermann H, Braun N, Ecto-nucleotidases molecular structures, catalytic properties, and functional roles in the nervous system. Prog Brain Res 1999;12:371-385.

6. Khakh BS. Molecular physiology of P2X receptors and ATP signalling at synapses. Nat Rev Neurosci 2001b; 2:165-174.
7. North RA, Barnard EA. Nucleotide receptors. Curr Opin Neurobiol 1997; 7:346-357.
8. Khakh BS, Burnstock G, Kennedy C et al. International union of pharmacology. XXIV. Current status of the nomenclature and properties of P2X receptors and their subunits. Pharmacol Rev 2001a; 53:107-118.
9. Burnstock G. Current status of purinergic signalling in the nervous system. Prog Brain Res 1999; 120:3-10.
10. Burnstock G. The past, present and future of purine nucleotides as signalling molecules. Neuropharmacology 1997; 36(9):1127-1139.
11. Amadio S, D'Ambrosi N, Cavaliere F et al. P2 receptor modulation and cytotoxic function in cultured CNS neurones. Neuropharmacology 2002; 42:489-501.
12. Volonté C, Merlo D. Selected P2 purinoceptor modulators prevent glutamate-evoked cytotoxicity in cultured cerebellar granule neurons. J Neurosci Res 1996; 45(2):183-193.
13. Volonté C, Ciotti MT, D'Ambrosi N et al. Neuroprotective effects of modulators of P2 receptors in primary culture of CNS neurones. Neuropharmacology 1999; 38(9):1335-1342.
14. Fields RD, Stevens B. ATP: an extracellular signaling molecule between neurons and glia. Trends Neurosci 2000; 23:625-633.
15. Burnstock G, Williams M. P2 purinergic receptors: modulation of cell function and therapeutic potential. J Pharmacol Exp Ther 2000; 295:862-869.
16. Cavaliere F, D'Ambrosi N, Sancesario G et al. Hypoglycaemia-induced cell death: features of neuroprotection by the P2 receptor antagonist basilen blue. Neurochem Int 2001a; 38(3):199-207.
17. Cavaliere F, D'Ambrosi N, Ciotti MT et al. Glucose deprivation and chemical hypoxia: neuroprotection by P2 receptor antagonists. Neurochem Int 2001b; 38(3):189-197.
18. Burnstock G. Potential therapeutic targets in the rapidly expanding field of purinergic signalling. Clin Med 2002; 2:45-53.
19. Neary JT, Rathbone MP, Cattabeni F et al. Trophic actions of extracellular nucleotides and nucleosides on glial and neuronal cells. Trends Neurosci 1996; 19:13-18.
20. D'Ambrosi N, Murra B, Cavaliere F et al. Interaction between ATP and nerve growth factor signalling in the survival and neuritic outgrowth from PC12 cells. 2001; Neurosci 108:527-534.
21. Mulryan K, Gitterman DP, Lewis CJ et al. Reduced vas deferens contraction and male infertility in mice lacking $P2X_1$ receptors. Nature 2000; 403(6765):86-89.
22. Prasad M, Fearon IM, Zhang M et al. Expression of $P2X_2$ and $P2X_3$ receptor subunits in rat carotid body afferent neurones: role in chemosensory signalling. J Physiol 2001; 537:667-677.
23. Thomas T, Ralevic V, Bardini M et al. Evidence for the involvement of purinergic signalling in the control of respiration. Neuroscience 2001; 107:481-490.
24. Torres GE, Egan TM, Voigt MM. Hetero-oligomeric assembly of P2X receptor subunits. Specificities exist with regard to possible partners. J Biol Chem 1999; 274(10):6653-6659.
25. D'Ambrosi N, Cavaliere F, Merlo D et al. Antagonists of P2 receptor prevent NGF-dependent neuritogenesis in PC12 cells. Neuropharmacology 2000; 39(6):1083-1094.
26. Petruska JC, Cooper BY, Gu JG et al. Distribution of P2X1, P2X2, and P2X3 receptor subunits in rat primary afferents: relation to population markers and specific cell types. J Chem Neuroanat 2000; 20(2):141-162.
27. Pankratov YuV, Lalo UV, Dashkin AN et al. Heterogeneity of the functional expression of P2X3 and P2X2/3 receptors in the primary nociceptive neurons of rat. Neurochem Res 2001; 26(8-9):993-1000.
28. Ueno S, Tsuda M, Iwanaga T. Cell type-specific ATP-activated responses in rat dorsal root ganglion neurons. Br J Pharmacol 1999; 126(2):429-436.
29. Vulchanova L, Riedl MS, Shuster SJ et al. Immunohistochemical study of the P2X2 and P2X3 receptor subunits in rat and monkey sensory neurons and their central terminals. Neuropharmacology 1997; 36(9):1229-1242.
30. Le KT, Babinski K, Seguela P. Central P2X4 and P2X6 channel subunits coassemble into a novel heteromeric ATP receptor. J Neurosci 1998a; 18(18):7152-7159.
31. Cunha RA, Ribeiro JA. ATP as a presynaptic modulator. Life Sci 2000; 68(2):119-137.

32. Virginio C, MacKenzie A, Rassendren FA et al. Pore dilation of neuronal P2X receptor channels. Nat Neurosci 1999; 2(4):315-321.
33. North RA, Surprenant A. Pharmacology of cloned P2X receptors. Annu Rev Pharmacol Toxicol 2000;40:563-580.
34. Nakazawa K. ATP-activated current and its interaction with acetylcholine-activated current in rat sympathetic neurons. J Neurosci 1994;14(2):740-750.
35. Zhou X, Galligan JJ. Non-additive interaction between nicotinic cholinergic and P2X purine receptors in guinea-pig enteric neurons in culture. J Physiol 1998; 513 (Pt 3):685-697.
36. Salter MW, De Koninck Y. An ambiguous fast synapse: a new twist in the tale of two transmitters. Nat Neurosci 1999; 2(3):199-200.
37. Edwards FA, Robertson SJ, Gibb AJ. Properties of ATP receptor-mediated synaptic transmission in the rat medial habenula. Neuropharmacology 1997; 36(9):1253-1268.
38. Rogers M, Colquhoun LM, Patrick JW et al. Calcium flux through predominantly indipendent purinergic ATP and nicotinic acetylcholine receptors. J. Neurophysiol. 1997; 77(3):1407-1417.
39. Nakatsuka T, Ataka T, Kumamoto E et al. Alteration in synaptic inputs through C-afferent fibers to substantia gelatinosa neurons of the rat spinal dorsal horn during postnatal development. Neurosci 2000; 99(3):549-556.
40. Sun XP, Stanley EF. An ATP-activated, ligand-gated ion channel on a cholinergic presynaptic nerve terminal. Proc Natl Acad Sci USA 1996; 93(5):1859-1863.
41. Merlo D, Volonté C. Binding and functions of extracellular ATP in cultured cerebellar granule neurons. Biochem Biophys Res Commun 1996; 225(3):907-914.
42. Jo YH, Schlichter R. Synaptic corelease of ATP and GABA in cultured spinal neurons. Nat Neurosci 1999; 2(3):241-245.
43. Chittajallu R, Alford S, Collingridge GL. Ca2+ and synaptic plasticity. Cell Calcium 1998; 24(5-6):377-385.
44. Kennedy Mary B, Mu-ming Poo. Signalling mechanisms. Current Opinion in Neurobiology 1999; 9:263-266.
45. Rose CR, Konnerth A. Stores not just for storage. intracellular calcium release and synaptic plasticity. Neuron 2001; 31:519-522.
46. Zucker RS, Regehr WG. Short-term synaptic plasticity. Annu Rev Physiol 2002;64:355-405.
47. Pozzo-Miller LD, Connor JA et al. Microheterogeneity of calcium signalling in dendrites. J Physiol 2000; 525:53-61.
48. McAllister AK. Cellular and molecular mechanisms of dendrite growth. Cereb Cortex 2000; 10(10):963-973.
49. Yuste R, Bonhoeffer T. Morphological changes in dendritic spines associated with long-term synaptic plasticity. Annu Rev Neurosci 2001;24:1071-1089.
50. Gazzaley A, Kay S, Benson DL. Dendritic spine plasticity in hippocampus. Neuroscience 2002; 111(4):853-862.
51. Harris KM. Structure, development, and plasticity of dendritic spines. Curr Opin Neurobiol 1999; 9(3):343-348.
52. Matus A. Actin-based plasticity in dendritic spines. Science 2000; 290(5492):754-758.
53. Li GH, Lee EM, Blair D et al. The distribution of P2X receptor clusters on individual neurons in sympathetic ganglia and their redistribution on agonist activation. J Biol Chem 2000; 275:29107-112.
54. Dutton JL, Poronnik P, Li GH, Holding CA et al. P2X(1) receptor membrane redistribution and down-regulation visualized by using receptor-coupled green fluorescent protein chimeras. Neuropharmacol 2000; 39:2054-2066.
55. Khakh BS, Smith WB, Chiu CS et al. Activation-dependent changes in receptor distribution and dendritic morphology in hippocampal neurons expressing P2X2-green fluorescent protein receptors. Proc Natl Acad Sci 2001c; 98:5288-5293.
56. Song HJ, Poo MM. Signal transduction underlying growth cone guidance by diffusible factors. Curr Opin Neurobiol 1999; 9(3):355-363.
57. Bernstein M, Lichtman JW. Axonal atrophy: the retraction reaction. Curr Opin Neurobiol 1999; 9(3):364-370.
58. Kanjhan R, Housley GD, Burton LD et al. Distribution of the P2X2 receptor subunit of the ATP-gated ion channels in the rat central nervous system. J Comp Neurol 1999; 407 :11-32.

59. Yao ST, Barden JA, Finkelstein DI et al. Comparative study on the distribution patterns of P2X(1)-P2X(6) receptor immunoreactivity in the brainstem of the rat and the common marmoset (Callithrix jacchus): association with catecholamine cell groups. J Comp Neurol 2000; 427:485-507.
60. Collo G, North RA, Kawashima E et al. Cloning of P2X5 and P2X6 receptors and the distribution and properties of an extended family of ATP-gated ion channels. J Neurosci 1996;16(8):2495-2507.
61. Vulchanova L, Arvidsson U, Riedl M et al. Differential distribution of two ATP-gated channels (P2X receptors) determined by immunocytochemistry. Proc Natl Acad Sci USA 1996; 93(15):8063-8067.
62. Yao ST, Barden JA, Lawrence AJ. On the immunohistochemical distribution of ionotropic P2X receptors in the nucleus tractus solitarius of the rat. Neuroscience 2001;108(4):673-685.
63. Ergene E, Dunbar JC, O'Leary DS et al. Activation of P2-purinoceptors in the nucleus tractus solitarius mediate depressor responses. Neurosci Lett 1994; 174(2):188-192.
64. Barraco RA, O'Leary DS, Ergene E et al. Activation of purinergic receptor subtypes in the nucleus tractus solitarius elicits specific regional vascular response patterns. J Auton Nerv Syst 1996; 59(3):113-124.
65. Bradbury EJ, Burnstock G, McMahon SB. The expression of P2X3 purinoreceptors in sensory neurons: effects of axotomy and glial-derived neurotrophic factor. Mol Cell Neurosci 1998; 12:256-268.
66. Ueno T, Ueno S, Kakazu Y et al. Bidirectional modulation of P2X receptor-mediated response by divalent cations in rat dorsal motor nucleus of the vagus neurons. J Neurochem 2001; 78(5):1009-1018.
67. Loesch A, Burnstock G. Immunoreactivity to P2X(6) receptors in the rat hypothalamo-neurohypophysial system: an ultrastructural study with extravidin and colloidal gold-silver labelling. Neuroscience 2001; 106(3):621-631.
68. Le KT, Villeneuve P, Ramjaun AR et al. Sensory presynaptic and widespread somatodendritic immunolocalization of central ionotropic P2X ATP receptors. Neuroscience 1998b; 83(1):177-190.
69. Rubio ME, Soto F. Distinct Localization of P2X receptors at excitatory postsynaptic specializations. J Neurosci 2001; 21:641-653.
70. West AE, Chen WG, Dalva MB et al. Calcium regulation of neuronal gene expression. Proc Natl Acad Sci USA 2001; 98(20):11024-11031.
71. Gu X, Olson EC, Spitzer NC. Spontaneous neuronal calcium spikes and waves during early differentiation. J Neurosci 1994; 14:6325-6335.
72. Gu X, Spitzer NC. Distinct aspects of neuronal differentiation encoded by frequency of spontaneous Ca2+ transients. Nature 1995; 375:784-787.
73. Komuro H, Rakic P. Intracellular Ca2+ fluctuations modulate the rate of neuronal migration. Neuron 1996; 17:275-285.
74. Owens DF, Kriegstein AR. Patterns of intracellular calcium fluctuation in precursor cells of the neocortical ventricular zone. J Neurosci 1998; 18:5374-5388.
75. Cheung KK, Burnstock G. Localization of P2X3 receptors and coexpression with P2X2 receptors during rat embryonic neurogenesis. J Comp Neurol 2002; 443:368-382.
76. Meyer MP, Groschel-Stewart U, Robson T et al. Expression of two ATP-gated ion channels, P2X5 and P2X6, in developing chick skeletal muscle. Dev Dyn 1999; 216:442-449.
77. Jarlebark LE, Housley GD, Thorne PR. Immunohistochemical localization of adenosine 5'-triphosphate-gated ion channel P2X(2) receptor subunits in adult and develo rat cochlea. J Comp Neurol 2000; 421:289-301.
78. Nikolic P, Housley GD, Luo L et al. Transient expression of P2X(1) receptor subunits of ATP-gated ion channels in the developing rat cochlea. Brain Res Dev Brain Res 2001; 126:173-182.
79. Jang IS, Rhee JS, Kubota H et al. Developmental changes in P2X purinoceptors on glycinergic presynaptic nerve terminals projecting to rat substantia gelatinosa neurones. J Physiol 2001; 536(Pt 2):505-519.
80. Kidd EJ, Miller KJ, Sansum AJ et al. Evidence for P2X3 receptors in the developing rat brain. Neurosci 1998; 87:533-539.
81. Rathbone MP, Middlemiss PJ, Gysbers JW et al. Trophic effects of purines in neurons and glial cells. Prog Neurobiol 1999; 59:663-690.

82. Kukley M, Barden JA, Steinhauser C et al. Distribution of P2X receptors on astrocytes in juvenile rat hippocampus. Glia 2001; 36:11-21.
83. Verderio C, Matteoli M. ATP mediates calcium signaling between astrocytes and microglial cells: modulation by IFN-gamma. J Immunol 2001; 166(10):6383-6391.
84. Guthrie PB, Knappenberger J, Segal M et al. ATP released from astrocytes mediates glial calcium waves. J Neurosci 1999; 19:520-528.
85. Cotrina ML, Lin JH, Lopez-Garcia JC, Naus CC, Nedergaard M. ATP-mediated glia signaling. J Neurosci 2000; 20(8):2835-2844.
86. Scemes E. Components of astrocytic intercellular calcium signaling. Mol Neurobiol 2000; 22(1-3):167-179.
87. Stevens B, Fields RD. Response of Schwann cells to action potentials in development. Science 2000; 287(5461):2267-2271.
88. Abbracchio MP, Ceruti S, Langfelder R et al. Effects of ATP analogues and basic fibroblast growth factor on astroglial cell differentiation in primary cultures of rat striatum. Int J Dev Neurosci 1995; 13(7):685-693.
89. Abe K, Saito H. Effect of ATP on astrocyte stellation is switched from suppressive to stimulatory during development. Brain Res 1999; 850(1-2):150-157.
90. Franke H, Krugel U, Illes P. P2 receptor-mediated proliferative effects on astrocytes in vivo. Glia 1999; 28(3):190-200.
91. Franke H, Grosche J, Schadlich H et al. P2X receptor expression on astrocytes in the nucleus accumbens of rats. Neurosci 2001a; 108:421-429.
92. Franke H, Krugel U, Schmidt R et al. P2 receptor-types involved in astrogliosis in vivo. Br J Pharmacol 2001b; 134(6):1180-1189.
93. Ciccarelli R, Ballerini P, Sabatino G et al. Involvement of astrocytes in purine-mediated reparative processes in the brain. Int J Dev Neurosci 2001; 19(4):395-414.
94. Aldskogius H, Kozlova EN. Central neuron-glial and glial-glial interactions following axon injury. Prog Neurobiol 1998; 55(1):1-26.
95. Aldskogius H. Microglia in neuroregeneration. Microsc Res Tech 2001; 54:40-46.
96. Walz W, Ilschner S, Ohlemeyer C et al. Extracellular ATP activates a cation conductance and a K+ conductance in cultured microglial cells from mouse brain. J Neurosci 1993; 13(10):4403-4411.
97. Haas S, Brockhaus J, Verkhratsky A et al. ATP-induced membrane currents in ameboid microglia acutely isolated from mouse brain slices. Neuroscience 1996; 75(1):257-261.
98. Wang X, Franciosi S, Bae JH et al. Expression of P2y and P2x receptors in cultured human microglia. Proc West Pharmacol Soc 1999;42:79-81.
99. Visentin S, Renzi M, Frank C et al. Two different ionotropic receptors are activated by ATP in rat microglia. J Physiol 1999; 519 Pt 3:723-736.
100. Wollmer MA, Lucius R, Wilms H et al. ATP and adenosine induce ramification of microglia in vitro. J Neuroimmunol 2001; 115(1-2):19-27.
101. Hide I, Tanaka M, Inoue A et al. Extracellular ATP triggers tumor necrosis factor-alpha release from rat microglia. J Neurochem 2000; 75(3):965-972.
102. Chakfe Y, Seguin R, Antel JP et al. ADP and AMP induce interleukin-1beta release from microglial cells through activation of ATP-primed P2X7 receptor channels. J Neurosci 2002; 22(8):3061-3069.
103. Mattson MP, Chan SL, Duan W. Modification of brain aging and neurodegenerative disorders by genes, diet, and behavior. Physiol Rev 2002; 82(3):637-672.
104. Barden JA, Bennett MR. Distribution of P2X purinoceptor clusters on individual rat dorsal root ganglion cells. Neurosci Lett 2000; 287(3):183-186.
105. Tsuzuki K, Kondo E, Fukuoka T et al. Differential regulation of P2X(3) mRNA expression by peripheral nerve injury in intact and injured neurons in the rat sensory ganglia. Pain 2001; 91: 351-360.
106. Bennett DL, Michael GJ, Ramachandran N et al. A distinct subgroup of small DRG cells express GDNF receptor components and GDNF is protective for these neurons after nerve injury. J Neurosci 1998; 18:3059-3072.
107. Ramer MS, Bradbury EJ, McMahon SB. Nerve growth factor induces P2X(3) expression in sensory neurons. J Neurochem 2001; 77:864-875.

108. Novakovic SD, Kassotakis LC, Oglesby IB et al. Immunocytochemical localization of P2X3 purinoceptors in sensory neurons in naive rats and following neuropathic injury. Pain 1999; 80:273-282.
109. Xu GY, Huang LY. Peripheral inflammation sensitizes P2X receptor-mediated responses in rat dorsal root ganglion neurons. J Neurosci 2002; 22(1):93-102.
110. Goldberg JL, Barres BA. The relationship between neuronal survival and regeneration. Annu Rev Neurosci 2000; 23:579-612.
111. McMahon SB, Priestley JV. Peripheral neuropathies and neurotrophic factors: animal models and clinical perspectives. Curr Opin Neurobiol 1995; 5:616-624.
112. Apfel SC. Neurotrophic factors in peripheral neuropathies: therapeutic implications. Brain Pathol 1999; 9:393-413.
113. Priestley JV, Ramer MS, King VR et al. Stimulating regeneration in the damaged spinal cord. J Physiol Paris 2002; 96:123-133.
114. Cavaliere F, Florenzano F, Amadio S et al. Up regulation of $P2X_2$, $P2X_4$ receptors and ischemic cell death: prevention by P2 antagonists. Neurosci 2003; 120(1):85-98.
115. Bringmann A, Pannicke T, Moll V et al. Upregulation of P2X(7) receptor currents in Muller glial cells during proliferative vitreoretinopathy. Invest Ophthalmol Vis Sci 2001; 42(3):860-867.
116. Moller T, Kann O, Verkhratsky A et al. Activation of mouse microglial cells affects P2 receptor signaling. Brain Res 2000; 853(1):49-59.
117. Jabs R, Paterson IA, Walz W. Qualitative analysis of membrane currents in glial cells from normal and gliotic tissue in situ: down-regulation of Na+ current and lack of P2 purinergic responses. Neuroscience 1997; 81(3):847-860.
118. James G, Butt AM. Changes in P2Y and P2X purinoceptors in reactive glia following axonal degeneration in the rat optic nerve. Neurosci Lett 2001; 312:33-36.
119. Privat A, Valat J, Fulcrand J. Proliferation of neuroglial cell lines in the degenerating optic nerve of young rats. A radioautographic study. J Neuropathol Exp Neurol 1981; 40(1):46-60.
120. Thomas T, Spyer KM. ATP as a mediator of mammalian central CO2 chemoreception. J. Physiol. 2000; 523:441-447.
121. Bahr M, Bonhoeffer F. Perspectives on axonal regeneration in the mammalian CNS. Trends Neurosci 1994;17(11):473-479.
122. Stichel CC, Muller HW. Experimental strategies to promote axonal regeneration after traumatic central nervous system injury. Prog Neurobiol 1998; 56(2):119-148.
123. Lieberman AR. The axon reaction: a review of the principal features of perikaryal responses to axon injury. Int Rev Neurobiol 1971;14:49-124.
124. Koeppen AH, Barron KD. The neuropathology of olivopontocerebellar atrophy. Adv Neurol 1984;41:13-38.
125. Aldskogius H, Kozlova EN. Central neuron-glial and glial-glial interactions following axon injury. Prog Neurobiol 1998; 55(1):1-26.
126. Bahr M. Target-specific guidance cues for regenerating axons are reexpressed in the lesioned adult mammalian central nervous system. Adv Neurol 1997; 73:83-90.
127. Bahr M. Live or let die—retinal ganglion cell death and survival during development and in the lesioned adult CNS. Trends Neurosci 2000; 23(10):483-490.
128. Strata P, Rossi F. Plasticity of the olivocerebellar pathway. Trends Neurosci 1998; 21(9):407-413.
129. Fry EJ. Central nervous system regeneration: mission impossible? Clin Exp Pharmacol Physiol 2001; 28(4):253-258.
130. Kwon BK, Liu J, Messerer C et al. Survival and regeneration of rubrospinal neurons 1 year after spinal cord injury. Proc Natl Acad Sci USA 2002; 99(5):3246-3251.
131. Vaillant AR, Mazzoni I, Tudan C et al. Depolarization and neurotrophins converge on the phosphatidylinositol 3-kinase-Akt pathway to synergistically regulate neuronal survival. J Cell Biol 1999; 146(5):955-966.
132. Shen S, Wiemelt AP, McMorris FA et al. Retinal ganglion cells lose trophic responsiveness after axotomy. Neuron 1999; 23(2):285-295.
133. Goldberg JL, Espinosa JS, Xu Y et al. Retinal ganglion cells do not extend axons by default: promotion by neurotrophic signaling and electrical activity. Neuron 2002; 33(5):689-702.

134. Raisman G. Olfactory ensheathing cells—another miracle cure for spinal cord injury? Nat Rev Neurosci 2001; 2(5):369-375.
135. Blesch A, Lu P, Tuszynski MH. Neurotrophic factors, gene therapy, and neural stem cells for spinal cord repair. Brain Res Bull 2002; 57(6):833-838.
136. Herdegen T, Skene P, Bahr M. The c-Jun transcription factor—bipotential mediator of neuronal death, survival and regeneration. Trends Neurosci 1997; 20(5):227-231.
137. Wehrle R, Caroni P, Sotelo C et al. Role of GAP-43 in mediating the responsiveness of cerebellar and precerebellar neurons to axotomy. Eur J Neurosci 2001; 13:857-870.
138. Rossi F, Bravin M, Buffo A et al. Intrinsic properties and environmental factors in the regeneration of adult cerebellar axons. Prog Brain Res 1997; 114:283-296.
139. Strata P, Tempia F, Zagrebelsky M et al. Reciprocal trophic interactions between climbing fibres and Purkinje cells in the rat cerebellum. Prog Brain Res 1997; 114:263-282.
140. Brodal A, Kawamura K. Olivocerebellar projection: a review. Adv Anat Embryol Cell Biol 1980; 64:1-140.
141. Florenzano F, Viscomi MT, Cavaliere F et al. Cerebellar lesion up-regulates $P2X_1$ and $P2X_2$ purinergic receptors in precerebellar nuclei. Neurosci 2002; 115:425-434.
142. Ruigrok TJ, de Zeeuw CI, Voogd J et al. Hypertrophy of inferior olivary neurons: a degenerative, regenerative or plasticity phenomenon. Eur J Morphol 1990; 28:224-239.
143. Buffo A, Fronte M, Oestreicher, AB et al. Degenerative phenomena and reactive modifications of the adult rat inferior olivary neurons following axotomy and disconnection from their targets. Neurosci 1998; 85:587-604.
144. Chen S, Aston-Jones G. Cerebellar injury induces NADPH diaphorase in Purkinje and inferior olivary neurons in the rat. Exp Neurol 1994; 126(2):270-276.
145. Saxon DW, Beitz AJ. Induction of NADPH-diaphorase/nitric oxide synthase in the brainstem trigeminal system resulting from cerebellar lesions. J Comp Neurol 1996; 371(1):41-71.
146. Abbracchio MP, Cattabeni F. Brain adenosine receptors as targets for therapeutic intervention in neurodegenerative diseases. Ann N Y Acad Sci 1999; 890:79-92.
147. Jurkowitz MS, Litsky ML, Browning MJ et al. Adenosine, inosine, and guanosine protect glial cells during glucose deprivation and mitochondrial inhibition: correlation between protection and ATP preservation. J Neurochem 1998; 71(2):535-548.
148. Phillis JW, O'Regan MH, Perkins LM. Adenosine 5'-triphosphate release from the normoxic and hypoxic in vivo rat cerebral cortex. Neurosci Lett 1993;151(1):94-96.
149. Juranyi Z, Sperlagh B, Vizi ES. Involvement of P2 purinoceptors and the nitric oxide pathway in [3H]purine outflow evoked by short-term hypoxia and hypoglycemia in rat hippocampal slices. Brain Res 1999; 823(1-2):183-190.
150. Moroni F, Meli E, Peruginelli F et al. Poly(ADP-ribose) polymerase inhibitors attenuate necrotic but not apoptotic neuronal death in experimental models of cerebral ischemia. Cell Death Differ 2001; 8(9):921-932.
151. Laake JH, Haug FM, Wieloch T et al. A simple in vitro model of ischemia based on hippocampal slice cultures and propidium iodide fluorescence. Brain Res Brain Res Protoc 1999; 4(2):173-184.
152. Liu PK, Robertson CS, Valadka A. The association between neuronal nitric oxide synthase and neuronal sensitivity in the brain after brain injury. Ann N Y Acad Sci 2002; 962:226-241.
153. Vizi ES, Sperlagh B. Receptor- and carrier-mediated release of ATP of postsynaptic origin: cascade transmission. Prog Brain Res 1999;120:159-169.
154. Morigiwa K, Quan M, Murakami M et al. P2 Purinoceptor expression and functional changes of hypoxia-activated cultured rat retinal microglia. Neurosci Lett 2000; 282(3):153-156.

Chapter 6

Lesion-Induced Axonal Sprouting in the Central Nervous System

Thomas Deller, Carola A. Haas, Thomas M. Freiman, Amie Phinney, Mathias Jucker and Michael Frotscher

Summary

Injury or neuronal death often come about as a result of brain disorders. Inasmuch as the damaged nerve cells are interconnected via projections to other regions of the brain, such lesions lead to axonal loss in distal target areas. The central nervous system responds to deafferentation by means of plastic remodeling processes, in particular by inducing outgrowth of new axon collaterals from surviving neurons (collateral sprouting). These sprouting processes result in a partial reinnervation, new circuitry, and functional changes within the deafferented brain regions. Lesioning of the entorhinal cortex is an established model system for studying the phenomenon of axonal sprouting. Using this model system, it could be shown that the sprouting process respects the pre-existing lamination pattern of the deafferented fascia dentata, i. e., it is layer-specific. A variety of different molecules are involved in regulating this reorganization process (extracellular matrix molecules, cell adhesion molecules, transcription factors, neurotrophic factors, growth-associated proteins). It is proposed here that molecules of the extracellular matrix define the boundaries of the laminae following entorhinal lesioning and in so doing limit the sprouting process to the deafferented zone. To illustrate the role of axonal sprouting in disease processes, special attention is given to its significance for neurodegenerative disorders, particularly Alzheimer's disease (AD), and temporal lobe epilepsy. Finally, we discuss both the beneficial as well as disadvantageous functional implications of axonal sprouting for the injured organism in question.

Introduction

The central nervous system (CNS) of a healthy adult organism is distinguished by a large number of coherently interconnected nerve cells. At the conclusion of their developmental phase, these nerve cells are highly specialized and, as a general rule, post-mitotic. As such, it was assumed until well into the middle of the last century that nerve cell fibers as well, which conduct information and signals between neurons, were incapable of modification anatomically. The CNS was construed as a "hard-wired" system of nerve cells and their axons. In the past forty years, this view of the brain has undergone fundamental change. It could be shown that plastic reorganization can, and does, occur at any point in time even in the fully developed brain and that the ability to remodel synaptic connections appears to be one of the physiological properties of neurons.[1] The most recent findings even indicate that the brain is capable of

Brain Repair, edited by Mathias Bähr. ©2006 Eurekah.com
and Kluwer Academic / Plenum Publishers.

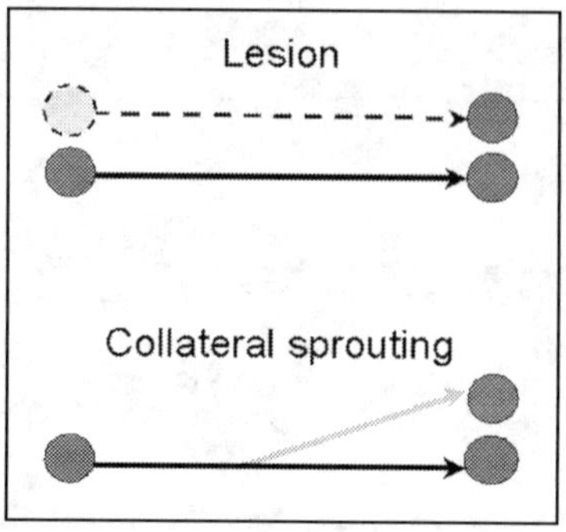

Figure 1. Collateral sprouting after CNS lesion. Injuries and diseases of the CNS can result in the death of involved nerve cells (upper portion of figure). Such a lesion also leads to the destruction of synapses in the target region of the neurons, i.e., to deafferentiation of subsequent connected brain regions. In response to the loss, surviving nerve cells send out new axon collaterals and reinnervation takes place (lower portion of the figure).

producing new nerve cells well into old age and these neurons, in turn, are interconnected with existing nerve cells.[2-4] Such plastic processes enable the adult organism to learn and to adapt to a constantly changing environment.

The physiological ability to reorganize synaptic connections is retained by the adult brain even following injury and in the course of neurodegenerative diseases. Even under these conditions nerve tissue possesses the capacity to modify nerve fiber connectivities and produce new synapses.[5-7] However, different parts of the nervous system differ in their ability to reorganize themselves following lesion: Whereas neurons in the peripheral nervous system (PNS) are able to regenerate their axon over long distances, neurons located in the CNS are incapable to do so. After a central fiber tract has been severed, fiber connectivities are permanently interrupted with resulting axonal and synaptic loss in the distally located regions of the brain. Although the affected nerve fibers cannot reconnect to their original target region, the central nervous system reacts to this deafferentiation with remodeling processes of a plastic nature within the denervated brain regions. Surviving nerve fibers sprout new axon collaterals (collateral sprouting) and form new synapses (reactive synaptogenesis), thereby reinnervating the denervated brain region (Fig. 1). However, this sprouting process does not restore original anatomical connections. It leads to, in the final analysis, a partial reinnervation of the deafferented brain regions, with many new interconnectivities, and considerable functional changes within the involved brain areas.

Collateral sprouting is a common response of the CNS to injury. It was first described in the spinal cord[8] and shortly thereafter in the brain.[9] Since then, lesion-induced forms of neuronal plasticity have been demonstrated in numerous species and brain regions.[7,10-16] Possible triggers of axonal sprouting in the adult brain are, in effect, every form of injury and any disease that lead to the deafferentiation of a given brain area. Accordingly, sprouting processes occur following head traumata[15] where a singular massive loss of nerve cells takes place but also during the course of neurodegenerative disorders[17]and certain forms of epilepsy,[18,19] in which slowly increasing neuronal loss in particular regions of the brain is at the forefront. In effect, axonal sprouting is a widespread plastic response on the part of the CNS to a lesion.[20] As such, a detailed neuroanatomical profile of axonal sprouting, identification of the cellular and molecular mechanisms involved, as well as elucidation of their respective functional implications for the organism as a whole would seem to be both logical and of significance for the ongoing development of therapeutic interventional strategies.

Reorganization of the CNS after Entorhinal Lesioning

Subsequent to a lesion, the reorganization of the CNS entails a very complicated series of events that cannot be simulated in vitro. One classic model system for studying axonal sprouting is the lesioning of the entorhinal cortex in small rodents. Following removal of the entorhinal cortex, one encounters degeneration of the entorhinal fibers in the hippocampus and loss of up to 90% of all synapses within the outer molecular layer of the fascia dentata. The entorhinal lesioning model is regarded as well characterized and easily accessible experimentally.[5,16,21-23] Moreover, it offers enormous advantages in comparison with other lesioning models: first, the lengthy distance between the site of the lesion and the deafferented region facilitates the accurate distinction between neurodegenerative and sprouting processes. Second, the layered termination of the entorhinal fibers in the fascia dentata permits clean delineation of the deafferented zone from the adjacent regions of the CNS (Fig. 2).

The entorhinal lesioning model was described for the first time in the 1970s and initially characterized employing histochemical and electron microscopy techniques.[24-31] At that time, however, the direct demonstration of sprouting fibers was not technically possible and, thus, it was unclear until well into the 1990s precisely in what manner the morphology of individual axons changes within the context of the sprouting phenomenon. Nonetheless, insofar as these morphological data are crucial for gauging and assessing the effects of experimental invasive methods on sprouting axons, we began our studies first of all with a detailed description of axonal sprouting at the level of single, isolated fibers.

For this purpose, we marked the sprouting fibers with the sensitive anterograde tracer *Phaseolus vulgaris* leucoagglutinin (PHAL).[32] This procedure enabled us to visualize single axons from projection fibers in control animals as well as in those with entorhinal lesions and compare them. For purposes of quantifying the changes, we determined the degree of branching on the part of the axons and the number of boutons within a clearly defined axon segment. In addition, we characterized the sprouting-associated changes in fiber morphology. As paradigmatic for the sprouting behavior of anatomically heterologue and homologue fiber systems, the GABAergic commissural[33,34] and the glutamatergic crossed entorhinal[35] fiber systems were studied. In comparison with control animals, we observed post-lesion a significant increase of the degree of branching and total fiber length for single fibers as well as an increase in the number of boutons contained within a defined axon segment. Further, we also found alterations in the normal fiber morphology. Whereas some sprouting axons developed a large number of short, truncated axonal processes (axonal extensions), other fibers gave rise to conspicuous glomerular or tangle-like formations.[33,35]

Remarkably, these neuroanatomical analyses revealed that the axonal sprouting adapted itself strictly to the existing lamination pattern in the dentate gyrus. Thus, after entorhinal lesioning, the results were deafferentiation of the outer molecular layer and sprouting by surviving axons within the layer (Figs. 3a and b). Fiber systems of adjacent intact layers, e.g., the commissural fibers belonging to the inner molecular layer, do not succeed in penetrating the deafferented zone (Figs. 3c and d). As such, the ordered, laminated structure of these brain regions remains intact even after extensive deafferentiation of the gyrus dentatus.[34,36]

Molecular Regulation of Axonal Sprouting

The neuroanatomical description of axonal sprouting poses the question as to which cells and molecules are involved in the regulation of the sprouting process. Since an axonal growth process is involved, one's thoughts quite logically turn, primarily, to those classes or groups of molecules that likewise play an important role in axonal growth processes during development. Here, a basic distinction is made between so-called "intrinsic" molecules that are expressed by growing neurons and "extrinsic" molecules that reside in the local microenvironment of the growing axon. Expression of intrinsic growth-associated molecules (e.g., growth-associated

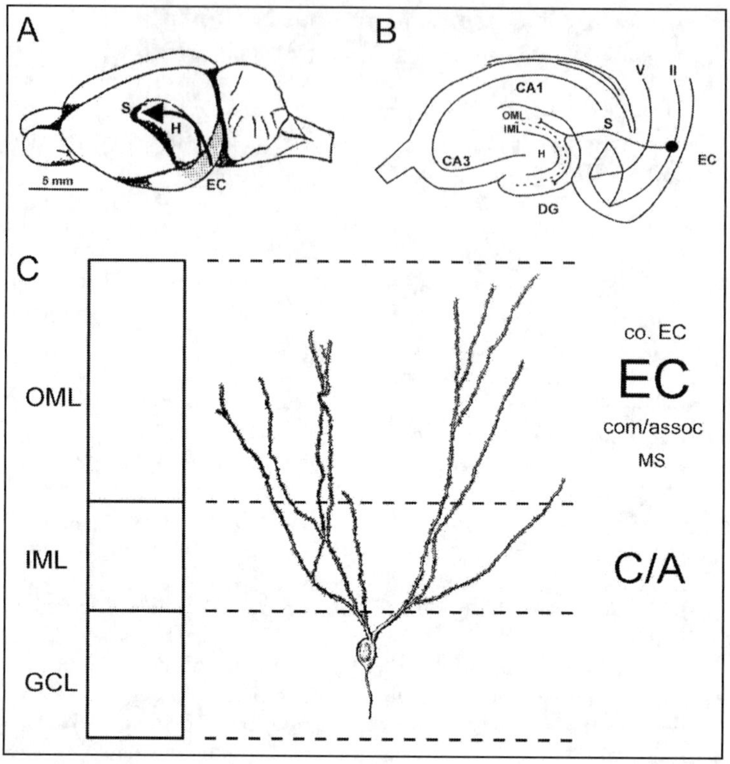

Figure 2. The entorhinal projection to the rat fascia dentata. A) The entorhinal cortex (EC) projects to the hippocampus (H) in a topographically ordered manner. Dorsal portions of the entorhinal cortex project to septal (S) portions of the hippocampus, whereas more ventrally located portions of the entorhinal cortex project to temporal levels of the hippocampus. Thus, the lesion site in the ECL model is far away from the denervated septal portion of the hippocampus. B) Schematic diagram of a slice through the hippocampal formation. Entorhino-hippocampal fibers originate from layer II neurons in the EC. They fasciculate, form the perforant pathway, and terminate in the outer molecular layer (OML) of the dentate gyrus (DG). H, hilus; IML, inner molecular layer; S, subiculum. C) A typical granule cell of the fascia dentata is illustrated. Afferents terminate on the dendritic arbor of this cell in either a lamina-specific or a diffuse fashion. In the outer molecular layer (OML) fibers of the ipsilateral entorhinal cortex (EC), contralateral entorhinal cortex (CO. EC), and GABAergic commissural and associational fibers (com/assoc) terminate. The inner molecular layer (IML) is occupied by glutamatergic commissural and associational fibers (C/A). Fibers from the medial septum (MS) terminate throughout all layers. GCL, granule cell layer. Reprinted from Deller et al; Restor Neurol Neurosci 2001; 19:159-167. ©2001 IOS Press.

protein 43, GAP-43) is required for a nerve cell even to acquire the capacity for growth in the first place, its "growth competence," so to speak. What is involved is a genetically determined growth program that becomes activated by the neuron during axonal growth.[37-44] Extrinsic molecules are found in the target area of the growing axon where they are important for neuronal pattern formation. These homing-molecules point the way for the growing axon to the correct target region and the target cell. They regulate the degree of axonal branching, the number of synapses that are formed, and, presumably, they also play a role in synaptic stabilization (Fig. 4). The extrinsic molecules frequently are subdivided further into those which stimulate axonal growth and those inhibiting it. To the extrinsic molecules belong a variety of cell adhesion molecules (e.g., N-CAM), extracellular matrix molecules (e.g., tenascin-C,

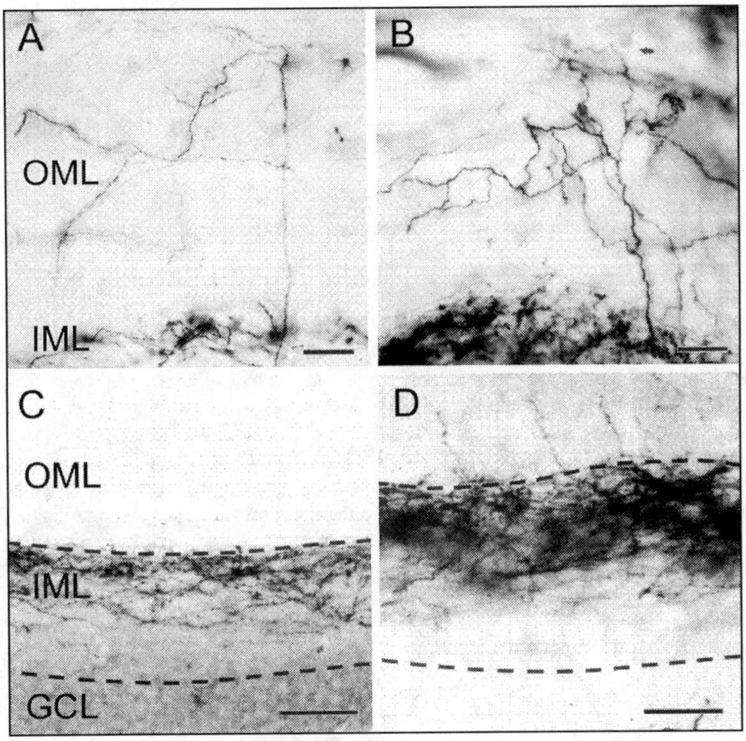

Figure 3. Sprouting in the fascia dentata after ECL is layer-specific. A) Commissural fiber to the outer molecular layer in a control animal. These fibers do not give off branches to the inner plexus, and the number of collaterals in the outer molecular layer is low. IML= inner molecular layer; OML= outer molecular layer. B) Commissural fiber to the outer molecular layer four weeks after ECL. The number of axonal collaterals has considerably increased in the OML. Note that this axon does not branch in the inner molecular layer similar to the axon shown in A. C) Section of the fascia dentata immunostained for PHAL in a control animal. The commissural projection to the inner molecular layer terminates within 70 μm from the granule cell layer. D) Section of the fascia dentata of an animal four weeks after entorhinal cortex lesion. The termination field of commissural fibers to the inner molecular layer has expanded by 30-40 μm (only very few fibers grow beyond this layer). Scale bars: a, b= 30μm; c, d= 50 μm. Reprinted from Deller et al; Restor Neurol Neurosci 2001; 19:159-167. ©2001 IOS Press.

neurocan), and numerous diffusible molecules (e.g., growth factors and cytokines), which can be produced by nerve cells or glial cells within the target region.[45-52]

What are the molecular bases for axonal sprouting? To identify the molecular and cellular players involved in the sprouting process, the entorhinal lesion model is once again a useful experimental paradigm. Using this model system we and others have analyzed suitable molecular candidates subsequent to entorhinal lesioning and have correlated both the spatial and temporal changes on the part of the expression pattern seen in these molecules with axonal sprouting. Where clear correlations exist, genetically altered mice (transgenic mice, knock-out mice) are now being used to demonstrate causal relationships between certain molecules and the phenomenon of axonal sprouting. In our previous studies, we could identify several molecules that might be of regulatory significance for the axonal sprouting that occurs after entorhinal lesioning. Among the molecules we have examined are neurotrophic factors (nerve growth-factor, NGF; brain-derived neurotrophic factor, BDNF; ciliary neurotrophic factor, CNTF),[53,54] transcription factors (c-Fos, c-Jun, Jun B, and STAT3),[55-57] cell adhesion molecules (neural

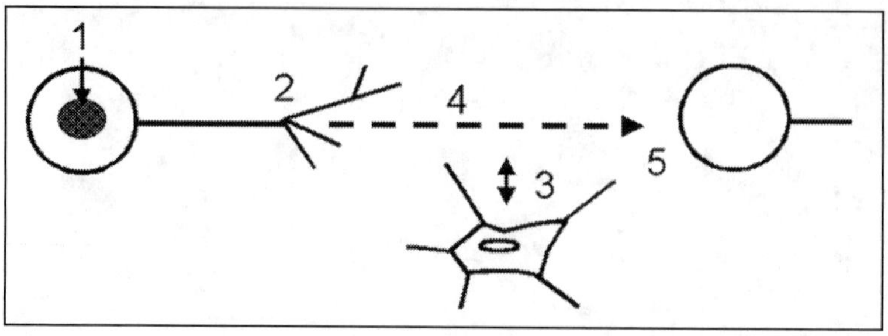

Figure 4. Molecular regulation of axonal growth. Both intrinsic (1) as well as extrinsic (2-5) molecules are involved in the molecular regulation of axonal growth processes. Intrinsic molecules (e.g., GAP-43) are expressed by growing and regenerating neurons. These molecules are part of a nerve cell's genetic growth program that is activated within the context of axonal growth processes and which are essential for the growth competence of nerve cells. Extrinsic molecules are found along the axon's route to the target region as well as within the target region. They provide local guidance cues for the growing axons. Extrinsic molecules can intervene at various points in the axonal growth process both in a growth promoting as well as growth inhibiting manner. Thus, extrinsic molecules regulate the outgrowth of growth cones (2), interactions with glia cells (3), elongation of the growing axon (4), and the formation and stabilization of synapses (5). The extrinsic molecules include cell surface molecules, extracellular matrix molecules, and diffusible molecules (growth factors and cytokines), which can be produced by nerve cells but also by glia cells.

adhesion molecule L1, neural cell adhesion molecule N-CAM),[58,59] together with extracellular matrix molecules (tenascin-C; DSD-1-proteoglycan; reelin; brevican; neurocan).[60-63] Because it is beyond the scope of the present review to discuss all molecular changes that have been reported following entorhinal lesion, and several recent reviews exist on this subject matter,[16,23,64] we will focus here on extracellular matrix molecules and their role in the reorganization process after entorhinal lesioning.

Extracellular Matrix Molecules Direct Axonal Growth Processes

Over the past years, it has become increasingly clear that the extracellular matrix (ECM) is of great importance for axonal growth processes during development and following injury to the CNS.[49,65-75] The ECM consists of structurally and functionally different molecules, among them many proteoglycans and glycoproteins, which are organized around a hyaluronan backbone into a complicated three-dimensional structure.[69] As such, it constitutes the growth substrate for axons in the brain and is capable of exerting influence via its molecular composition on the growth process itself. Of particular functional significance is the fact that the molecular composition of the ECM during development displays regional differences in the brain. Its diversified components, which include growth-permissive, growth-stimulating, and growth-inhibitory molecules, finally accounts for the ECM's effect on the growing axon. It is assumed, therefore, that brain regions which are well supplied with inhibitory ECM molecules are avoided by growing axons, whereas brain areas that are richly endowed with growth-stimulating ECM molecules provide a favorable substrate for the growing axon. In this manner, growth boundaries could emerge on the basis of the local distribution of the ECM molecules in the brain during development.[46,76]

It has been suggested that ECM molecules could assume a role similar to the one they play during the developmental phase in the injured brain. And, in fact, long-term changes in the ECM's composition occur within the area surrounding a lesion, including, in particular, an increase in the concentration of chondroitinsulfate-proteoglycans (CSPG). These molecules inhibit axonal growth, and it has been assumed that their presence constitutes a major barrier to axonal regeneration within the CNS.[66,71,72,74] Based upon these findings, we postulated that CSPG molecules might also be of significance for the regulation of axonal sprouting in deafferented brain regions and, as such, we studied the molecular composition of the ECM in the deafferented hippocampus at various time points following entorhinal cortex lesioning. To visualize each of the ECM molecules, we employed antibodies, and we identified the cells that synthesize theses molecules with the help of in situ hybridization combined with immunocytochemistry. Out of these experiments emerged evidence showing that in deafferented brain regions the molecular composition of the ECM is considerably altered. Thus, tenascin-C,[60] DSD-1-proteoglycan,[60] brevican,[62] and neurocan[61] are synthesized by reactive astrocytes in the fascia dentata following entorhinal lesioning, are released into the ECM, and accumulate there. As a result, the composition of the ECM in the deafferented region is altered in relation to intact neighboring layers, and clearly-defined molecular boundaries exist separating the deafferented regions of the CNS from those areas that are still intact (Fig. 5).

What are the functional implications of such a regional alteration within the ECM following lesioning? We are operating under the assumption that it regulates the local sprouting process within the deafferented zone. It appears that the layer-specific alteration in the ECM produces, in the process, different effects on the axons that sprout within the deafferented layer than on those of the adjacent layers. The sprouting axons encounter within the deafferented external molecular layer a milieu that is favorable to growth. This environment consists of ECM molecules, cell adhesion molecules, growth factors, and cytokines, which in toto regulate and nurture the axonal sprouting process and in so doing the reinnervation of the deafferented region as well. In their attempt to extend their growth into the deafferented zone, the axons of the adjacent layers (e.g., the fibers belonging to the undamaged internal molecular layer) would be confronted by a molecular barrier that harbors growth-inhibiting ECM molecules, such as neurocan. The fact that these kinds of molecular barriers are capable of hindering the ingrowth of fibers into a given brain area could account for why, subsequent to entorhinal lesioning, ingrowth of fibers originating in the adjacent layers does not take place (Fig. 3).[36] Theoretically, the altered composition of the ECM could, in this manner, confine the sprouting process to the axons of the deafferented zone (Fig. 6).[77,78]

Axonal Sprouting in Alzheimer's Disease

In the course of neurodegenerative disorders, such as Alzheimer's disease (AD), a gradual but increasing degree of neuronal loss occurs. As a result of cell death, the synapses in subsequent, connected regions of the brain also die, and surviving nerve cells generate new axon collaterals. Because, initially, only a few cells belonging to a single projection tract become affected in neurodegenerative illnesses, collateral sprouting at the time of disease onset occurs in those types of nerve fibers that can functionally compensate for the neuronal loss. Viewed in this light, axonal sprouting appears to be a welcome compensatory mechanism for such patients and it is believed that its occurrence delays the appearance of clinical symptoms.[17,79]

In the case of AD, however, there are signs that axonal sprouting may not always be advantageous. In this neurodegenerative disorder, characterized, in part, by the extracellular deposition of amyloid (plaques), one finds indications of aberrant axonal growth processes in the vicinity of amyloid deposits.[79] These observations suggest that not only compensatory but also pathological sprouting processes occur in AD.

Initial signs of pathological sprouting phenomena in AD were found in the brains of AD patients, i.e., in human autopsy specimens.[79-81] Since postmortal tissue is only of limited use in experimental studies, it was not possible initially to supply any further explanation for the aberrant sprouting processes in the vicinity of the plaques. It was only with the advent and development of the transgenic mouse model for AD that new experimental perspectives presented themselves. Accordingly, it became possible to apply modern experimental methods to

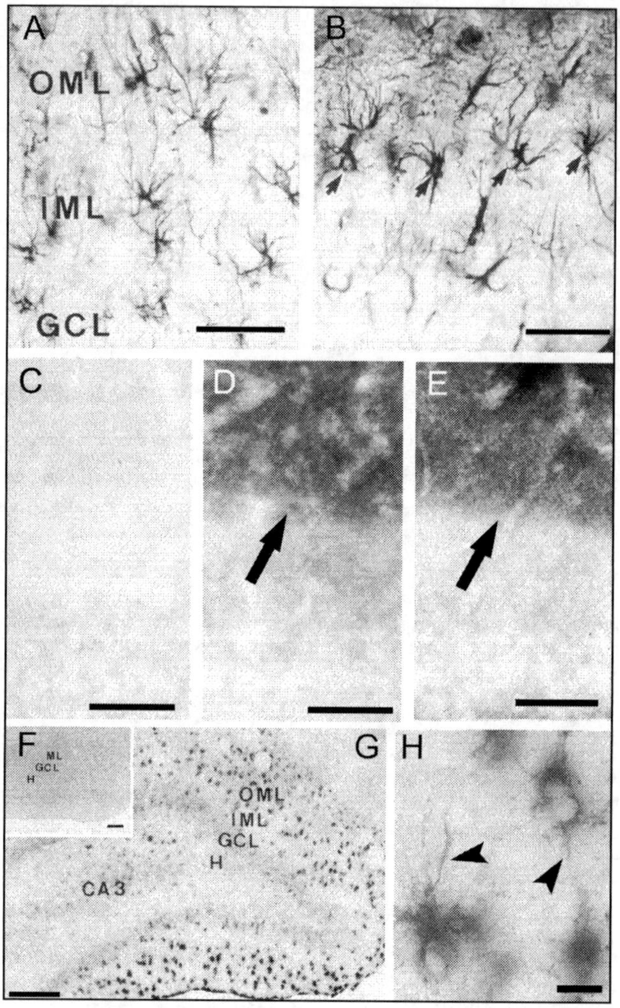

Figure 5. Reactive astrocytes change the molecular composition of the extracellular matrix in the denervated outer molecular layer A, B. Immunoreactivity for GFAP in a control animal (A) and in an animal 14 days post lesion (B). In the control (A), astrocytes are fairly evenly distributed throughout the molecular layer. After ECL (B), GFAP immunoreactivity is increased in the outer molecular layer (OML), and astrocytes form a row of cells at the border between the inner molecular layer (IML) and the OML (arrows). C, D) Immunoreactivity for the chondroitin sulfate proteoglycan neurocan in a control animal (C) and in an animal 14 days post lesion (D). In the control (C), neurocan immunoreactivity is not above background. After ECL (D), a distinct neurocan immunoreactive band is visible in the OML (arrow). E) Immunoreactivity for tenascin-C in an animal 14 days post lesion. Note that the tenascin-C enriched outer molecular layer forms a sharp molecular boundary towards the non-denervated IML (arrow). F G) Neurocan mRNA expression in a control animal (F) and in an animal four days postlesion (G). In the control (F) no neurocan mRNA expressing cells can be detected in the molecular layer. After ECL (G), many cellular profiles are observed in the outer molecular layer of the fascia dentata. (H) Fascia dentata four days after entorhinal cortex lesion. This section was double labeled for GFAP (immunohistochemistry) and neurocan mRNA (in situ hybridization). Several double labeled astroglial cells are visible. Immunostaining for GFAP identifies astrocytes and labels their somata and proximal processes (arrowheads). In situ hybridization for neurocan mRNA identifies neurocan mRNA expressing cells. Scale bars: A-E= 50 μm; F, G= 40 μm; H=10 μm.

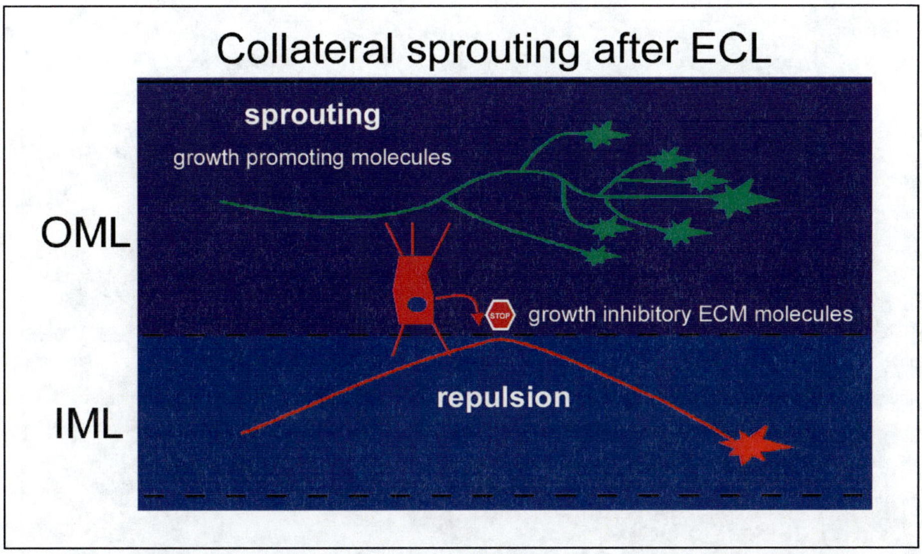

Figure 6. Collateral sprouting after entorhinal cortex lesion. Molecules of the ECM may guide the sprouting fibers after entorhinal cortex lesion (ECL): Fibers that sprout within the denervated outer molecular layer (OML) grow within a growth promoting environment. This environment contains a variety of growth promoting ECM molecules, cell adhesion molecules, and cytokines. In contrast, fibers that attempt to grow into the denervated OML from the neighboring inner molecular layer (IML) meet a barrier of inhibitory ECM molecules which are secreted by reactive astrocytes. This molecular barrier could prevent ingrowth of fibers from non-denervated adjacent zones into the denervated OML.

the problem of axonal sprouting for the purpose of characterizing axonal growth processes at the level of single nerve fibers. We used a transgenic mouse (APP23 transgenic mouse)[82,83] for our investigations. The APP23 mouse overexpresses a mutated form of the human amyloid precursor protein (APP) that was discovered in the brains of patients, in whose families AD occurs with a marked frequency. At six months of age, the transgene in the brains of the mice results in the pronounced formation of amyloid deposits that resemble very closely the plaques seen in AD. To determine what influence amyloid deposition has on neighboring axons, we marked the entorhinal projection to the hippocampus with the anterograde tracer PHAL.[84] Inasmuch as this projection tract becomes involved very early in AD, it seemed an especially suitable candidate for our studies pertaining to axonal alterations in the vicinity of amyloid plaques.[85-87]

The entorhinal fiber projection to the hippocampus in aged control cases and in young APP23 transgenic animals displayed a normal morphology (Figs. 7A and C). In these animals, we were able to locate the entorhinal fibers in their usual layer, namely the middle and outer molecular layer of the fascia dentata. Isolated fibers displayed a long, drawn-out course and developed numerous small boutons. In contradistinction to these, we found in older transgenic animals accentuated axonal changes near the amyloid deposits (Figs. 7B, D and E). Most noticeable were the large swellings (dystrophic boutons) formed by the entorhinal fibers in the immediate vicinity of the plaques. Under the electron microscope, these boutons exhibited the properties that are typical for dystrophic neurites, such as they also are found in AD.[84]

Evidence, however, of dystrophic changes within the confines of the normal termination zone of the entorhinal fibers does not mean that these structures originate via axonal sprouting. In effect, it is conceivable that such changes come about both by means of degenerative as well as axonal growth processes. Accordingly, we sought additional indications of axonal growth in

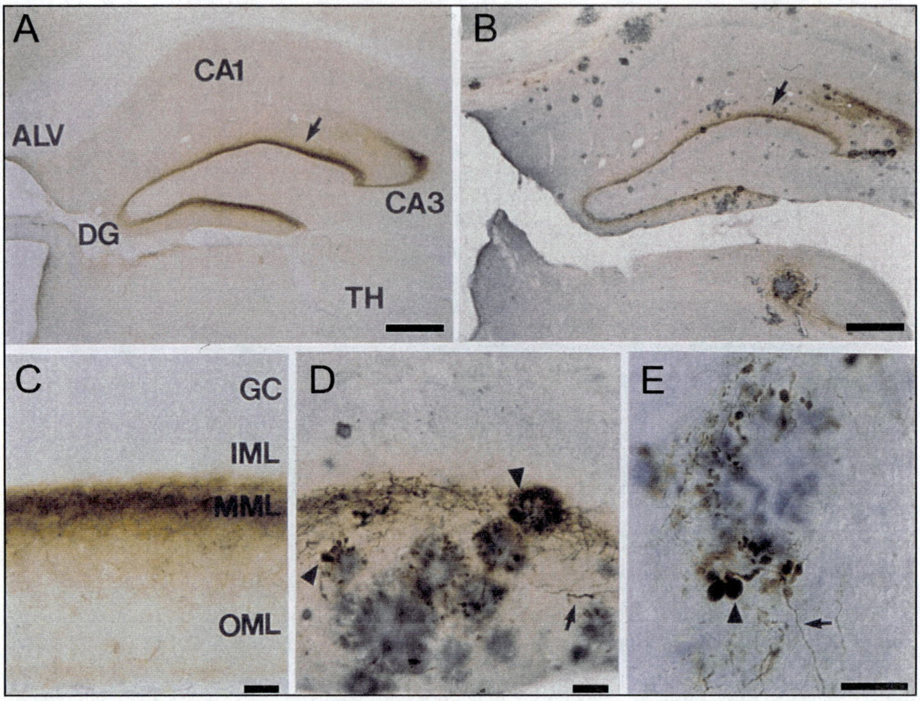

Figure 7. Entorhinal axonal projections in APP23 mice. A, B) Anterogradely Phaseolus vulgaris-Leucoagglutinin (PHAL)-labeled projection (arrow) form the entorhinal cortex to the molecular layer of the dentate gyrus (DG) in an 18-months-old nontransgenic control mouse (A) and an 18-months-old APP23 mouse (B). Note the amyloid deposits (blue-gray reaction product) throughout the hippocampus, the thalamus (TH), and within the alveus (ALV). CA1, CA3, hippocampal subfields. C, D) Higher magnification of the inferior balde of DG from a control mouse (C) and an APP23 transgenic mouse. Note the thickened PHAL-labeled axons (arrow) and ballooned, spheroidal axon terminals (arrowheads) in the vicinity of amyloid plaques. GC= granule cell layer; IML= inner molecular layer; MML= middle molecular layer; OML= outer molecular layer. E) High magnification of PHAL-labeled axons with the characteristic dystrophic terminals around an amyloid plaque in CA3. Many PHAL-labeled axons appear normal until directly adjacent to the amyloid (arrow), then typically curve around the amyloid periphery, forming several small swellings followed by a large terminal balloon-shaped swelling that turns away from the plaque (arrowhead). Scale bars: A, B= 300 μm; C-E= 25 μm. Reprinted from Phinney et al. J Neurosci 1999; 19:8552-8559. ©1999 Society for Neuroscience.

the vicinity of amyloid deposits. Here, we were aided by the fact that entorhinal fibers in control cases were found only in specific regions and layers of the brain and that the presence of entorhinal fibers at ectopic sites would be decisive evidence for the existence of reactive axonal growth processes in the mutant animals. We found such evidence for reactive axonal growth, once again, within the fascia dentata of the hippocampal formation. In this brain area, entorhinal fibers terminate in a layer-specific manner in the outer molecular layer and commissural/associational fibers terminate layer-specifically in the inner molecular layer (Fig. 2). In control cases, this laminar organization is respected by both projections systems, and the fiber systems are confined to their respective layers. In APP23 transgenic animals, however, we observed that this fiber stratification is missing close to the plaques. As entorhinal fibers occur within the "wrong" inner molecular layer (Fig. 8A), commissural fibers are found ectopically within the

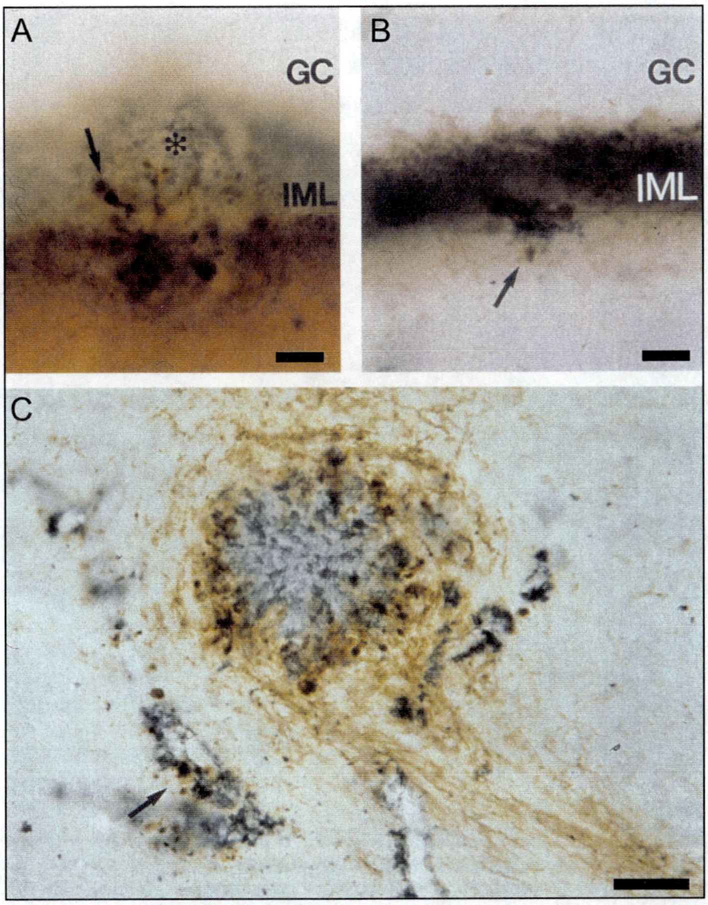

Figure 8. Plaque-associated ectopic axon terminals in APP23 transgenic mice. A) Disruption of the layer-specific termination of PHAL-labeled entorhinal axons in the vicinity of an amyloid plaque (asterisk) is evidenced by the invasion of these fibers (brown reaction product) into the inner molecular layer (IML) of the dentate gyrus, which is also labeled for calretinin (blue-gray reaction product). Calretinin is specific for the commissural/associational fibers in the mouse and labels specifically the IML. GC= granule cell layer of the dentate gyrus. B) Calretinin-labeled commissural/associational fibers (brown) also form dystrophic axon terminals (arrow) and deviate from their normally specific termination in the IML when in vicinity of amyloid (blue-gray). C) High magnification of the amyloid plaque (blue-gray) shown in the thalamic region in Figure 7B. Entorhinal fibers curve around the amyloid and form dystrophic boutons directly adjacent to the amyloid. Interestingly, PHAL-labeled entorhinal axons also formed dystrophic boutons around vascular amyloid deposits (arrow). Scale bars: A, B= 10 μm; C= 50 μm. Reprinted from Phinney et al. J Neurosci 1999; 19:8552-8559. ©1999 Society for Neuroscience.

"wrong" outer molecular layer (Fig. 8B), and, on these ectopic sites, both fiber systems give rise to typically dystrophic boutons.

We found a further piece of evidence for aberrant axonal growth in the thalamus of the APP23 transgenic mouse. There, amyloid deposits can be visualized routinely in the vicinity of blood vessels (vascular amyloid).[88] Whereas, in controls, entorhinal fibers display no particular affinity to blood vessels, we found in APP23 transgenic animals countless dystrophic entorhinal

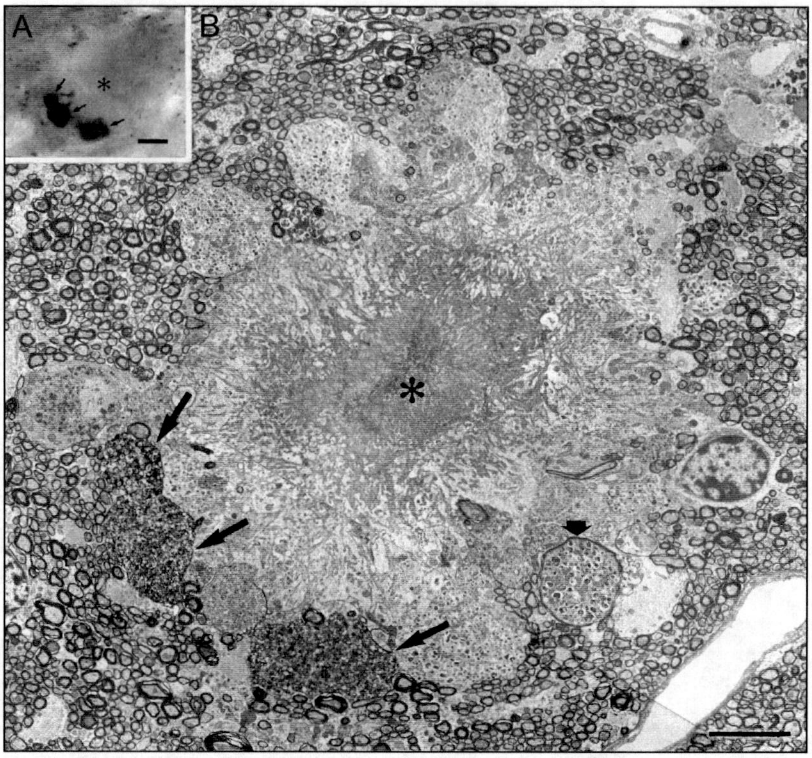

Figure 9. Entorhinal fibers form dystrophic boutons. A) Light micrograph of an amyloid plaque (asterisk) in the alveus of the hippocampus. Three PHAL-labeled entorhinal boutons surround the plaque (arrows). B) Electron micrograph of the plaque illustrated in A. The core amyloid (asterisk) is surrounded by numerous dystrophic neurites. The three large entorhinal boutons illustrated in A are indicated by arrows. One of the unlabeled dystrophic neurites is myelinated (short arrow). Scale bars: A= 10 μm; B= 5 μm. Reprinted from Phinney et al. J Neurosci 1999; 19:8552-8559. ©1999 Society for Neuroscience.

boutons in the immediate vicinity of vascular amyloid (Fig. 8C). In the final analysis, the entorhinal fibers could only have reached these locations via axonal growth.

Presumably, the best example that we found for abnormal axonal sprouting in the APP23 transgenic mice was near amyloid plaques in the hippocampal white matter, in the alveus. There, entorhinal fibers make their way en-route to their target cells in the hippocampus. At this point, they are myelinated and do not produce any synapses. Nonetheless, we observed numerous entorhinal boutons in the alveus surrounding the plaques (Fig. 9A) and could identify them electron microscopically as dystrophic neurites (Fig. 9B).[84] Many of these entorhinal boutons are immunoreactive for the growth-associated protein GAP-43, which is typically in evidence as a feature of axonal growth during development, axonal regeneration, and occurs in the course of sprouting processes. Moreover, the dystrophic boutons contain the synaptic protein synaptophysin and show ultrastructural changes that point in the direction of a partial synaptic differentiation.[84]

The molecular mechanisms that underlie this form of abnormal axonal sprouting are the focus of our recent investigative efforts. It is conceivable that ECM alterations in the plaque periphery, such as the accumulation of perlecan,[82] to name but one example, have an important role to play. Furthermore, amyloid deposition in the brain activates astrocytes and microglia in the vicinity of the plaques. Activated glial cells could, for their part, synthesize growth factors and cytokines, thereby exerting an influence on the sprouting process.[89,90]

At present, the following picture emerges: axonal alterations appear only in the neighborhood of amyloid; entorhinal axons grow close to amyloid deposits and into the "wrong" areas of the CNS; and entorhinal axons are immunopositive for the growth-associated protein GAP-43 and display morphological as well as molecular characteristics of synaptic differentiation. These findings indicate that dystrophic entorhinal boutons come into existence by way of axonal growth processes. Furthermore, inasmuch as axonal sprouting can be shown to occur exclusively in the vicinity of amyloid deposits, the conclusion can be reached that amyloid deposition is responsible for these axonal growth processes either directly or, more likely, indirectly via the activation of glial cells.

Axonal Sprouting in Temporal Lobe Epilepsy

Sprouting processes have also been described in other diseases of the nervous system in which neuron loss occurs. In these disorders, neuronal damage leads to a plastic reaction of the brain and a reorganization of the central nervous circuitry. A prominent example in this context is human temporal lobe epilepsy (TLE). This disease is linked strongly to severe neuronal loss in a characteristic pattern known as hippocampal sclerosis (HS). It is assumed that following preferential loss of hippocampal neurons, such as those located in the hilus and hippocampal subfield CA1, surviving neurons sprout new axon collaterals and reinnervate the partially denervated hippocampus. The sprouting fibers form a new hippocampal circuitry that is thought to contribute to seizure pathogenesis as well as seizure propagation.[18,19,91]

Because TLE frequently is drug-resistant, hippocampal surgery is often necessary to achieve seizure control. These human hippocampi could be studied, and morphological evidence for the reorganization of hippocampal circuits was found. The most prominent changes were observed in the dentate gyrus, where granule cells sprout new collaterals into the inner molecular layer and establish recurrent synapses with granule cell dendrites. Histochemical staining methods, such as the Timm's stain,[92,93] visualize ectopic supragranular mossy fibers in sclerotic hippocampi (Fig. 10 A, B). Furthermore, based on Timm's staining and dynorphin immunoreactivity, mossy fibers sprout across subfield boundaries into the hippocampal subfield CA1 and subiculum.[94,95] Finally, evidence was obtained for the sprouting of neuropeptide Y, somatostatin, and interneurons containing gamma-aminobutyric acid (GABA).[96,97] Within other regions of the hippocampus considerable network reorganization has also been reported.[98] Thus, anatomic circuit reorganization occurs to a considerable extent in this disorder, and numerous aberrant connectivities are formed by the sprouting fibers.

At present, only very few studies have been published, in which the connectivity of the human hippocampus is analyzed in detail, e.g., at the level of single axons or that of single identified cells.[99,100] As such, we sought a method for studying the reorganization of hippocampal pathways at the level of local fiber tracts. Because anterograde tracing techniques are used for this purpose in experimental animals in vivo[101] as well as in vitro,[102] we adapted known anterograde tracing methods to human hippocampal slices and employed this technique to analyze mossy fiber sprouting in the human brain.[103] Hippocampal slices were obtained from the hippocampi of patients with HS or from those with mass lesions located in the temporal lobe. All of the patients gave their written consent for the use of their brain tissue for scientific research purposes. All of these studies were approved by the University of Freiburg's Ethics Committee. Hippocampal slices were kept alive in a slice chamber, and the anterograde tracer neurobiotin was iontophoretically injected into the granule cell layer of the dentate gyrus. Single injections resulted in the labeling of small groups of granule cells (Fig. 10 C, D). The axonal arbor of these cells could be partially reconstructed, and single mossy fibers could be followed from the soma to the inner molecular layer of the sclerotic dentate gyrus (Fig. 10 E,

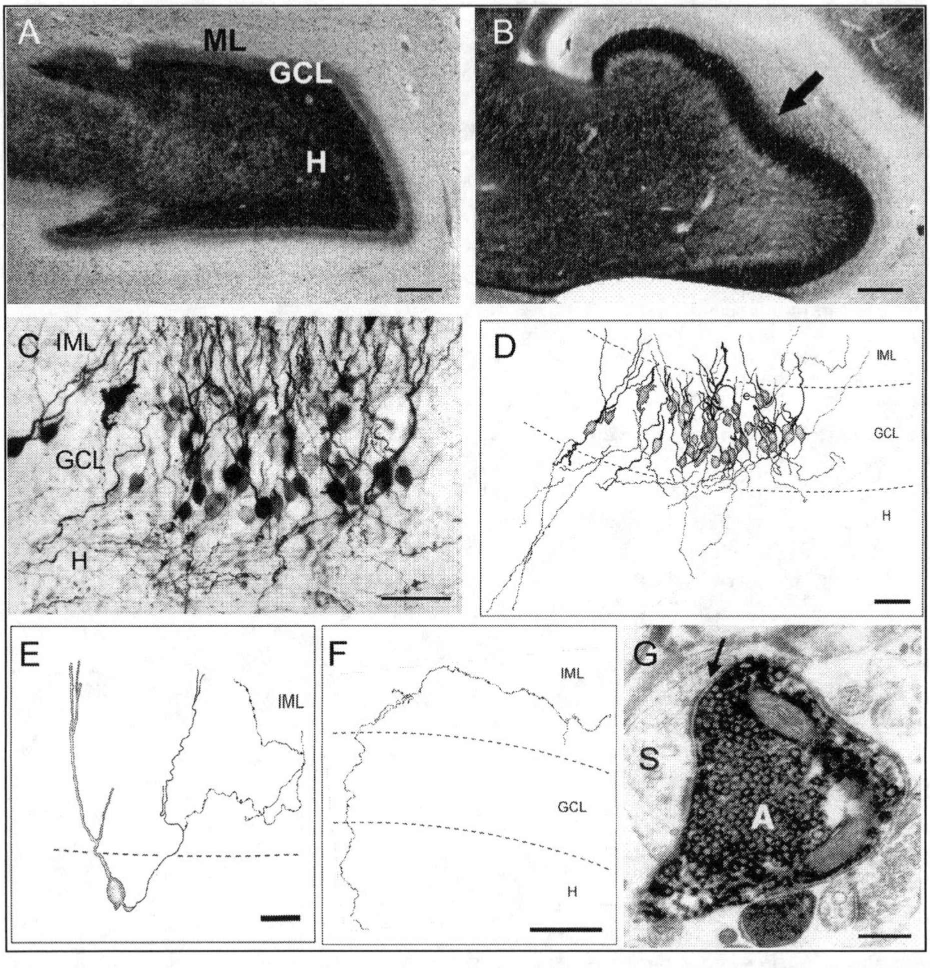

Figure 10. Mossy fiber sprouting in human hippocampus. A) Timm-stained section of the dentate gyrus. Patient with tumor-associated TLE (astrocytoma; WHO II°). The cytoarchitecture of the granule cell layer (GCL) appears to be normal. Mossy fibers are darkly stained. They are abundant in the hilus (H) and virtually absent from the molecular layer (ML). B) Timm-stained section of the dentate gyrus. Patient with Ammon's horn sclerosis. The granule cell layer is less compact compared to A. Mossy fibers form a dense fiber plexus in the inner portion of the molecular layer (arrow). C, D) Medium sized injection of the anterograde tracer neurobiotin (NB) in the granule cell layer (GCL) of the dentate gyrus. A group of granule cells has taken up the tracer. Single granule cells can be identified and their axonal projections can be reconstructed using a camera lucida (D, E, F). H, hilus; IML, inner molecular layer. E, F) Camera lucida reconstructions of a granule cell (E; c.f., D) and a granule cell axon (F) in the inner molecular layer. G) Neurobiotin-labeled mossy fiber synapse in the IML of the dentate gyrus. Note that the anterogradely labeled bouton forms an asymmetric synapse with a small postsynaptic element, most likely a granule cell spine (S). The arrow points at the synaptic cleft. A= axon terminal. Scale bars: A, B= 500 μm; C, D= 50 μm; E, F= 25 μm; G= 0.25 μm.

F). Electron microscopy revealed asymmetric mossy fiber synapses on spiny neurons in the inner molecular layer, presumably granule cells (Fig. 10 G).

The description of the sprouting process in human HS is only the first step towards unlocking the molecular and electrophysiological processes that underlie the reorganization of the hippocampus and seizure pathogenesis. In this context, we have learned a great deal from animal models of TLE. By using these model systems, the supragranular sprouting response of mossy fibers was studied and changes in neurotrophic factors,[104] extracellular matric molecules, transcription factors, and other sprouting-related molecules[18] were analyzed. Very likely, the most widely accepted model for the recurrent mossy fiber sprouting response involves seizure-induced alterations in neurotrophic factors. In short, it has been demonstrated that seizures alter the levels of nerve growth factor (NGF), brain-derived neurotrophic factor (BDNF), and fibroblast growth factor (FGF) in the dentate gyrus. This may induce the granule cells to sprout new axon collaterals, which then grow into the inner molecular layer and reinnervate proximal granule dendrites. Although neurotrophic factors are likely to play a crucial role in the *induction* of axonal outgrowth, extracellular matrix (ECM) molecules and cell adhesion molecules may be just as important for the *guidance* of sprouting fibers. Along this line, several cell adhesion and ECM molecules have been described in animal models of TLE, among them neural cell adhesion molecule[105] and tenascin-C,[106,107] and it has been proposed that these molecules guide the sprouting axons.

In human hippocampi, similar changes were observed, which in turn suggests that comparable mechanisms may be operative in human TLE. For instance, BDNF as well as NT-3 were also found to be associated with increased mossy fiber Timm's staining,[108] and the distribution of tenascin-C in HS indicates that ECM molecules may have an important role to play in the rewiring of the hippocampus.[109] In this context, it should be taken into account that the absence of individual trophic molecules may be compensated for by others. As an example, in mice deficient in BDNF, we still observed the sprouting of mossy fibers in the hippocampus.[110]

What Is the Functional Significance of Axonal Sprouting?

The pressing question arises as to how we should judge the functional significance of axonal sprouting for the affected organism. Is it an essentially positive process that can be viewed as a natural means of repair and a compensatory mechanism that leads to the functional improvement of an organism, or should it be classified as an undesirable process that is accompanied by detrimental consequences for an injured organism? Our experimental data and those reported by other research groups suggest that the functional significance of axonal sprouting for an injured organism can not be judged in a sweeping manner, and that both the degree of axonal sprouting as well as its functional consequences depend on the age of the organism, the brain regions injured, and the type of lesion involved. Depending on these variables, the phenomenon of axonal sprouting brings in its train both favorable as well as unfavorable consequences for the subsequent functioning of an organism that has been subjected to injury or trauma. Several excellent reviews and commentaries are recommended to the reader in which the functional significance of axonal sprouting processes have been discussed at greater length for lesion-induced reorganization in the neocortex,[15,17,111] the spinal cord,[112] and the hippocampal formation following entorhinal lesion.[15,113,114]

The following examples should serve to illustrate the two-faced aspect of axonal sprouting: Young children, by whom the cortex had to be removed from one side of the brain, compensate their motor deficits via axon collateral sprouting processes. What is probably involved here are the intact motor axons of the other hemisphere, which are capable of crossing over the midline at spinal cord level. These fibers sprout into the deafferented spinal chord and can thereby regulate movements on both sides of the body.[115,116] In contrast to these favorable aspects of axonal sprouting behavior, one also finds instances of detrimental connectivities that can result in the CNS via axonal sprouting. So the report by Woolf and coworkers[117] describing how, following injury of a peripheral nerve, collateral sprouting occurs on the part of sensible fibers in the spinal chord. In the process, these fibers gain access to the spinal chord's pain pathways, which in turn can trigger painful sensations at a light touch . In this case, axonal sprouting is not only disadvantageous for the injured organism but is involved at the most fundamental level in the pathogenesis of a chronically painful disorder.

With regard to AD, the difficulty of assessing axonal sprouting becomes clear even within a single case. Accordingly, whereas compensatory collateral sprouting on the part of surviving nerve fibers delays the clinical presentation of symptoms, on the one hand,[17] aberrant sprouting behavior in the vicinity of amyloid deposits could well contribute to the pathogenesis of dementia on the other.[84] Particularly the changes seen in the white matter could result in pronounced dysfunction since even minute damage to a fiber tract is capable of interrupting the connectivity between two brain regions. It can be assumed, therefore, that aberrant axon growth in the vicinity of plaques contributes to cognitive deficits in AD.[84]

In view of this situation, it does not appear reasonable to evaluate the significance of axonal sprouting in a generalized manner. In the event that sprouting processes play a role in certain disorders, they initially should be classified morphologically, molecularly, and by means of behavioral experiments. Only by using this approach can the functional importance of axon sprouting for a given disease be assessed successfully, regulatory molecules positively identified, and, finally, the means provided for intervening in that process. In the long run and in this way, new therapeutic strategies could be developed for treating specific neurological disorders.

Acknowledgments

The authors thank all those who contributed of their time and talents to the studies reviewed in this chapter. We thank Dr. Kelly Del Tredici for help with the manuscript. This work was supported by the Deutsche Forschungsgemeinschaft (SFB 505; SFB 269; TR-3), the Swiss National Science Foundation, and Alzheimer Forschung International.

References

1. Singer W. Activity-dependent self-organization of synaptic connections as a substrate for learning. In: Changeux JP, Konishi M, eds. The neural and molecular bases of learning. London: John-Wiley, 1987:301-336.
2. Gage FH, Kempermann G, Palmer TD et al. Multipotent progenitor cells in the adult dentate gyrus. J Neurobiol 1998; 36:249-266.
3. Kempermann G, van Praag H, Gage FH. Activity-dependent regulation of neuronal plasticity and self repair. Prog Brain Res 2000; 127:35-48.
4. Kempermann G, Kuhn HG, Gage FH. More hippocampal neurons in adult mice living in an enriched environment. Nature 1997; 386:493-495.
5. Lynch G, Cotman C. The hippocampus as a model for studying anatomical plasticity in the adult brain. In: Isaacson RL, Pribram KH, eds. The Hippocampus. Vol. 1. New York: Plenum Press, 1975:123-154.
6. Cotman CW, Nadler JV. Reactive synaptogenesis in the hippocampus. In: Cotman CW, ed. Neuronal Plasticity. New York: Raven Press, 1978:227-271.
7. Cotman CW, Nieto-Sampedro M, Harris EW. Synapse replacement in the nervous system of adult vertebrates. Physiol Rev 1981; 61:684-784.
8. Liu CN, Chambers WW. Intraspinal sprouting of dorsal root axons. Arch Neurol Psych 1958; 79:46-61.
9. Raisman G. Neuronal plasticity in the septal nuclei of the adult rat. Brain Res 1969; 14:25-48.
10. Raisman G, Field PM. A quantitative investigation of the development of collateral reinnervation after partial deafferentation of the septal nuclei. Brain Res 1973; 50:241-264.
11. Chen S, Hillman DE. Plasticity of the parallel fiber-purkinje cell synapse by spine takeover and new synapse formation in the adult rat. Brain Res 1982; 240:205-220.
12. Rubel EW, Smith ZDG, Steward O. Sprouting in the avian brainstem auditory pathway: dependence on dendritic integrity. J Comp Neurol 1981; 202:397-414.
13. Wells J, Tripp LN. Time course of reactive synaptogenesis in the subcortical somatosensory system. J Comp Neurol 1987; 255:466-475.

14. Schwegler G, Schwab ME, Kapfhammer JP. Increased collateral sprouting of primary afferents in the myelin-free spinal cord. J Neurosci 1995; 15:2756-2767.
15. Steward O. Reorganization of neuronal circuitry following central nervous system trauma: naturally occurring processes and opportunities for therapeutic intervention. In: Salzman SK, Faden AI, eds. The Neurobiology of Central Nervous System Trauma. New York: Oxford University Press, 1994:266-287.
16. Deller T, Frotscher M. Lesion-induced plasticity of central neurons: sprouting of single fibers in the rat hippocampus after unilateral entorhinal lesion. Prog Neurobiol 1997; 53:687-727.
17. Moore RY, Zigmond MJ. Compensatory mechanisms in central neurodegenerative disease. In: Calne DB, ed. Neurodegenerative Diseases. Philadelphia: Saunders, 1994:355-370.
18. Chevassus-Au-Louis N, Niquet J, Ben-Ari Y et al. Cellular plasticity. In: Engel J Jr, Pedley TA, eds. Epilepsy: A comprehensive textbook. Philadelphia: Lippincott-Raven, 1997:387-396.
19. Mathern GW, Babb TL, Armstrong DL. Hippocampal sclerosis. In: Engel J Jr, Pedley TA, eds. Epilepsy: A comprehensive textbook. Philadelphia: Lippincott-Raven, 1997:133-155.
20. Steward O. Cholinergic sprouting is blocked by repeated induction of electroconvulsive seizures, a manipulation that induces a persistent reactive state in astrocytes. Exp Neurol 1994; 129:103-111.
21. Gall C, Ivy G, Lynch G. Neuroanatomical plasticity. Its role in organizing and reorganizing the central nervous system. Human Growth 1986; 2:411-436.
22. Steward O. Synapse replacement on cortical neurons following denervation. Cereb Cortex 1991; 9:81-132.
23. Savaskan NE, Nitsch R. Molecules involved in reactive sprouting in the hippocampus. Rev Neurosci 2001; 12:195-215.
24. Lynch G, Matthews DA, Mosko S et al. Induced acetylcholinesterase-rich layer in rat dentate gyrus following entorhinal lesions. Brain Res 1972; 42:311-318.
25. Cotman CW, Matthews DA, Taylor D et al. Synaptic rearrangement in the dentate gyrus: histochemical evidence of adjustments after lesions in immature and adult rats. Proc Natl Acad Sci U S A 1973; 70:3473-3477.
26. Lynch G, Stanfield B, Cotman CW. Developmental differences in postlesion axonal growth in the hippocampus. Brain Res 1973; 59:155-168.
27. Steward O, Cotman CW, Lynch GS. Re-establishment of electrophysiologically functional entorhinal cortical input to the dentate gyrus deafferented by ipsilateral entorhinal lesions: innervation by the contralateral cortex. Exp Brain Res 1973; 18:396-414.
28. Paranavelas JG, Lynch G, Brecha N et al. Spine loss and regrowth in hippocampus following deafferentation. Nature 1974; 248:71-73.
29. Zimmer J. Extended commissural and ipsilateral projections in postnatally de-entorhinated hippocampus and fascia dentata demonstrated in rats by silver impregnation. Brain Res 1973; 64:293-311.
30. Zimmer J. Proximity as a factor in the regulation of aberrant axonal growth in postnatally deafferented fascia dentata. Brain Res 1974; 72:137-142.
31. Zimmer J, Hjorth-Simonsen A. Crossed pathways from the entorhinal area to the fascia dentata.II: Provokable in rats. J Comp Neurol 1975; 161:71-102.
32. Gerfen CR, Sawchenko PE. An anterograde neuroanatomical tracing method that shows the detailed morphology of neurons, their axons and terminals: immunohistochemical localization of an axonally transported plant lectin, Phaseolus vulgaris-Leucoagglutinin (PHAL). Brain Res 1984; 290:219-238.
33. Deller T, Frotscher M, Nitsch R. Morphological evidence for the sprouting of inhibitory commissural fibers in response to the lesion of the excitatory entorhinal input to the rat dentate gyrus. J Neurosci 1995; 15:6868-6878.
34. Deller T, Nitsch R, Frotscher M. Layer-specific sprouting of commissural fibers to the rat fascia dentata after unilateral entorhinal cortex lesion: A Phaseolus vulgaris-Leucoagglutinin tracing study. Neuroscience 1996; 71:651-660.
35. Deller T, Frotscher M, Nitsch R. Sprouting of crossed entorhinodentate fibers after a unilateral entorhinal lesion: anterograde tracing of fiber reorganization with phaseolus vulgaris-leucoagglutinin (PHAL). J Comp Neurol 1996; 365:42-55.
36. Frotscher M, Heimrich B, Deller T. Sprouting in the hippocampus is layer-specific. Trends Neurosci 1997; 20:218-223.

37. Fawcett JW. Intrinsic neuronal determinants of regeneration. Trends Neurosci 1992; 15:5-8.
38. Caroni P. Intrinsic neuronal determinants that promote axonal sprouting and elongation. Bio Essays 1997; 19:767-775.
39. Aigner L, Arber S, Kapfhammer JP et al. Overexpression of the neural growth-associated protein GAP-43 induces nerve sprouting in the adult nervous system of transgenic mice. Cell 1995; 83:269-278.
40. Caroni P. Driving the growth cone. Science 1998; 281:1465-1466.
41. Laux T, Fukami K, Thelen M et al. GAP43, MARCKS, and CAP23 modulate PI(4,5)P(2) at plasmalemmal rafts, and regulate cell cortex actin dynamics through a common mechanism. J Cell Biol 2000; 149:1455-1472.
42. Frey D, Laux T, Xu L et al. Shared and unique roles of CAP23 and GAP43 in actin regulation, neurite outgrowth, and anatomical plasticity. J Cell Biol 2000; 149:1443-1454.
43. Benowitz LI, Routtenberg A. GAP-43: an intrinsic determinant of neuronal development and plasticity. Trends Neurosci 1997; 20:84-91.
44. Oestreicher AB, DeGraan PNE, Gispen WH et al. B-50, the growth associated protein-43: modulation of cell morphology and communication in the nervous system. Prog Neurobiol 1997; 53:627-686.
45. Goodman CS. Mechanisms and molecules that control growth cone guidance. Annu Rev Neurosci 1996; 19:341-377.
46. Faissner A, Götz B, Joester A et al. Tenascin-C glycoproteins in neural pattern formation and plasticity. The neurosci 1996; 8:347-356.
47. Tessier-Lavigne M, Goodman CS. The molecular biology of axon guidance. Science 1996; 274:1123-1133.
48. Kapfhammer JP. The Axon. In: Ferretti P, Géraudie J, eds. Cellular and molecular basis of regeneration: from invertebrates to humans. Chichester: John Wiley, 1998:355-378.
49. Bovolenta P, Fernaud-Espinosa I. Nervous system proteoglycans as modulators of neurite outgrowth. Prog Neurobiol 2000; 61:113-132.
50. Van Vactor D, Lorenz LJ. Introduction: invertebrate axons find their way. Cell Mol Life Sci 1999; 55:1355-1357.
51. Pires Neto MA, Braga-de-Souza S, Lent R. Extracellular matrix molecules play diverse roles in the growth and guidance of central nervous system axons. Braz J Med Biol Res 1999; 32:633-638.
52. Van Vactor DV, Lorenz LJ. Neural development: The semantics of axon guidance. Curr Biol 1999; 9:201-204.
53. Förster E, Naumann T, Deller T et al. Cholinergic sprouting in the rat fascia dentata after entorhinal lesion is not linked to early changes in neurotrophin mRNA expression. Neuroscience 1997; 80:731-739.
54. Lee MY, Deller T, Kirsch M et al. Differential regulation of CNTF and CNTF receptor alpha expression in astrocytes and neurons of the fascia dentata following entorhinal cortex lesion. J Neurosci 1997; 17:1137-1146.
55. Haas CA, Deller T, Frotscher M. Basal expression, subcellular distribution, and upregulation of the proto-oncogene c-jun in the rat dentate gyrus after unilateral entorhinal cortex lesion. Neuroscience 1997; 81:33-45.
56. Haas CA, Frotscher M, Deller T. Differential induction of c-Fos, c-Jun and Jun B in the rat central nervous system following unilateral entorhinal cortex lesion. Neuroscience 1999; 90:41-51.
57. Xia XG, Hofmann HD, Deller T et al. Induction of STAT3 signaling in activated astrocytes and sprouting septal neurons following entorhinal cortex lesion in adult rats. Mol Cell Neurosci 2002; 21:379-39.:
58. Jucker M, Mondadori C, Mohajeri H et al. Transient upregulation of NCAM mRNA in astrocytes in response to entorhinal cortex lesions and ischemia. Mol Brain Res 1995; 28:149-156.
59. Jucker M, D'Amato F, Mondadori C et al. Expression of the neural adhesion molecule L1 in the deafferented dentate gyrus. Neuroscience 1996; 75:703-715.
60. Deller T, Haas CA, Naumann T et al. Upregulation of astrocyte-derived tenascin-c correlates with neurite outgrowth in the rat dentate gyrus after unilateral entorhinal cortex lesion. Neuroscience 1997; 81:829-846.

61. Haas CA, Rauch U, Thon N et al. Entorhinal cortex lesion in adult rats induce the expression of the neuronal chondroitin sulfate proteoglycan neurocan in reactive astrocytes. J Neurosci 1999; 19:9953-9963.
62. Thon N, Haas CA, Rauch U et al. The chondroitin sulphate proteoglycan brevican is upregulated by astrocytes after entorhinal cortex lesions in adult rats. Eur J Neurosci 2000; 12:2547-2558.
63. Haas CA, Deller T, Krsnik Z et al. Entorhinal cortex lesion does not alter reelin mRNA expression in the dentate gyrus of young and adult rats. Neuroscience 2000; 97:25-31.
64. Collazos-Castro JE, Nieto-Sampedro M. Developmental and reactive growth of dentate gyrus afferents: Cellular and molecular interactions. Restor Neurol Neurosci 2001; 19:169-187.
65. Pearlman AL, Sheppard AM. Extracellular matrix in early cortical development. Prog Brain Res 1996; 108:117-134.
66. Höke A, Silver J. Proteoglycans and other repulsive molecules in glial boundaries during development and regeneration of the nervous system. Prog Brain Res 1996; 108:149-163.
67. Yamada H, Fredette B, Shitara K et al. The brain chondroitin sulfate proteoglycan brevican associates with astrocytes ensheathing cerebellar glomeruli and inhibits neurite outgrowth from granule neurons. J Neurosci 1997; 17:7784-7795.
68. Margolis RU, Margolis RK. Chondroitin sulfate proteoglycans as mediators of axon growth and pathfinding. Cell Tiss Res 1997; 290:343-348.
69. Rauch U. Modeling an extracellular environment for axonal pathfinding and fasciculation in the central nervous system. Cell Tiss Res 1997; 290:349-356.
70. Faissner A. The tenascin gene family in axon growth and guidance. Cell Tiss Res 1997; 290:331-341.
71. Stichel CC, Müller HW. The CNS lesion scar: new vistas on an old regeneration barrier. Cell Tiss Res 1998; 294:1-9.
72. Davies SJA, Goucher DA, Doller C et al. Robust regeneration of adult sensory axons in degenerating white matter of the adult rat spinal cord. J Neurosci 1999; 19:5810-5822.
73. Moon LD, Asher RA, Rhodes K et al. Regeneration of CNS axons back to their target following treatment of adult brain with chondroitinase ABC. Nature Neurosci 2001; 4:465-466.
74. Asher RA, Morgenstern DA, Moon LD et al. Chondroitin sulphate proteoglycans: inhibitory components of the glial scar. Prog Brain Res 2001; 132:611-619.
75. Fawcett JW, Asher RA. The glial scar and central nervous system repair. Brain Res Bull 1999; 49:377-391.
76. Powell EM, Fawcett JW, Geller HM. Proteoglycans provide neurite guidance at an astrocyte boundary. Mol Cell Neurosci 1997; 10:27-42.
77. Deller T, Haas CA, Frotscher M. Reorganization of the rat fascia dentata after a unilateral entorhinal cortex lesion: Role of the extracellular matrix. Ann NY Acad Sci 2000; 911:207-220.
78. Deller T, Haas CA, Frotscher M. Sprouting in the hippocampus after entorhinal cortex lesion is layer-specific but not translaminar: Which molecules may be involved? Restor Neurol Neurosci 2001; 19:159-167.
79. Geddes JW, Monaghan DT, Cotman CW et al. Plasticity of hippocampal circuitry in alzheimer's disease. Science 1985; 230:1179-1181.
80. Geddes JW, Anderson KJ, Cotman CW. Senile plaques as aberrant sprout-stimulating structures. Exp Neurol 1986; 94:767-776.
81. Masliah E, Fagan AM, Terry RD et al. Reactive synaptogenesis assessed by synaptophysin immunoreactivity is associated with GAP-43 in the denate gyrus of the adult rat. Exp Neurol 1991; 113:131-142.
82. Sturchler-Pierrat C, Abramowski D, Duke M et al. Two amyloid precursor protein transgenic mouse models with alzheimer disease-like pathology. Proc Natl Acad Sci U S A 1997; 94:13287-13292.
83. Calhoun ME, Wiederhold KH, Abramowski D et al. Neuron loss in APP transgenic mice. Nature 1998; 395:755-756.
84. Phinney AL, Deller T, Stalder M et al. Cerebral amyloid induces aberrant axonal sprouting and ectopic terminal formation in amyloid precursor protein transgenic mice. J Neurosci 1999; 19:8552-8559.
85. Hyman BT, Kromer LJ, VanHoesen GW. Reinnervation of the hippocampal perforant pathway zone in alzheimer's disease. Ann Neurol 1987; 21:259-267.

86. Hyman BT, VanHoesen GW, Kromer LJ et al. Perforant pathway changes and the memory impairment of alzheimer,s disease. Ann Neurol 1986; 20:472-481.
87. Braak H, Braak E. Neuropathological stageing of Alzheimer-related changes. Acta Neuropathol 1991; 82:239-259.
88. Calhoun ME, Burgermeister P, Phinney AL et al. Neuronal overexpression of mutant APP results in prominent deposition of cerebrovascular amyloid. Proc Natl Acad Sci U S A 1999; 96:14088-14093.
89. Stalder M, Phinney AL, Probst A et al. Association of microglia with amyloid plaques in brains of APP23 transgenic mice. Am J Pathol 1999; 154:1673-1684.
90. Stalder M, Deller T, Staufenbiel M et al. 3D-reconstruction of microglia and amyloid in APP23 transgenic mice: no evidence of intracellular amyloid. Neurobiol Aging 2001; 22:427-434.
91. Sloviter RS. The functional organization of the hippocampal dentate gyrus and its relevance to the pathogenesis of temporal lobe epilepsy. Ann Neurol 1994; 35:640-654.
92. Sloviter RS. A simplified Timm stain procedure compatible with formaldehyde fixation and routine paraffin embedding of rat brain. Brain Res Bull 1982; 8:771-774.
93. Sutula T, Cascino G, Cavazos J et al. Mossy fiber synaptic reorganization in the epileptic human temporal lobe. Ann Neurol 1989; 26:321-330.
94. Babb TL, Pretorius JK, Mello LE et al. Synaptic reorganizations in epileptic human and rat kainate hippocampus may contribute to feedback and feedforward excitation. Epilepsy Res Suppl 203; 9:193-202.
95. Babb TL, Kupfer WR, Pretorius JK et al. Synaptic reorganization by mossy fibers in human epileptic fascia dentata. Neuroscience 1991; 42:351-363.
96. de Lanerolle NC, Kim JH, Robbins RJ et al. Hippocampal interneuron loss and plasticity in human temporal lobe epilepsy. Brain Res 1989; 495:387-395.
97. Mathern GW, Babb TL, Pretorius JK et al. Reactive synaptogenesis and neuron densities for neuropeptide Y, somatostatin, and glutamate decarboxylase immunoreactivity in the epileptogenic human fascia dentata. J Neurosci 1995; 15:3990-4004.
98. Lehmann TN, Gabriel S, Kovacs R et al. Alterations of neuronal connectivity in area CA1 of hippocampal slices from temporal lobe epilepsy patients and from pilocarpine-treated epileptic rats. Epilepsia 2000; 41:190-194.
99. Isokawa M, Levesque MF, Babb TL et al. Single mossy fiber axonal systems of human dentate granule cells studied in hippocampal slices from patients with temporal lobe epilepsy. J Neurosci 1993; 13:1511-1522.
100. Franck JE, Pokorny J, Kunkel DD et al. Physiologic and morphologic characteristics of granule cell circuitry in human epileptic hippocampus. Epilepsia 1995; 36:543-558.
101. van Haeften T, Wouterlood FG. Neuroanatomical tracing at high resolution. J Neurosci Meth 2000; 103:107-116.
102. Boulton CL, Haebler DV, Heinemann U. Tracing of axonal connections by rhodamine-dextran amine in the rat hippocampal-entorhinal cortex slice preparation. Hippocampus 1992; 2:99-106.
103. Freiman TM, Gimbel K, Honegger J et al. Anterograde tracing of human hippocampus in vitro - a neuroanatomical tract tracing technique for the analysis of local fiber tracts in human brain. J Neurosci Meth 2002; 120:95-103.
104. Gall CM, Isackson PJ. Limbic seizures increase neuronal production of messenger RNA for nerve growth factor. Science 1989; 245:758-761.
105. Niquet J, Jorquera J, Ben-Ari Y et al. N-CAM immunoreactivity on mossy fibers and reactive astrocytes in the hippocampus of epileptic rats. Brain Res 1993; 626:106-116.
106. Niquet J, Jorquera J, Faissner A et al. Gliosis and axonal sprouting in the hippocampus of epileptic rats are associated with an increase of tenascin-C immunoreactivity. J Neurocytol 1995; 24:611-624.
107. Nakic M, Mitrovic N, Sperk G et al. Kainic acid activates transient expression of tenascin-C in the adult rat hippocampus. J Neurosci Res 1996; 44:355-362.
108. Mathern GW, Babb TL, Micevych PE et al. Granule cell mRNA levels for BDNF, NGF, and NT-3 correlate with neuron losses or supragranular mossy fiber sprouting in the chronically damaged and epileptic human hippocampus. Mol Chem Neuropathol 1997; 30:53-76.
109. Scheffler B, Faissner A, Beck H et al. Hippocampal loss of tenascin boundaries in ammon's horn sclerosis. Glia 1997; 19:35-46.

110. Bender R, Heimrich B, Meyer M et al. Hippocampal mossy fiber sprouting is not impaired in brain-derived neurotrophic factor-deficient mice. Exp Brain Res 1998; 120:399-402.
111. Schneider GE. Is it really better to have your brain lesion early? A revision of the "kennard principle." Neuropyschol 1979; 17:557-583.
112. Schwab ME, Bartholdi D. Degeneration and regeneration of axons in the lesioned spinal cord. Physiol Rev 1996; 76:319-370.
113. Ramirez JJ. The functional significance of lesion-induced plasticity of the hippocampal formation. Brain Plas Adv Neurology 1997; 73:61-82.
114. Ramirez JJ. The role of axonal sprouting in functional reorganization after CNS injury: Lessons from the hippocampal formation. Restor Neurol Neurosci 2001; 19:237-262.
115. Cao Y, Vikingstad EM, Huttenlocher PR et al. Functional magnetic resonance studies of the reorganization of the human hand sensorimotor area after unilateral brain injury in the perinatal period. Proc Natl Acad Sci USA 1994; 91:9612-9616.
116. Carr LJ, Harrison LM, Evans AL et al. Patterns of central motor reorganization in hemiplegic cerebral palsy. Brain 1993; 116:1223-1247.
117. Woolf CJ, Shortland P, Coggeshall RE. Peripheral nerve injury triggers central sprouting of myelinated afferents. Nature 1992; 355:75-78.

CHAPTER 7

A Kinase with a Vision:

Role of ERK in the Synaptic Plasticity of the Visual Cortex

Gian Michele Ratto and Tommaso Pizzorusso

Introduction

We look at these written words with two eyes, their neuronal representations are elaborated separately in the two retinae and they are conveyed to two separate zones of the thalamus. The segregation in eye specific structures is broken only in the primary visual cortex, where neurons responsive to both eyes can be finally found. The cortical circuitry that brings together information from the two eyes is exquisitely tuned during the early post natal life, in a critical period in which synaptic changes are driven by the electrical activity evoked by the visual stimulation. Though the plasticity of binocular vision has served as a model for the study of synaptic plasticity for over 40 years, the identity of the molecular mechanisms involved in this process has remained elusive. Recently, we have offered evidences, gathered both in vivo and in vitro, suggesting that the Extracellular-signal Regulated Kinase 1/2 (ERK1/2) plays a crucial role in the control of this form of plasticity.

In this chapter we will at first describe ocular dominance in visual cortex and how it is shaped by visual activity during development. Then we will analyze the factors that are known to influence visual plasticity and their transduction pathway. Finally, we will show how the activation of this intracellular machinery is necessary for visual cortical plasticity in vitro and in vivo.

Critical Period for Ocular Dominance

Most of the richness and complexity of the external world is conveyed to us by the sense of vision. The architecture of the visual system complicates gradually from lower mammalian to primates and man, in a way corresponding to the increasing importance of vision in respect of the other sensorial modalities. Still, many functional features of the visual system are shared at all levels of phylogenetic development, both in terms of the neuronal architecture and in terms of the developmental processes leading to the mature system. Among these conserved themes, one of the most striking feature is the fact that the visual system gradually learns to see during a definite period of development. During this time window most parameters determining the performances of the visual system improve gradually, in correspondence of a progressive tuning of cortical circuitry. This period, aptly named critical period, begins roughly at eye opening and lasts for a few weeks or months depending on the species, and culminating to a duration of a few years in humans. At this time the circuitry of the visual system is prone to changes according to the visual experience impinging on the subject, as it is exemplified by the

Brain Repair, edited by Mathias Bähr. ©2006 Eurekah.com
and Kluwer Academic / Plenum Publishers.

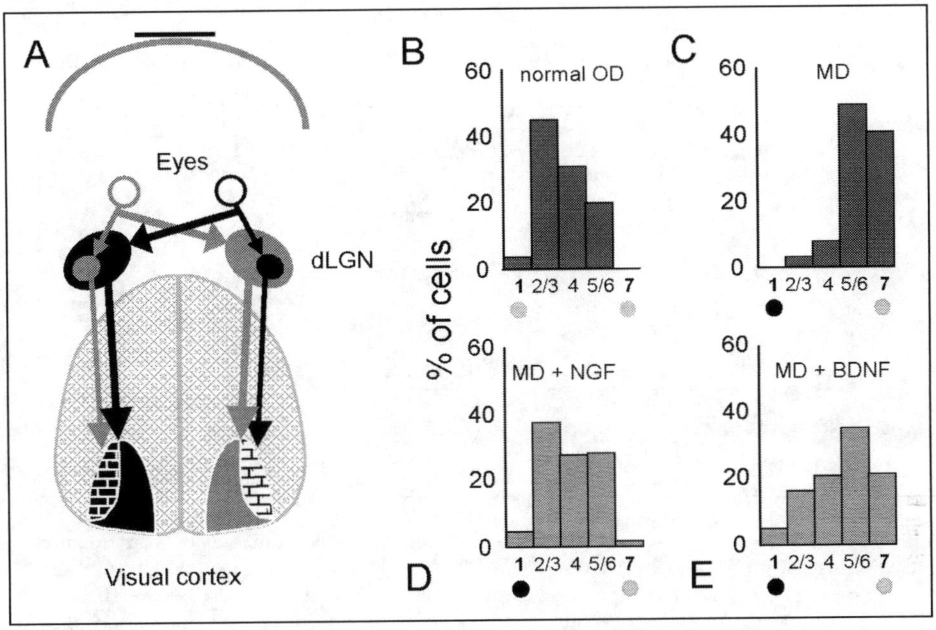

Figure 1. Organisation of the mammalian visual system, and definition of ocular dominance. A) Each eye projects to both sides of the thalamus with unequal weight. In rodents, 95% of the retinal fibres occupy most of the controlateral thalamus, while the remaining fibers project to a small volume of the ipsilateral thalamus. These two zones are still segregated, in the sense that post-synaptic thalamic neurons respond to either one or the other eye. The thalamo-cortical projections converge on the binocular portion of the visual cortex, and thus cortical neurons are responsive to stimuli presented to both eyes. B) The histogram is used to classify the ocular dominance of cortical neurons, from a score of 1 (responsive only to the controlateral eye) to 7 (responsive only to the ipsilateral eye). C) The monocular deprivation during the critical period (post natal day 21 to about 32 in rats and mice) cause a dramatic shift of ocular dominance, which can be partially rescued with exogenous infusion of NGF or BDNF (D and E respectively).

maturation of the ocular dominance of cortical neurons. The information incoming from the two eyes is initially segregated in two separate pathways that converge only at the level of the binocular portion of the primary visual cortex (Fig. 1A). Not until this stage we find neurons responsive to visual stimuli presented to both eyes. Accordingly, cortical neurons can be classified in base of their ocular dominance, with a score that defines the degree of control of each eye in determining the cell responsiveness (Fig. 1B). A score of 1 or 7 indicates that the eye is responsive only to the contralateral or ipsilateral eye respectively, score 4 indicates equal responsiveness to both eyes. The ocular dominance in a mature rodent is described by the histogram 1B. Clearly the contralateral eye is somewhat dominant, but there is a significative contribution by the ipsilateral eye also. If during the critical period one eyelid is sutured (monocular deprivation) the ocular dominance distribution shifts drastically toward the non deprived eye (Fig. 1C).[1]

Factors Critical for Ocular Dominance: Electrical Activity and Neurotrophins

In the quest for the mechanisms controlling the plasticity of ocular dominance, we must understand first how monocular deprivation is translated into the neuronal alphabet of spike trains. The suture of the eyelid does not sink the retina in uniform darkness since light changes

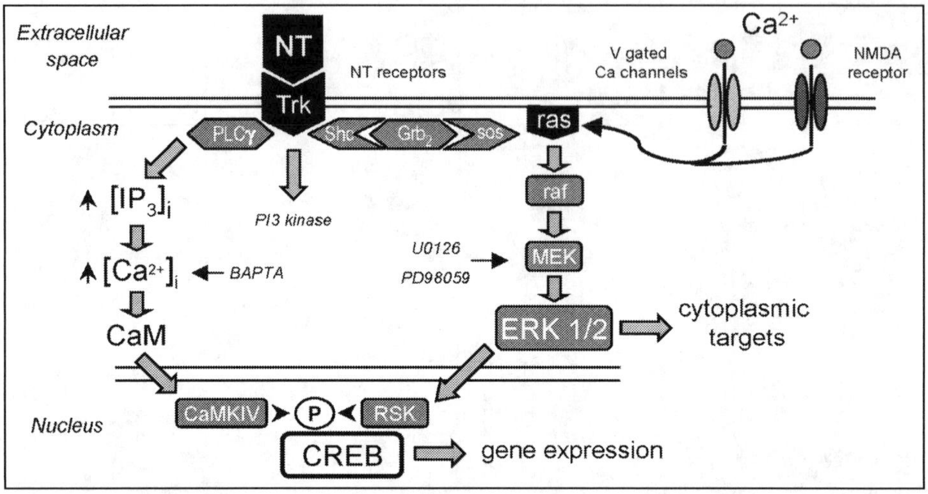

Figure 2. Signal transduction pathway associated with neurotrophins and electrical activity. Neurotrophins transduction is started at the Tyrosine Kinase receptors. There are three flavours of receptors (TrkA, B and C) with different preferential affinity for the various neurotrophins: NGF (TrkA), BDNF (TrkB), NT3 (TrkC) and NT4 (TrkB). The phosphatidylinositol 3-kinase pathway mediates the control of neuronal survival, and it is not considered here. Electrical activity triggers the ras-ERK pathway in a Ca-dependent way. After its phosphorylation ERK 1/2 can control gene expression and exert local effects on cytoplasmatic or membrane-bound substrates.

are attenuated but still discernible; what is taken away from the visual world is the presence of spatial contrast. The thalamic fibers carrying the input originating from the deprived eye miss the high frequency trains characteristic of the perception of edges. Thus, the strengthening of the normal eye at expenses of the deprived eye, can be interpreted as an expansion of the synaptic target of the active fibers against the terminals with depressed electrical activity. Another factor that contributes to regulate this competitive process between the two eyes, is the availability of neurotrophins in the visual cortex.[2-4] Indeed, the involvement of neurotrophins on monocular deprivation is directly demonstrated by the fact that their exogenous administration during deprivation prevents the shift of ocular dominance caused by the unbalance of electrical activity[5-8] (Fig. 1D-E). Furthermore, the duration of the critical period for monocular deprivation is shortened in mice overexpressing BDNF.[9]

These observations reveal that monocular deprivation is a play with two actors on stage: electrical activity and neurotrophins, with a central role played by BDNF. Therefore it is reasonable to assume that the mechanisms controlling plasticity of ocular dominance must sit somewhere between the intracellular signaling pathways activated by neurotrophins and those which convert visually driven activity into long-lasting changes of cortical circuitry. A possible scheme of molecular interactions between neurotrophin signaling and activity dependent signaling is represented in Figure 2. Most of the physiological functions of neurotrophins begin with their binding with the tyrosine kinase receptors (Trk receptor).[10] Consequently, the receptor self-phosphorylates and exposes consensus sites for various Trk substrates. Two pathways that have been well characterized in vitro are especially interesting, because they converge on the transcription factor CREB (cAMP Response Element Binding protein) that, upon activation, begins CRE-mediated transcription. The offspring of this episode of gene expression has an important role for memory and learning in Drosophila, Aplysia, and mouse.[11-13]

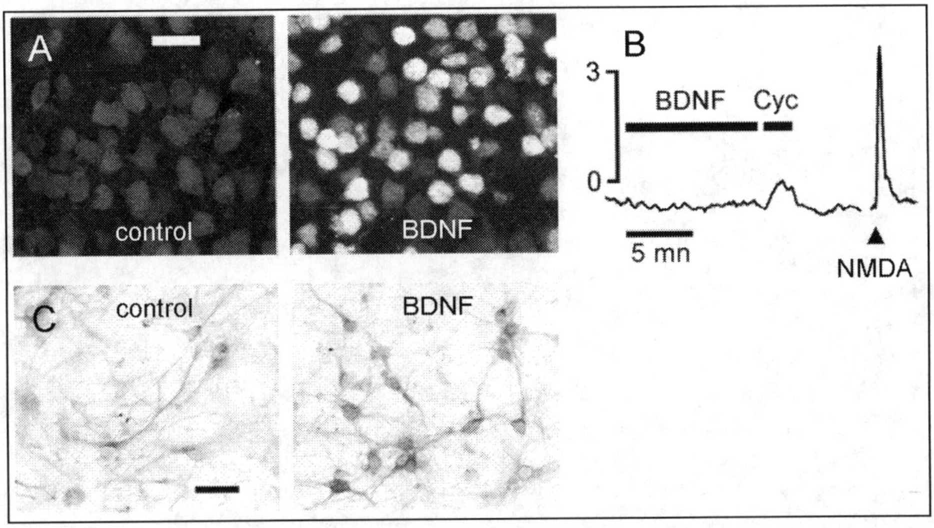

Figure 3. Effects of BDNF on neurons of the visual cortex. A) A 1 hr exposure of a slice of visual cortex to BDNF (200 ng/ml) caused a strong phosphorylation of the response element CREB (bar 20 μm). B) In the same conditions Ca-imaging at the confocal microscope failed to show any response to BDNF, but the cells were responsive to brief pulses of Cyclopiazonic acid (50 μM) and of NMDA (20 μM). C) BDNF (50 ng/ml for 30 min) is a powerful activator of ERK 1/2, as shown by the strong pERK immunofluorescence in cell body and dendrites (bar 50 μm).

The first pathway leading from Trk to CREB passes through the activation of phospholipase Cγ, the production of Inositol 3-phosphate (IP_3) and the consequent release of calcium from internal stores.[14] At the end of this cascade, CREB phosphorylation would be brought about by the Ca-dependent activation of CaMK IV.[15] The second pathway does not involve a Ca change, and the Trk receptors are linked by the complex of adapter proteins shc-Grb-sos to the ras-ERK 1/2 cascade. Eventually, these molecular stepping-stones lead to the cell nucleus since phosphorylated ERK causes CREB activation by means of the intermediate kinase RSK. In these schemes coalesce experimental evidences gathered mainly in vitro, and do not reflect necessarily what occurs in vivo. For example, the data regarding a direct action of neurotrophins on intracellular calcium are rather controversial, and have been obtained mainly in culture as discussed elsewhere.[16] In the next section we will show that, in the visual cortex, BDNF causes CREB activation by means of the Ca–indepent cascade only.

Activation of ERK 1/2 Is Necessary for BDNF-Induced Phosphorylation of CREB

Our initial goal is to show what pathways are involved in mediating BDNF action in the visual cortex. Simply put, we must answer to the following questions:

1. Does BDNF activate CREB in neurons of the visual cortex?
2. Does BDNF cause an increase of intracellular calcium and/or activate ERK 1/2 to mediate the activation of CREB?

Experiments executed on both cultured neurons and acute slices obtained from the visual cortex show that BDNF activates CREB, strongly arguing for an involvement of CREB in mediating the action of BDNF in the visual system. This is demonstrated by immunoreaction with an antibody raised against CREB phosphorylated at the Ser 133 residue (Fig. 3A). Exposing cortical slices to BDNF in similar conditions, we performed Ca-imaging experiments to

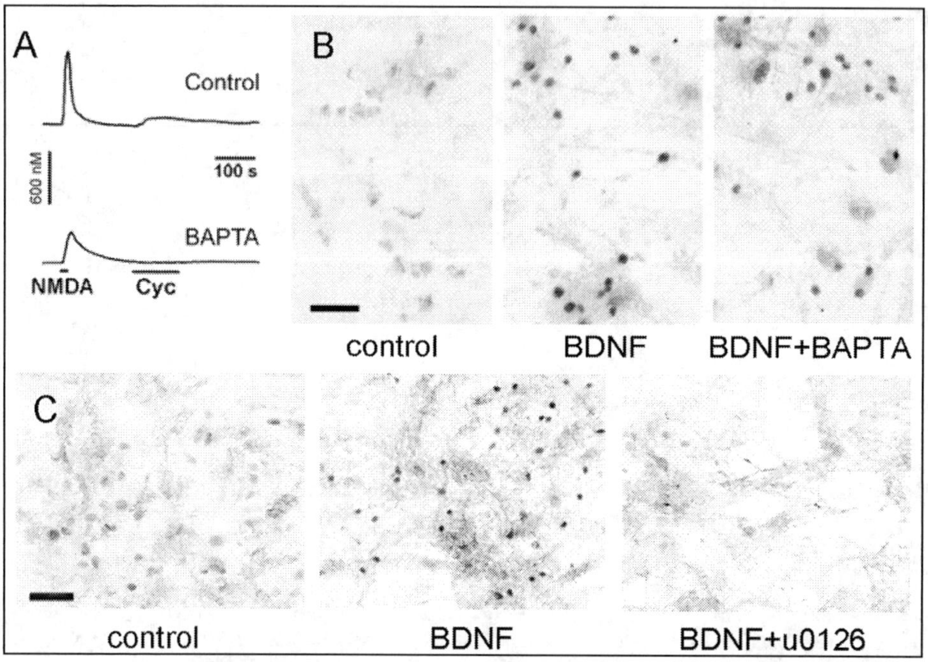

Figure 4. BDNF causes CREB phosphorylation in a Ca-independent way but requires ERK activation. A) Incubation with the cell permeable form of the calcium chelator BAPTA quenched the Ca transient caused by release from the intracellular stores (cyclopiazonic acid 50 μM), and strongly reduced the size of the NMDA response (20 μM). B) In the same conditions, BDNF (50 ng/ml for 15 min) still caused a strong phosphorylation of CREB in cultured neurons (bar 50 μm). C) In contrast, treatment with the MEK inhibitor U0126 completely blocked CREB phosphorylation (bar 50 μm).

detect any calcium increase triggered by the neurotrophin. In a set of remarkably uneventful experiments, we recorded hundreds of cells and we observed a Ca-change only in an handful of neurons (13 out of over 1000 cells), even if the imaging system was sensitive enough to reveal the tiny Ca transient unmasked by the SERCA inhibitor Cyclopiazonic acid (Fig. 3B). More was to be found on the other branch of the cascade departing from the Trk receptor and leading to the nucleus, since BDNF proved to be a powerful activator of ERK 1/2. This is shown by the strong increase in phospho-specific immunostaining of cultured neurons, after exposure to BDNF (Fig. 3C).

The necessity of either pathways can be demonstrated by blocking them with specific agents and looking at the effects on CREB phosphorylation. The two tools in our hands were the Ca chelator BAPTA, to interrupt the Ca-dependent pathway, and the molecule U0126, a specific blocker for the ERK kinase MEK.[17,18] In neurons loaded with a cell-permeant form of BAPTA, the calcium changes due to the release from intracellular stores were virtually suppressed, and even the much larger changes caused by the influx of external calcium through the NMDA receptors were heavily depressed (Fig. 4A). Therefore, given the high affinity and binding speed of BAPTA with Ca, this treatment should effectively quench any Ca change that might have been left undetected in the imaging experiments. In these conditions the pCREB staining after exposure to BDNF was virtually unaffected (Fig. 4B). This data, together with the Ca imaging experiments, prove that Ca is not a key effector in the pathway between TrkB activation and CREB phosphorylation in visual neuron. The opposite situation holds true for

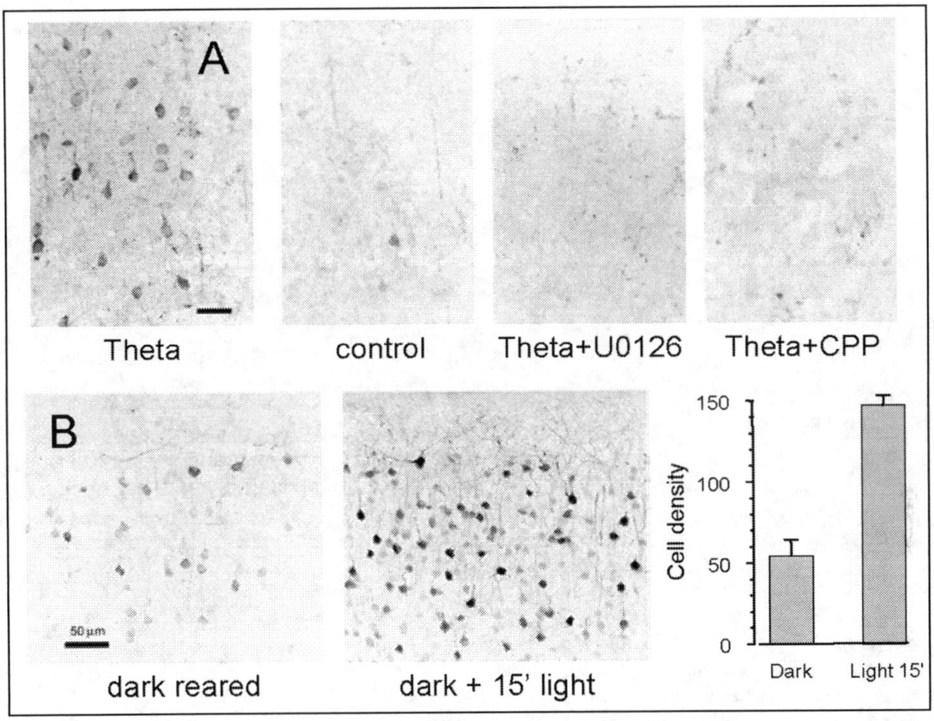

Figure 5. Neuronal activity causes ERK phosphorylation. A) A theta burst delivered to the white matter activated ERK in the visual cortex 5 min after stimulation. A similar control slice that did not receive the theta stimulation was much more weakly stained. ERK activation by the theta burst was blocked by U0126 (20 µM) and by a similar concentration of the NMDA antagonist CPP (bar 40 µm). B) ERK phosphorylation was also induced by a 15 min exposure to light after a period or dark rearing.

the ras-ERK pathway: pre-incubation of cortical neurons with the MEK blocker U0126, completely suppressed the pCREB induction operated by stimulation with BDNF (Fig. 4C). Therefore, BDNF dependent CREB phosphorylation requires ERK activation and it is independent on changes of intracellular Ca.[16]

ERK 1/2 Is Phosphorylated by Activity and Visual Stimulation in the Cortex

The surprise of meeting a Ca-independent process does not last long, since Ca comes back on scene when we study the pathway linking electrical activity to CREB phosphorylation. Indeed, activity-mediated CREB phosphorylation is quantitatively dependent on intracellular calcium increases caused by influx through voltage sensitive channels and NMDA receptors.[19-22] While the details of this regulation are still elusive, it is likely that the Ca influx acts on the ras-ERK pathway at the level of ras and /or raf. The involvement of NMDA receptors on activity dependent ERK phosphorylation is demonstrated in Figure 5: slices from the visual cortex received a special pattern of stimuli from an extracellular electrode. The pattern, named theta burst, consists of short bursts at high frequency (100 Hz), repeated at 5Hz frequency. Fifteen minutes after stimulation, staining with pERK antibody revealed a strong level of phosphorylation that required Ca influx through the NMDA receptor, since ERK activation was completely blocked by the NMDA receptor antagonist CPP (Fig. 5A). It could be argued that

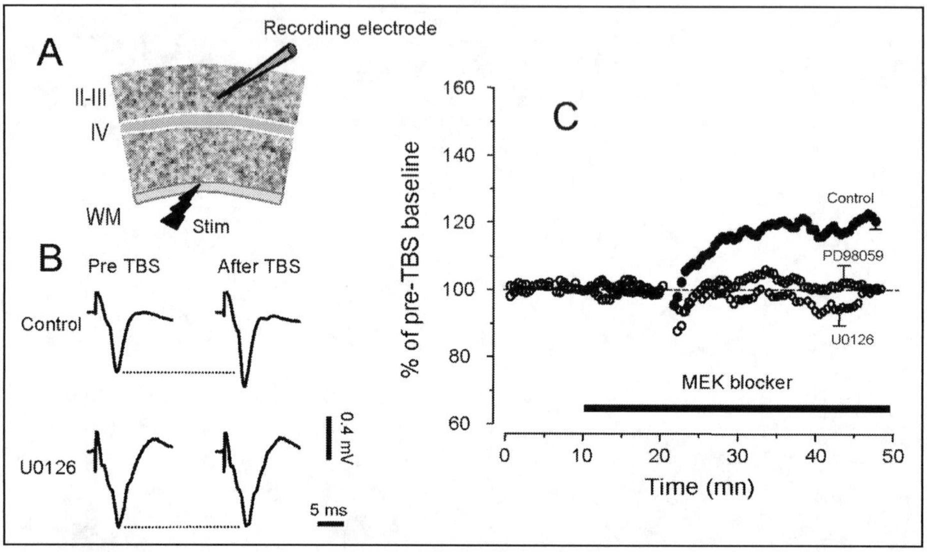

Figure 6. ERK activation is necessary for LTP induction in visual cortex. A) Experiments were performed on slices from the visual cortex of rats within the critical period. The stimulating electrode is placed in the white matter (WM), and the recordings are obtained from neurons in the superficial layers. B) The artefact caused by the stimulation is followed by the downward field response. After theta burst the response shows potentiation in control conditions, but not after incubation with U0126. C) The amplitude of the field response is tested every 30 s before and after the TBS and plotted in function of time. Each point represents averages from 18, 5 and 7 different slices recorded in control, PD98059 and U0126 respectively.

the in vitro stimulation is different from the pattern of electrical activity that is evoked in the cortex by visual stimuli. Thus, we desired to test whether visual stimulation would bring about ERK phosphorylation in behaving animals. Rats were kept in darkness for 3 days, and then were exposed to light. Twenty minutes afterward, the animals were sacrificed and processed for immunohistochemistry with the pERK antibody. Figure 5B shows that exposure to visual environment caused a robust activation of neurons in the visual cortex.[23,23a]

In conclusion, since ERK acts as a convergence point between electrical activity and neurotrophins, it appears to be in a strategic position to translate the regulatory actions of activity and neurotrophins into changes of cortical circuitry. Of course, it remains to be demonstrated a direct link between ERK activation and the regulation of synaptic strength.

ERK Activation Is Required for Synaptic Plasticity in Vitro and in Vivo

In models of synaptic plasticity, such as long term potentiation (LTP) or depression (LTD), it has been shown that the changes of neuronal connectivity are the result of a complex chain of events involving calcium entry through NMDA receptors and voltage-gated calcium channels, activation of protein kinases, gene expression and protein synthesis (see refs. 24,25 for recent reviews). In acute slices of visual cortex it is possible to stimulate the incoming excitatory fibers by means of extracellular electrodes and the evoked post-synaptic activity can be recorded from the superficial layers (Fig. 6A). LTP of the thalamo-cortical circuit can be induced by applying a theta burst stimulation in the white matter. Since this form of LTP is present only during the critical period, it has been suggested that this form of synaptic plasticity might be implicated in the activity-dependent refinement of cortical circuitry occurring during the critical period.[26]

Test stimuli were delivered every 30 sec by an electrode placed on the white matter, while recording the evoked field response from layer II/III. The presentation of a theta burst induced a potentiation of the response amplitude of about 20% in control conditions (Fig. 6B,C). If the theta was preceded by a 10 min incubation with the MEK inhibitors U0126 or PD98059, the potentiation was completely suppressed. It is interesting to notice that the time course of the response amplitude of slices treated with the inhibitors, differs from controls immediately after TBS delivery (Fig. 6C). The rapidity of the onset of the effect of MEK blockage suggests that ERK action is required for some mechanism of potentiation that, at least initially, is independent on gene transcription. Separate experiments have shown that the inhibitory effect of U0126 on LTP were not caused by reduced responses to the theta burst or by reduced activation of the NMDA receptors that are required for the induction of this form of LTP.[23]

To test the role of ERK pathway in visual plasticity, we needed a way of delivering the MEK inhibitors to the cortex during monocular deprivation. The drugs can be supplied by osmotic minipumps with their outlet placed immediately in front of the binocular visual cortex. Treatment was provided for one week at the peak of the critical period, as schematized in Figure 7A. After seven days of treatment the MEK inhibitor was still effective. This was proven in an experiment in which electrical activity was strongly increased by acute infusion with picrotoxin, a blocker of $GABA_A$ receptor. This treatment caused a robust ERK phosphorylation in control cortex but not in the cortex treated with U0126 (Fig. 7B). At this point we are finally ready to study whether activation of the ERK cascade is necessary for the experience-dependent plasticity occurring during monocular deprivation. Indeed, block of ERK activation by U0126 or PD98059 prevented MD effects (Fig. 7C). In normal P28 rat the overwhelming majority of visual cortical cells are binocular, with a clear dominance of the controlateral eye, and one week of deprivation at this age should cause a massive shift of responsiveness to the normal eye. This plastic change is clearly prevented by the blockage of the ERK pathway.

Beyond ERK: Mechanisms Controlling Synaptic Plasticity in the Visual Cortex

What are the cellular mechanisms critical for the plasticity of ocular dominance and that are blocked by the inhibition of the ras-ERK pathway? From the molecular point of view, the possible targets of ERK after its visually driven activation are at two different levels: at the nucleus and at the cytoplasm. In the first case activated ERK translocates to the nucleus where it can start CRE-mediated gene expression, with the consequent production of gene transcripts essential for establishment or maintenance of plastic changes.[12,27-29a] Indeed, recent observations by us and others, suggest that protein synthesis is necessary for ocular dominance plasticity, and that visual activity regulates CRE-mediated gene expression.[30,23a] The second scheme envisions a local action of ERK that, upon its activity/neurotrophin dependent activation, phosphorylates certain substrates that are critical for synaptic transmission or neuronal excitability. Such an intriguing possibility, is consistent with the observed distribution of the pERK staining that is very strong in dendrites and not only at the cell body, and with the rapid effect of ERK blockade after theta. Recent data show that there are at least three possible sites for acute ERK action: First, in Aplysia ERK is required for the downregulation and internalization of the adhesion molecule Ap-CAM, a key step in the induction of long term facilitation.[31] Second, ERK can act at synaptic level, since it has been shown that synapsin I has consensus sequences for ERK.[32] Furthermore, ERK phosphorylates synapsin I in response to neurotrophis,[33] with consequent effects on glutamate release.[34] Finally, ERK can phosphorylate the potassium channel Kv4.2,[35] which is one of the main determinant of the rapidly inactivating potassium current. Given the importance of this conductance in determining spike duration, neuronal excitability and spike back-propagation, it is conceivable that this might be

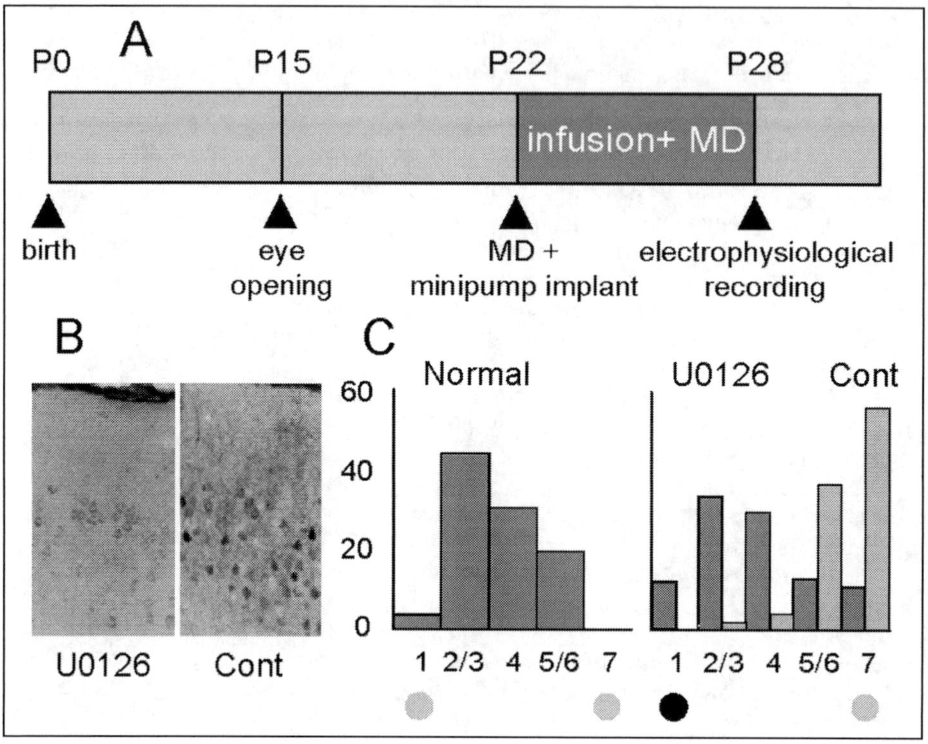

Figure 7. ERK activation is necessary for ocular dominance shift after monocular deprivation. A) Time schedule of experimental treatment. The period of monocular deprivation coincides with the duration of infusion and it is located at the peak of the critical period. B) At the end of the treatment the cortex was exposed and perfused with 1 mM of picrotoxin. At this concentration the electrical activity increased of about a factor four, and this was reflected by ERK induction, which was still blocked on the cortex treated with U0126 (250 μM). C) Average distributions of the ocular dominance for 6 control rats (histogram at left), and 11 rats that were monocularly deprived (right). Seven of the deprived rats were implanted with a minipump delivering U0126. Similar results were obtained with the second MEK inhibitor, PD98059.

another key target for ERK actions on synaptic connectivity. Further details of the mechanisms controlling the acute and long term effects of ERK on neuronal plasticity are still no more than hypotheses.

References

1. Wiesel TN, Hubel DH. Comparison of the effects of unilateral and bilateral eye closure on cortical unit responses in kittens. J Neurophysiol 1995; 28(6):1029-1040
2. Katz LC, Shatz CJ. Synaptic activity and the construction of cortical circuits. Science 1996; 274:1133-1138.
3. Pizzorusso T, Maffei L. Plasticity in the developing visual system. Curr Opin Neurol 1996; 9:122-125.
4. McAllister AK, Katz LC, Lo DC. Neurotrophins and synaptic plasticity. Annu Rev Neurosci 1999; 22:295-318.
5. Maffei L, Berardi N, Domenici L et al. Nerve growth factor (NGF) prevents the shift in ocular dominance distribution of visual cortical neurons in monocularly deprived rats. J Neurosci 1992; 12:4651-4662.

6. Riddle DR, Lo DC, Katz LC. NT-4-mediated rescue of lateral geniculate neurons from effects of monocular deprivation. Nature 1995; 378:189-191.
7. Galuske RA, Kim DS, Castren E et al. Brain-derived neurotrophic factor reversed experience-dependent synaptic modifications in kitten visual cortex. Eur J Neurosci 1996; 8:1554-1559.
8. Lodovichi C, Berardi N, Pizzorusso T et al. Effects of neurotrophins on cortical plasticity: same or different? J Neurosci 2000; 20:2155-2165.
9. Huang ZJ, Kirkwood A, Pizzorusso T et al. BDNF regulates the maturation of inhibition and the critical period of plasticity in mouse visual cortex. Cell 1999; 98:739-755.
10. Patapoutian A, Reichardt LF. Trk receptors: mediators of neurotrophin action. Curr Opin Neurobiol 2001; 11(3):272-280
11. Bourtchuladze R, Frenguelli B, Blendy J et al. Deficient long-term memory in mice with a targeted mutation of the cAMP-responsive element-binding protein. Cell 1994; 79:59-68.
12. Silva AJ, Kogan JH, Frankland PW et al. CREB and memory. Annu Rev Neurosci 1998; 21:127-148.
13. West AE, Chen WG, Dalva MB et al. Calcium regulation of neuronal gene expression. Proc Natl Acad Sci U S A 2001; 25:11024-11031.
14. Zirrgiebel U, Ohga Y, Carter B et al. Characterization of TrkB receptor-mediated signaling pathways in rat cerebellar granule neurons: Involvement of protein kinase C in neuronal survival. J Neurochem 1995; 65:2241-2250.
15. Finkbeiner S, Tavazoie SF, Maloratsky A et al. CREB: a major mediator of neuronal neurotrophin responses. Neuron 1997; 19:1031-1047.
16. Pizzorusso T, Ratto GM, Putignano E et al. Brain-derived neurotrophic factor causes cAMP response element-binding protein phosphorylation in absence of calcium increases in slices and cultured neurons from rat visual cortex. J Neurosci 2000; 15;20(8):2809-2816.
17. Favata MF, Horiuchi KY, Manos EJ et al. Identification of a novel inhibitor of mitogen-activated protein kinase kinase. J Biol Chem 1998; 273:18623-18632.
18. Roberson ED, English JD, Adams JP et al. The mitogen-activated protein kinase cascade couples PKA and PKC to cAMP response element binding protein phosphorylation in area CA1 of hippocampus. J Neurosci 1999; 19:4337-4348.
19. Ghosh A, Greenberg ME. Calcium signaling in neurons: molecular mechanisms and cellular consequences. Science 1995; 268:239-247.
20. Deisseroth K, Bito H, Tsien RW. Signaling from synapse to nucleus: postsynaptic CREB phosphorylation during multiple forms of hippocampal synaptic plasticity. Neuron 1996; 16:89-101.
21. Fields RD, Eshete F, Stevens B et al. Action potential-dependent regulation of gene expression: Temporal specificity in Ca^{2+}, cAMP-responsive element binding proteins, and mitogen-activated protein kinase signaling. J Neurosci 1997; 17:7252-7266.
22. Grewal SS, York RD, Stork PJ. Extracellular-signal-regulated kinase signalling in neurons. Curr Opin Neurobiol, 1999; 9:544-553.
23. Di Cristo G, Berardi N, Cancedda L et al. Requirement of ERK activation for visual cortical plasticity. Science. 2001; 22; 292:2337-2340.
23a. Cancedda L, Putignano E, Impey S et al. Patterned vision causes CRE-mediated gene expression in the visual cortex through PKA and ERK. Neurosci 2003; 23(18):7012-20.
24. Elgersma, Y, Silva AJ. Molecular mechanisms of synaptic plasticity and memory. Curr Opin Neurobiol 1999; 9:209-213.
25. Sweat JD. Toward a molecular explanation for long-term potentiation. Learning & Mem 1999; 6:399-416.
26. Kirkwood A, Lee HK, Bear MF. Co-regulation of long-term potentiation and experience-dependent synaptic plasticity in visual cortex by age and experience. Nature 1995; 375:328-31.
27. Sgambato V, Pages C, Rogard M et al. Extracellular signal-regulated kinase (ERK) controls immediate early gene induction on corticostriatal stimulation. J Neurosci 1998; 18:8814-25.
28. Impey S, Obrietan K, Storm DR. Making new connections: role of ERK/MAP kinase signaling in neuronal plasticity. Neuron 1999; 23:11-14.
29. Pham TA, Impey S, Storm DR et al. CRE-mediated gene transcription in neocortical neuronal plasticity during the developmental critical period. Neuron 1999; 22:63-72.

29a. Berardi N, Pizzorusso T, Ratto GM et al. Molecular basis of plasticity in the visual cortex. Trends Neurosci 2003; 26:369-378.
30. Taha SA, Stryker MP. Protein synthesis is required for ocular dominance plasticity. Soc for Neurosci Abs 2000; 322.12.
31. Bailey CH, Kaang BK, Chen M et al.Mutation in the phosphorylation sites of MAP kinase blocks learning- related internalization of apCAM in Aplysia sensory neurons. Neuron 1997; 18:913-924.
32. Matsubara M, Kusubata M, Ishiguro K et al. Site-specific phosphorylation of synapsin I by mitogen-activated protein kinase and Cdk5 and its effects on physiological functions. J Biol Chem 1996; 271:21108-21113.
33. Jovanovic JN, Benfenati F, Siow YL et al. Neurotrophins stimulate phosphorylation of synapsin I by MAP kinase and regulate synapsin I-actin interactions. Proc Natl Acad Sci U S A 1996; 93:3679-3683.
34. Jovanovic JN, Czernik AJ, Fienberg AA et al. Synapsins as mediators of BDNF-enhanced neurotransmitter release. Nat Neurosci 2000; 3(4):323-329
35. Adams JP, Anderson AE, Varga AW et al. The A-type potassium channel Kv4.2 is a substrate for the mitogen-activated protein kinase ERK. J Neurochem 2000; 75:2277-2287.

Chapter 8

Attempts to Restore Visual Function after Optic Nerve Damage in Adult Mammals

Tomomitsu Miyoshi, Takuji Kurimoto and Yutaka Fukuda

Abstract

Retinal ganglion cells (RGCs) and their axons, i.e., optic nerve (ON) fibers, provide a good experimental model for research on damaged CNS neurons and their functional recovery. After the ON transection most RGCs undergo retrograde and anterograde degeneration but they can be rescued and regenerated by transplantation of a piece of peripheral nerve (PN). When the nerve graft was bridged to the visual center, regenerating RGC axons can restore the central visual projection. Behavioral recovery of relatively simple visual function has been proved in such PN-grafted rodents. Intravitreal injections of various neurotrophic factors and cytokines to activate intracellular signaling mechanism of RGCs and electrical stimulation to the cut end of ON have promoting effects on their survival and axonal regeneration. Axotomized RGCs in adult cats are also shown to survive and regenerate their axons through the PN graft. Among the cat RGC types, Y cells, which function as visual motion detector, tend to survive and regenerate axons better than others. X cells, which are essential for acute vision, suffer from rapid death after ON transection but they can be rescued by intravitreal application of neurotrophins accompanied with elevation of cAMP. To restore visual function in adult mammals with damaged optic pathway, the comprehensive and integrative strategies of multiple approaches will be needed, taking care of functional diversity of RGC types.

Optic Nerve Regeneration and Functional Recovery of Vision in Rodents

During embryonic development the neural retina and optic nerve (ON) are formed as bilateral protrusions from a frontal part of the neural tube and thus these tissues are a part of the CNS. Basic studies on deterioration of the retina and the ON after the damage in the central visual pathway and various attempts to promote recovery of visual function will certainly benefit our knowledge on brain damage and its repair in general. The retinal ganglion cells (RGCs) send visual information, processed within the retina, to various visual centers of the brain through their axons in the ON. When the ON is severely injured, almost all RGCs die in adult mammals and there is no spontaneous axonal regeneration through the original optic pathway. However, these axotomized RGCs can regenerate axons when a segment of peripheral nerve (PN) is grafted to their cut ends.[1,2] Proportion of the RGCs with regenerated axons was, however, only 2-5% and even the best case approximately 10% in adult rats.[2] When

Brain Repair, edited by Mathias Bähr. ©2006 Eurekah.com and Kluwer Academic / Plenum Publishers.

the opposite end of the PN graft was bridged to the superior colliculus (SC), regenerating axons make synaptic contacts with proper target neurons.[2,3] The visual function of regenerated ON axons and re-established retino-collicular projection has been proved by electrophysiological recordings of single unit activities. Regenerated RGC axons, when recorded in the PN graft, revealed typical visual responses of ON, OFF or ON-OFF center with some surround.[4] Transsynaptic activation of re-innervated SC neurons was verified by electrical stimulation of the PN graft and visual stimulation to the operated eye.[5-7]

The behavioral evidence for the recovery of visual function in PN-grafted rodents has also been presented. Thanos et al[8] have first reported that pupillary light response could be recovered in ON-damaged rats 8 and 12 weeks after PN transplantation between the ON stump and the pretectum, though the response was weaker than in intact rats. After PN grafting into the SC in hamsters, Sasaki et al[9] have proved that these animals could learn avoidance behavior using light as conditioned stimulus. As shown in Figure 1A, the shuttle box was divided into two chambers by a partition which was low enough for the hamster to jump over. When the hamsters did not jump into the other chamber within ten seconds after the light on at the ceiling of shuttle box, they received electrical shocks to their foot. A session of thirty or fifty trials of such avoidance task was performed for 10 consecutive days. The hamsters with intact visual system acquired the avoidance behavior as the session proceeded, and achieved about 40% success trials after 10 sessions (Fig. 1B). Blind hamsters with bilateral ON transection did not show any improvement in the success rate of avoidance. On the other hand, the hamsters with PN grafts between the ON stump and the SC (the opposite ON was transected) showed a gradual increase in success rate although the extent was lower than that in normal hamsters (Fig. 1B). Sasaki et al[10,11] have further succeeded to show behavioral recovery of visual function in PN-grafted hamsters by using more natural paradigms such as counting spontaneous ambulating activity in light and dark conditions and measuring bodily movement coincided with EEG desynchronization induced by light. On the other hand, Thanos et al[12] have reported in PN-grafted rats that they could discriminate between vertical and horizontal stripes, with some morphological and electrophysiological evidence for restoration of retinotopic representation in reconstructed retino-collicular pathway.

Attempts to Promote RGC Survival and Their Axonal Regeneration in Rodents

As described above, at present only a small proportion of axotomized RGCs regenerate-their axons through the PN graft so that only a limited recovery of visual function can be expected. To restore higher visual function such as acute vision, shape or velocity discrimination, further efforts should be directed to the followings: (1) to increase the number of surviving RGCs, (2) to increase the number of regenerating RGC axons, and (3) to make retinotopic connection to the target neurons in visual centers.

After intraorbital ON transection, rat's RGCs start to die on day 3 and the survival proportion rapidly decreases to 10% of the normal on day 14[13] and then after the decrease was gradual.[14] The prevention of retrograde death of RGCs after ON transection is the first step for better recovery of visual function after PN grafting. Intravitreal injections of various neurotrophic factors have been shown to rescue axotomized RGCs from retrograde cell death in adult rats: NGF,[15] BDNF,[16,17] NT-4/5,[18,19] CNTF,[16] FGF,[20] IGF,[21] GDNF[22-24] and Neurturin.[24] The effect of peptidic neurotrophic factor such as BDNF and CNTF on RGC survival was dramatically enhanced by intracellular elevation of cAMP.[25] Other bioactive molecules such as TNF-α,[26] macrophage/microglia inhibitory factor (MIF),[27,28] caspase inhibitor[29] and Bax antisense oligonucleotide[30] were also reported to be effective. In addition, various molecules, cells and tissues derived from in vivo animal revealed promoting effect on RGC survival: PN,[31-33] Schwann cells,[34] activated macrophages,[35] collicular proteoglycan,[36] an artificial graft with

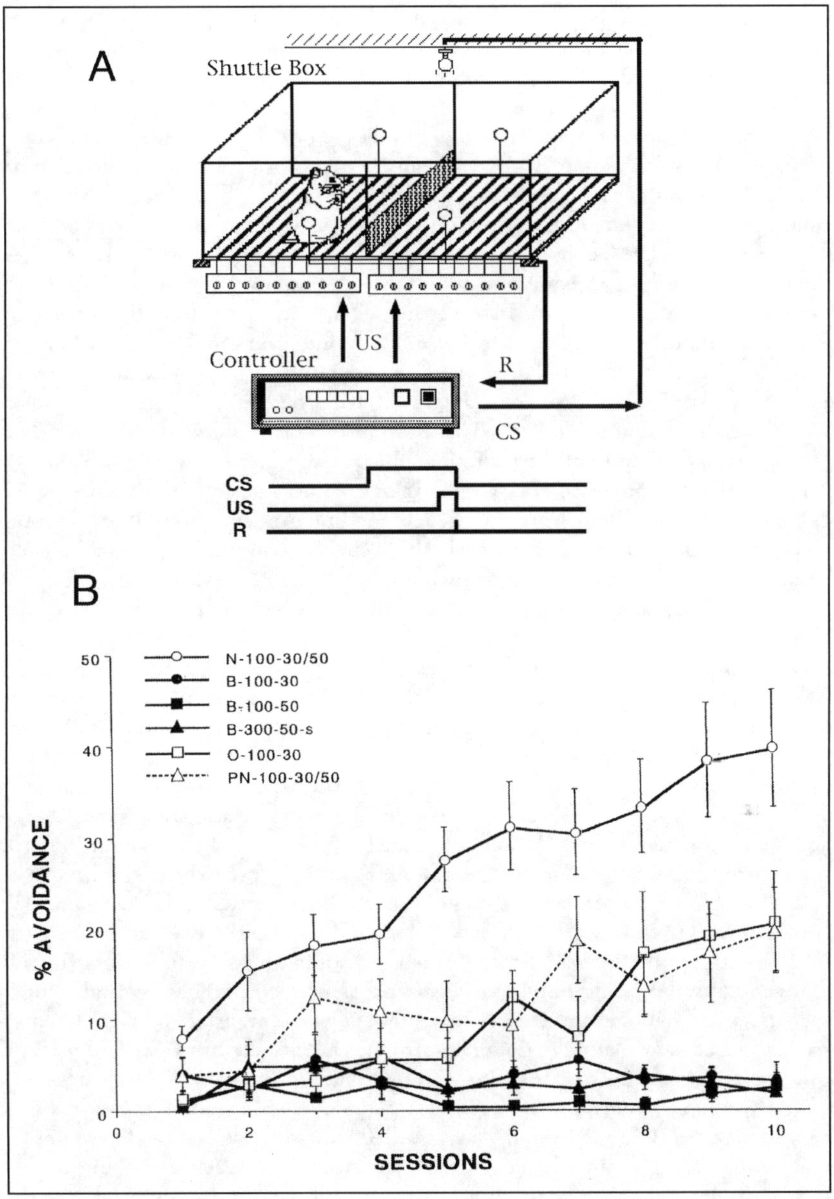

Figure 1. A) A schematic diagram of shuttle-box avoidance learning for hamsters. Turning the light on/off at the ceiling of shuttle box (conditioned stimulus; CS), electrical shocks to the foot (unconditioned stimulus; US) and movement of animal between the rooms (Response; R) were controlled and monitored. B) Increase of the rate of successful trials for avoidance response in normal (N), blind (B), one-eyed (O) and PN-grafted (PN) hamsters. The numbers of each group indicate the intensity of electrical shock (100 or 300 V) and the number of trials per day (30 or 50). The 's' in B-300-50-s indicates that small-sized shuttle box. PN-grafted hamsters showed statistically significant increase of avoidance scores to the same extent as one-eyed hamsters. Reprinted with permission from: Sasaki H, Inoue T, Iso H et al. Exp Neurol 1999; 159:377-390. ©1999 Elsevier Science.

Schwann cells and trophic factors,[37] and unknown molecule(s) from injured lens.[38,39] Most of these factors, however, become less effective as the day proceeded after the drug administration. Even prolonged administration of caspase inhibitor or NT-4/5 showed reduction of survival promoting effect at longer survival time.[40,41] A number of trials have been done to transfer genes for the purpose of long-term supplement of trophic factors with viral vector[42,43] or cDNA electroporation.[44] A combination of TrkB gene transfer to RGCs with adeno-associated virus and exogeneous BDNF supplement results in a marked enhancement of their survival in vivo.[45]

Sprouting and elongation of RGC axons is the next step for functional recovery of vision. Some of the neuroprotective factors administrated to the PN-grafted animals had also been reported to promote axonal regeneration in addition to survival of RCGs. The rate of axonal regeneration, however, could not be accurately estimated independently from the survival rate of RGCs, because improvement of the survival rate by some drugs may inevitably raise the regeneration rate after the PN transplantation. To evaluate the net effect of axonal regeneration, an experimental design that could separate the regeneration rate from the survival rate has been thought necessary. According to Yip and his associates,[46,47] intravitreal injections of CNTF did increase axonal regeneration of the rat RGCs which was induced by PN transplantation, whereas CNTF did not improve RGC survival in their experimental conditions. They showed that CNTF has promotive effect on axonal regeneration independently from its survival promotive effect. However, in the case of factors with both effects on survival and axonal regeneration, their net promoting effect on axonal regeneration was difficult to evaluate. Does the RGC survival by itself promote axonal regeneration? To answer this, *bcl*-2 overexpressed mice were useful. Bcl-2 is one of the anti-apoptotic factors and *bcl*-2 overexpressed mice have about twice RGCs as many as wild type mice because of the lack of apoptotic RGC death during development.[48,49] The relative survival of axotomized RGCs in *bcl*-2 overexpressed mice was enhanced to 65.3% as compared with 5.8% in wild type mice 4 weeks after ON transection.[49,50] Using these *bcl*-2 overexpressed mice, we recently assessed regeneration ability of axotomized RGCs separately from their survival ability. To examine whether the surviving RGCs in *bcl*-2 overexpressed mice can regenerate their axons, the surviving and regenerating RGCs were counted to evaluate the survival and regeneration rates separately.[49] Four weeks after axotomy and PN transplantation, the regeneration rate of surviving RGCs in *bcl*-2 overexpressed mice was not enhanced from that of wild type mice, although the survival rate of RGCs was enhanced more than ten times in *bcl*-2 overexpressed mice. The failure in promotion of RGC axonal regeneration in *bcl*-2 overexpressed mice was also confirmed by another in vivo experiment using ON crush.[51] These results clearly indicate that the survival is prerequisite, but not sufficient for axonal regeneration of RGCs. In a recent in vitro study using *bcl*-2-transfected RGCs, Goldberg et al.[52] have reported that many peptidic neurotrophic factors such as BDNF and CNTF promote the net axonal outgrowth. They also reported that intracellular cAMP elevation, as a result of depolarization by potassium or electrical stimulation via electrode allay on the culture dish, enhanced these peptidic neurotrophic factors' effect on axonal elongation. Given this, it is certain that various peptidic trophic factors stimulate net axonal regeneration into the PN graft, although in vitro neurite extension of CNS neurons may not be comparable to their axonal regeneration in vivo.

How we could promote axonal regeneration of once damaged RGCs other than or in addition to PN grafting and previously known neurotrophic factors? Lens injury has been shown to induce massive axonal regeneration of damaged retinal axons.[38,39,53] In the experiment of Fisher et al[53] almost 30% of damaged axons can regenerate without PN-grafting along the entire original visual pathway and reinnervate visual centers which has been verified by recording visual evoked responses from the visual cortex. Lens injury induced retinal infiltration of activated macrophages and these macrophages seemed to secrete factors that promote axonal regeneration.[38] Recently it was revealed that proteins under 30 kDa in the conditioned medium of activated macrophages have the regeneration-promoting activity on cultured RGCs.[88] Especially, the presence of 14 kDa-protein was parallel to the regeneration-promoting bioactivity on cultured RGCs. This protein was not identical to any of the tested several known trophic factors that can be secreted by activated macrophages. Further investigation for this protein is desired.

The final step is the reformation of specific synaptic connections with the target cells of the visual center. While Vidal-Sanz et al[2] have previously reported that RGC axons regenerating into the SC can find proper target neurons in upper visual layers of the SC, without growing downwards into the multisensory layers underneath, Zwimpfer et al[54] have reported that regenerating RGC axons, when guided into improper target neurons of the cerebellum, can also make synaptic connections. In a more recent study regenerating RGC axons could select their target nuclei, i.e., denervated retinorecipient nuclei, and eform persistent synapses for long time.[55] Thus, when regenerating RGC axons are allowed to select proper targets they prefer them and maintain proper synaptic connections.

Can regenerating RGC axons form synapses with target SC cells in a precise topographic manner as in normal animals? To answer this, Sauvé et al[7] systematically mapped receptive filed positions of single SC neurons throughout its rostrocaudal and mediolateral extent of the reinnervated SC. Although there was a tendency that the RGCs arising from more nasal retina innervate more caudal parts of the SC, this retinotopic preference was much weaker in PN-grafted hamsters than in intact ones. As to the ventral-dorsal axis of the retina, there was no significant tendency of retinotopic representation onto the SC after reinnervation. Sauvé et al[7] also noted some abnormal patterns of retinotopic projection such that single SC site is innervated by regenerating axons from a number of RGCs originating from multiple retinal loci and that a set of closely located RGCs innervate far distant parts on the SC surface. Thus, at present the precise retinotopic organization is not well established in reconstructed central visual pathway (see however, Thanos et al[12]). To improve this, at least two issues should be taken into account. First, by some additional means the survival of much more RGCs should be achieved so that more competitions occur among many regenerated axons to find out their own proper targets. Secondly, more extensive studies should be done on the expression of target finding molecules such as Eph receptors in regenerating RGC axons and ephrins on the surface of reinnervated SC.[56,57] Furthermore, in view of the recent studies that indicate the importance of rich environment or forced use of restored brain pathways for functional recovery, with such rehabilitational approach the reconstructed visual pathway in PN-grafted animals may regain specific synaptic reconnections of regenerating RGC axons with target cells in visual centers.[85-87]

RGC Survival and Their Axonal Regeneration in Adult Cats

To achieve the recovery of higher visual function such as acute vision and shape recognition, it is important to study on experimental animals having a well-developed visual system. In this section we will summarize our recent studies on axotomized RGCs and their axonal regeneration in adult cats. The cat's RGCs consist of three physiological types, termed Y, X and W cells, having different visual response properties.[58-61] Y and X cells were originally distinguished by their response summation properties of receptive fields; X cells summate linearly whereas Y cells nonlinearly.[74] Y cells have brisk-transient responses to stationary light stimulus, large receptive field center and highest axonal conduction velocity. The corresponding morphological type, α cells, have a largest soma, thick primary dendrites and a wide dendritic field.[62] X cells have a small receptive field center with brisk-sustained visual response and medium conduction velocity, and correspondingly β cells have a medium-sized soma and bushy dendrites which extend in a narrow area.[62] The W cell is a heterogeneous group of smaller RGCs and they are all characterized with sluggish visual responses to centered light stimuli and have relatively slow conduction velocities than Y and X cells. The W cell includes various

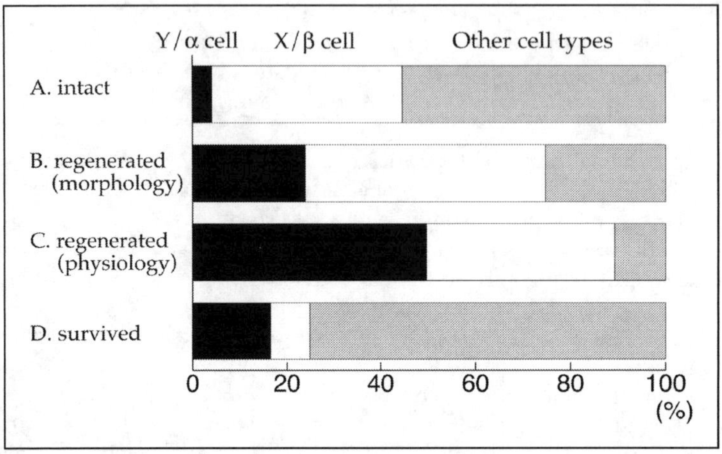

Figure 2. The relative proportion of each type of cat RGCs. A) Intact retinas. B) RGCs with regenerated axons identified morphologically after PN transplantation.[67] C) RGCs with regenerated axons identified electrophysiologically after PN transplantation.[69] D) RGCs surviving after ON transection, 2 months after surgery.[70] Both survival (D) and regeneration (C) rates of Y cells (α cells) were higher than those of X cells (β cells).

morphological types such as gamma,[62] delta[63] and epsilon cells[64] (see review ref. 65). Some W cells have been suggested to function in monitoring background light intensity.[58] In the area centralis of cat retina, the density of X cells (β cells) is highest and they contribute to the visual acuity of cat's vision. Because each cell type of the cat RGCs plays a specific role in visual information processing, it is important to investigate how different are the three types in the survival and axonal regeneration.

Since 1991 we have shown that adult cat's RGCs can regenerate their axons into the PN graft like rodent RGCs.[66-68] The type of RGCs with regenerated axons can be identified according to their dendritic morphology visualized by intracellular dye injection. All three major types (α, β and γ cells) were found to regenerate their axons. The overall regeneration rate of cat RGCs at two months after surgery was 3-4% of total RGCs. The relative proportions of α, β and other cell types were approximately 24, 50 and 24%, respectively (Fig. 2B). The relative proportion of α cells was about 6 times higher than the proportion in intact retina (4.2%, see Fig 2A), indicating that the α cell has the best regenerating capacity among the cat RGCs. Although some RGCs revealed signs of degenerative or regrowing changes of dendrites, the dendritic morphology of many RGCs with regenerated axons were comparable to those of normal ones.[67] Electron-microscopic analysis on regenerating axons revealed that about 20% of regenerated axons were myelinated but the rest were all unmyelinated two months after PN transplantation, thus axonal conduction velocity must be slower than normal.[68]

To reveal whether or not and how these regenerated axons transmit visual information through regenerated RGC axons in cats, we recorded single-unit activities from teased fibers within the PN graft about two months after surgery and analyzed their receptive field properties.[69] Many unit activities that responded to visual stimuli were recorded from the PN graft and almost all these unit activities preserved typical visual response properties of Y, X or W cell (Fig. 3). The highest sampling of Y cells confirmed the best regeneration ability of α cells in morphological study (Fig. 2B and C). The receptive field centers of Y and W cells were larger than those of X cells at corresponding eccentricities from the area centralis as in intact retinas.[58] The tendency that the receptive field centers of Y, X and W cells become larger at peripheral

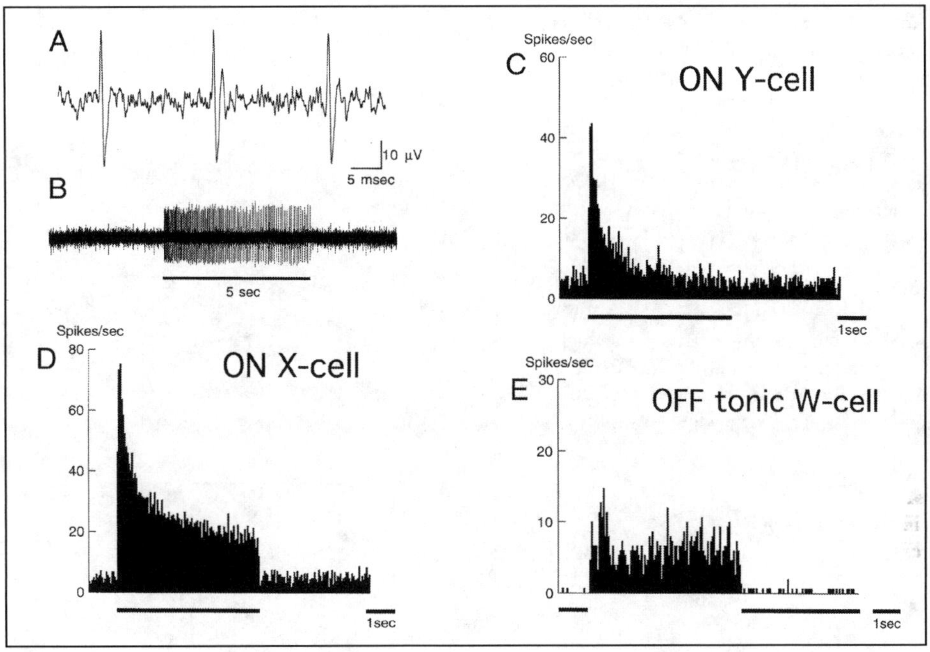

Figure 3. Physiological responses of cat RGCs recorded from their regenerated axons in PN graft. A, B) Examples of single-unit activity responded to a stationary light spot on the receptive field center. Expanded trace of spikes during light spot on (A) and the long trace of the same unit to a stationary light spot for 5 sec (B) are shown. C-E: Examples of peristimulus time histograms of ON-center Y cell with a brisk transient response (C), ON-center X cell with a brisk sustained response (D) and OFF-center W cell with sluggish sustained response (E) to stationary light spots centered on the receptive field center. The horizontal bar indicates the period of light on. The interruption in the bar of E indicates the period of light off. Reprinted with permission from: Miyoshi T, Watanabe M, Sawai H et al. Exp Brain Res 1999; 124:383-390. ©1999 Springer.

retina was also preserved in PN grafted cats, though the variation of receptive field center size in regenerated Y or X cells was larger than that of intact ones.[58] Within 10 degrees from the area centralis, however, the receptive field centers of X-cells with regenerated axons were larger than those in intact retinas. Another abnormalities in regenerating RGCs were low spontaneous activities and weak surround visual responses. These results suggest that some rearrangement of retinal circuitry and/or changes in membrane properties must have occurred as a consequence of degeneration and regeneration.

Does the difference of regeneration ability between Y cells (α cells) and X cells (β cells) depend on the ability of their survival? To clarify this, the survival of each RGC type was compared after ON transection.[70] Two months after ON transection, the same observation term as in regeneration study, α cells comprised 16% of surviving RGCs whereas β cells only 9% (Fig. 2D). Compared with the proportion in intact retina, α cells have better survival ability than β cells, whereas β cells are very vulnerable to axotomy. The two main types of cat RGCs were also quite different in the survival time course after axotomy.[71] Within 2 weeks after ON transection, the survival rate of β cells starts to decrease on day 3 and it rapidly falls to the level of about 20% on day 7 (Fig. 4C).[71] Then after, the survival rate of β cells gradually decreases to 12% on day 14. On the other hand, α cells do not have a rapid phase of death like β cells and the survival proportion gradually decreases to 64% on day 14 (Fig 4A and 4B). We

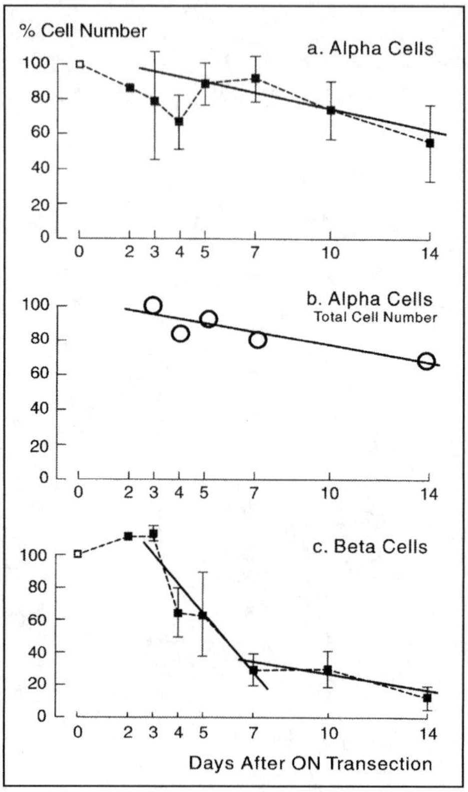

Figure 4. Changes of survival rates of α cells (a), and β cells (c), and ratios of estimated total numbers of α cells in ON-transected retinas to those in the intact retinas (b). Squares and vertical bars represent averages of normalized values of estimated cell number for the ON-transected retina to that in the intact retina with SD. Solid lines are regression lines calculated from original values between days 3-14 (a,b), or days 3-7 and days 7-15 (c). Broken lines (a,c) draw the averages. Note that survival ratios of β cells decrease rapidly from day 3 to day 7, then slowly after day 7, while those of surviving α cells decreases slowly. Reprinted with permission from: Watanabe M, Inukai N, Fukuda Y, Vis Neurosci 2001; 18:137-145. ©2001 Cambridge University Press.

have recently found that intraocular administration of caspase 3 inhibitor rescued β cells from rapid death after axotomy but did not affect the survival of α cells, indicating that the rapid cell death of β cells is due to apoptosis.[72] Because β cells contribute to the central vision, it is important to rescue β cells from rapid death for the recovery of higher visual function after the ON damage in cats.

Another question we asked was to what extent the cat's RGCs maintain their physiological properties after the ON transection. To answer this, we recorded single-unit activities from axotomized RGCs 5 and 14 days after ON transection.[73] On day 5 during which period β cells (X-cells) were dying rapidly, almost all recorded RGC activities kept their original visual response properties to light spot and also kept linear/non-linear response characteristics.[74] On day 14, when the rapid cell death period was over, the recording efficiency of X cells fell down. We also compared receptive filed center size as a function of eccentricity on 5 and 14 days after axotomy. On day 5 receptive field centers of approximately two third of Y cells and one third of X cells were smaller than those of respective types in intact retinas (Fig. 5).[58] On the other hand, on day 14 only a small number of Y cells had such shrunken receptive field centers. Spontaneous activity and response magnitude of axotomized RGC were significantly lower than those of intact RGCs. Taking the time course of X cells (β cells) death into consideration, X-cells' smaller receptive field centers on day 5 may have reflected some functional deterioration of axotomized X-cells just before their death.

Then we examined the effect of neurotrophic factors on the survival of α and β cells 14 days after ON transection.[75] Intravitreally injected BDNF, CNTF and GDNF were effective to promote the survival of β cells but not of α cells. Similar injections of NGF or NT-3 did not

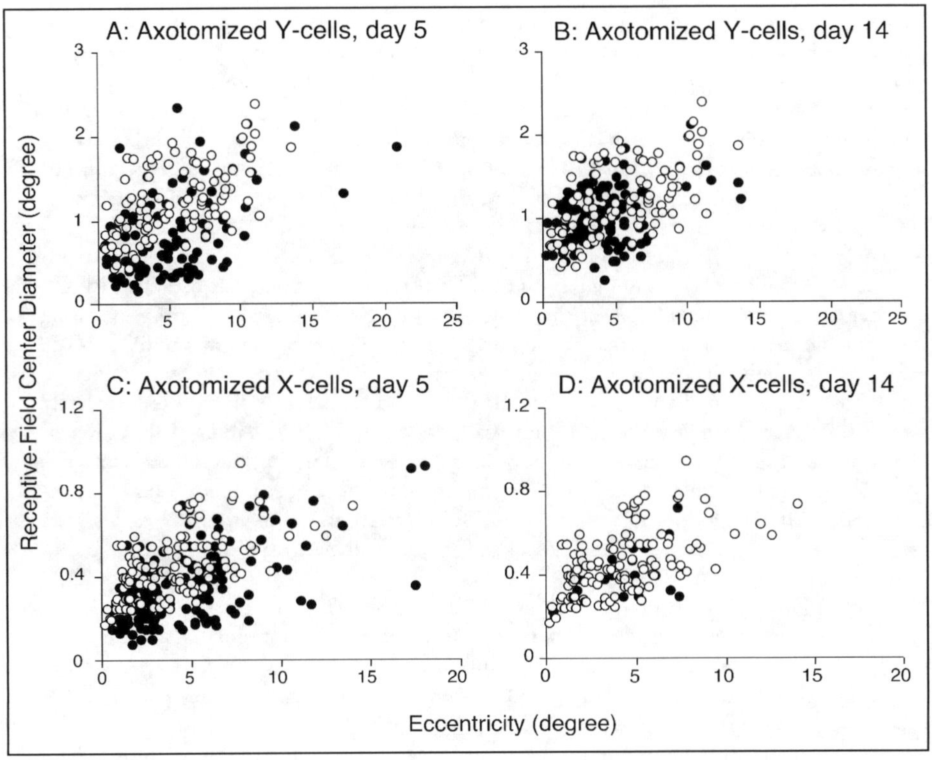

Figure 5. Changes of receptive field center diameters of axotomized cat RGCs on day 5 and 14 after ON transection. Open and filled circles indicate the receptive field centers of the intact and axotomized RGCs, respectively. A) day 5 axotomized Y-cells; B) day 14 axotomized Y-cells; C) day 5 axotomized X-cells; D) day 14 axotomized X-cells. In the axotomized retinas on day 5, Y- (A) and X-cells (C) tended to have smaller receptive field centers than those of intact RGCs throughout the eccentricity. By contrast, neither Y-cells nor X-cells on day 14 had such small receptive field centers (B and D), except for a few Y-cells. For each of the Y- and X-cells, ON-center and OFF-center subtypes were not distinguished because there was no consistent difference between these two subtypes. Reprinted with permission from: Takao M, Miyoshi T, Watanabe M et al. Exp Neurol 2002; 177:171-181. ©2002 Elsevier Science.

promote the survival of any type of cat RGCs. Intravitreal injections of a mixture of BDNF, CNTF and forskolin which enhances intracellular level of cAMP had the greatest effect, 4.7 folds increase, in promotion of β cell survival, whereas the same treatment had no effect on the survival of α cells.[75] In physiological experiment on axotomized cat RGCs, we found that intravitreal administration of this mixture preserved their spontaneous activities and prevented them from receptive field shrinkages.[76]

Neuroprotection of RGCs by Electrical Stimulation

In recent years it has been shown that electrical activity of CNS neurons plays an essential role in neural plasticity during development and even in adulthood. For example, to elicit long-term potentiation which is known to underlie learning and memory high frequency tetanic stimulation with electrical pulses is essential (see review ref. 77). Another example is that electrical activities of afferent inputs are crucial for fine adjustment of specific synaptic connections[78,79] and also for dendritic morphology of postsynaptic cells.[80]

During physiological studies on axotomized cat RGCs, we realized that the spontaneous discharge rate became quite low as the day proceeds after axotomy; on day 14 a majority of axotomized RGCs had no spontaneous activity.[73] The fall of spontaneous activity in these RGCs may reflect deafferentation of synaptic inputs from bipolar or amacrine cells within the retina besides some deterioration in their membrane properties. It is also possible that low or no spontaneous activity of axotomized RGCs contributes to curtail their survival. In support of this, we previously reported that after dark rearing of ON-transected and PN-grafted cats, RGCs revealed degenerative changes in their soma and dendritic shafts such as swelling and vacuolization,[81] indicating that light stimulus is necessary for the survival and axonal regeneration of RGCs. A similar observation has been made by Aigner et al[82] on axotomized rat RGCs.

Given these correlations between the RGC survival and the electrical activity or visual inputs, we hypothesized that electrical stimulation of the ON may prevent axotomized RGCs from retrograde cell death. To test this, we examined the in vivo effect of electrical stimulation of the ON on the survival of rat RGCs one week after axotomy.[83] The left ON was completely transected 5 days after applications of a fluorescent dye to bilateral SCs to label RGCs retrogradely. Just after the ON transection a train of monophasic electrical pulses (50 μsec in duration and 50 μA in amplitude) was applied for two hours to the ON stump via a pair of silver ball electrodes. One week after ON transection, the survival rates of axotomized RGCs were surveyed by evaluating the mean densities of retrogradely labeled RGCs in the treated and control retinas. As a result, 83% of RGCs survived after electrical stimulation of the ON, whereas only 54% of RGCs survived without ON stimulation (Fig. 6). The effect depends on the intensity of electrical currents and the sham stimulation did not show any effect. The enhancement of RGC survival by electrical stimulation of the ON was still clear 2 weeks after axotomy, though absolute survival rate became lower in both treated and non-treated retinas (Morimoto et al., unpublished observation).

It would be interesting to examine whether or not the activation of axotomized RGCs by electrical or light stimulation will promote in vivo axonal regeneration along the PN graft. As already mentioned in a preceding section, in a recent in vitro study, Goldberg et al[52] have shown a significant axonal growth from cultured RGCs after application of electrical currents in the culturing chamber in the medium with BDNF and CNTF. They further reported that the axonal growth promoting effect of electrical stimulation was mediated by elevation of intracellular level of cAMP and following activation of the downward pathway, i.e., protein kinase A (PKA)-mitogen activated protein kinase (MAPK) system.[52] Although in vivo experiments may not directly reflect the observations made in in vitro experiments, we can expect that a similar intracellular signaling pathway is activated during stimuli of RGC with electrical pulses or flashing light. In fact, the cAMP responsive element binding protein (CREB) has recently been implicated as a candidate for nuclear pathway that controls neuronal plasticity during retinogeniculate development.[84]

In conclusion, various factors and methods are required to overcome the obstacles hindering the RGC survival, regeneration and the precise retinotopic connection of their axons to the central visual target. Functional restoration of the damaged visual pathway may be accomplished by the combination of multiple strategies including PN transplantation, supplement of neurotrophic factors, controlling intracellular signaling, genetic alteration of cellular function and activation of RGCs by electrical or light stimulation. Another important issue that should be taken into account would be the diversity of visual functions among various types of RGCs especially in studies on animals with well-developed vision.

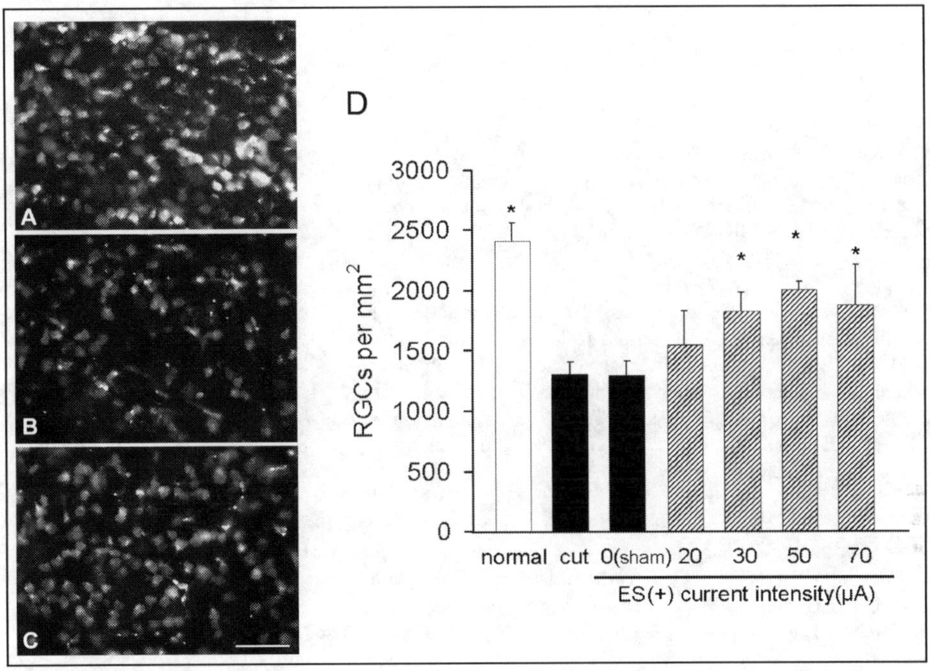

Figure 6. The effect of electrical stimulation on the survival of axotomized rat RGCs. A-C) The photomicrographs of FG-labeled RGCs in corresponding regions of an intact retina (A), 7 days after ON transection without electrical stimulation (B) and with electrical stimulation of 50μA (C). More RGCs survived in the retina with electrical stimulation than those without electrical stimulation. Scale bar: 50 μm. D) The mean RGC densities of the groups with different current intensities of electrical stimulation. Electrical stimulation of over 30μA significantly increased the surviving RGC densities. Reprinted with permission from: Morimoto T, Miyoshi T, Fujikado T et al. Neuroreport 2002; 13:227-230. ©2002 Lippincott Williams & Wilkins.

References

1. So K-F, Aguayo AJ. Lengthy regrowth of cut axons from ganglion cells after peripheral nerve transplantation into the retina of adult rats. Brain Res 1985; 328:349-354.
2. Vidal-Sanz MGM, Bray MP, Villegas-Pérez MP et al. Axonal regeneration and synapse formation in the superior colliculus by retinal ganglion cells in the adult rat. J Neurosci 1987; 7:2894-2909.
3. Carter DA, Bray GM, Aguayo AJ. Regenerated retinal ganglion cell axons can form well-differentiated synapses in the superior colliculus of adult hamsters. J Neurosci 1989; 9:4042-4050.
4. Keirstead SA, Vidal-Sanz M, Rasminsky M et al. Responses to light of retinal neurons regenerating axons into peripheral nerve grafts in the rat. Brain Res 1985; 359:402-406.
5. Keirstead SA, Rasminsky M, Fukuda Y et al. Electrophysiologic responses in hamster superior colliculus evoked by regenerating retinal axons. Science 1989; 246:255-257.
6. Sauvé Y, Sawai H and Rasminsky M. Functional synaptic connections made by regenerated retinal ganglion cell axons in the superior colliculus of adult hamsters. J. Neurosci 1995; 15:665-675.
7. Sauvé Y, Sawai H, Rasminsky M. Topological specificity in reinnervation of the superior colliculus by regenerated retinal ganglion cell axons in adult hamsters. J Neurosci 2001; 21:951-960.
8. Thanos S. Adult retinofugal axons regenerating through peripheral nerve grafts can restore the light-induced pupilloconstriction reflex. Eur J Neurosci 1992; 4:691-699.
9. Sasaki H, Inoue T, Fukuda Y et al. Light-dark discrimination after sciatic nerve transplantation to the sectioned optic nerve in adult hamsters. Vision Res 1993; 33:877-880.

10. Sasaki H, Coffey P, Villegas-Pérez MP et al. Light induced EEG desynchronization and behavioral arousal in rats with restored retinocollicular projection by peripheral nerve graft. Neurosci Lett 1996; 218:45-48.
11. Sasaki H, Inoue T, Iso H et al. Recovery of visual behaviors in adult hamsters with the peripheral nerve graft to the sectioned optic nerve. Exp Neurol 1999; 159:377-390.
12. Thanos S, Naskar R, Heiduschka P. Regenerating ganglion cell axons in the adult rat establish retinofugal topography and restore visual function. Exp Brain Res 1997; 114:483-491.
13. Barkelaar M, Clarke DB, Wang Y-C et al. Axotomy results in delayed death and apoptosis of retinal ganglion cells in adult rats. J Neurosci 1994; 14:4368-4374.
14. Villegas-Pérez MP, Vidal-Sanz M, Rasminsky M et al. Rapid and protracted phases of retinal ganglion cell loss follow axotomy in the optic nerve of adult rats. J Neurobiol 1993; 24:23-36.
15. Carmignoto G, Maffei L, Candeo P et al. Effect of NGF on the survival of rat retinal ganglion cells following optic nerve section. J Neurosci 1989; 9:1263-1672.
16. Mey J, Thanos S. Intravitereal injections of neurotrophic factors support the survival of axotomized retinal ganglion cells in adult rats in vivo. Brain Res 1993; 602:304-317.
17. Mansour-Robaey S, Clarke DB, Wang Y-C et al. Effects of ocular injury and administration of brain-derived neurotrophic factor on survival and regrowth of axotomized retinal ganglion cells. Proc Natl Acad Sci USA 1994; 91:1632-1636.
18. Cohen A, Bray GM, Aguayo AJ. Neurotrophin-4/5 (NT-4/5) increases adult rat retinal ganglion cell survival and neurite outgrowth in vitro. J Neurobiol 1994; 25:953-959.
19. Peinado-Ramon P, Salvador M, Villegas-Pérez MP et al. Effects of axotomy and intraocular administration of NT-4, NT-3, and brain-derived neurotrophic factor on the survival of adult rat retinal ganglion cells. A quantitative in vivo study. Invest Ophthalmol Vis Sci 1996; 37:489-500.
20. Sievers J, Hausmann B, Unsicker K et al. Fibroblast growth factors promote the survival of adult rat retinal ganglion cells after transection of the optic nerve. Neurosci Lett 1987; 76:157-162.
21. Kermer P, Klocker N, Labes M et al. Insulin-like growth factor-I protects axotomized rat retinal ganglion cells from secondary death via PI3-K-dependent Akt phosphorylation and inhibition of caspase-3 in vivo. J Neurosci 2000; 20:2-8.
22. Klocker N, Braunling F, Isenmann S et al. In vivo neurotrophic effects of GDNF on axotomized retinal ganglion cells. Neuroreport 1997; 8:3439-3442.
23. Koeberle PD, Ball AK. Effects of GDNF on retinal ganglion cell survival following axotomy. Vision Res 1998; 38:1505-1515.
24. Koeberle PD, Ball AK. Neurturin enhances the survival of axotomized retinal ganglion cells in vivo: combined effects with glial cell line-derived neurotrophic factor and brain-derived neurotrophic factor. Neuroscience 2002; 110:555-567.
25. Shen S, Wiemelt AP, McMorris FA et al. Retinal ganglion cells lose trophic responsiveness after axotomy. Neuron 1999; 23:285-295.
26. Diem R, Meyer R, Weishaupt JH et al. Reduction of potassium currents and phosphatidylinositol 3-kinase-dependent AKT phosphorylation by tumor necrosis factor-(alpha) rescues axotomized retinal ganglion cells from retrograde cell death in vivo. J Neurosci 2001; 21:2058-2066.
27. Thanos S, Mey J, Wild M. Treatment of the adult retina with microglia-suppressing factors retards axotomy-induced neuronal degradation and enhances axonal regeneration in vivo and in vitro. J Neurosci 1993; 13:455-466.
28. Thanos S, Mey J. Type-specific stabilization and target-dependent survival of regenerating ganglion cells in the retina of adult rats. J Neurosci 1995; 15:1057-1079.
29. Kermer P, Klocker N, Labes M et al. Inhibition of CPP32-like proteases rescues axotomized retinal ganglion cells from secondary cell death in vivo. J Neurosci 1998; 18:4656-4662.
30. Isenmann S, Engel S, Gillardon F et al. Bax antisense oligonucleotides reduce axotomy-induced retinal ganglion cell death in vivo by reduction of Bax protein expression. Cell Death Differ 1999; 6:673-682.
31. Turner JE, Blair JR, Chappel ET. Peripheral nerve implant effects on survival of retinal ganglion layer cells after axotomy initiated by a penetrating lesion. Brain Res. 1987; 419:46-54.
32. Villegas-Pérez MP, Visal-Sanz M, Bray GM et al. Influences of peripheral nerve grafts on the survival and regrowth of axotomized retinal ganglion cells in adult rats. J Neurosci 1988; 8:265-280.

33. Berry M, Carlile J, Hunter A. Peripheral nerve explants grafted into the vitreous body of the eye promote the regeneration of retinal ganglion cell axons severed in the optic nerve. J Neurocytol 1996; 25:147-170.
34. Maffei L, Carmignoto G, Perry VH et al. Schwann cells promote the survival of rat retinal ganglion cells after optic nerve section. Proc Natl Acad Sci U S A 1990; 87:1855-1859.
35. Lazarov-Spiegler O, Solomon AS, Zeev-Brann AB et al. Transplantation of activated macrophages overcomes central nervous system regrowth failure. FASEB J 1996; 10:1296-1302.
36. Huxlin KR, Dreher B, Schulz M et al. Effect of collicular proteoglycan on the survival of adult rat retinal ganglion cells following axotomy. Eur J Neurosci 1995; 7:96-107.
37. Negishi H, Dezawa M, Oshitari T et al. Optic nerve regeneration within artificial Schwann cell graft in the adult rat. Brain Res Bull 2001; 55:409-419.
38. Leon S, Yin Y, Nguyen J et al. Lens injury stimulates axon regeneration in the mature rat optic nerve. J Neurosci 2000; 20:4615-4626.
39. Fischer D, Pavlidis M, Thanos S. Cataractogenic lens injury prevents traumatic ganglion cell death and promotes axonal regeneration both in vivo and in culture. Invest Ophthalmol Vis Sci. 2000; 41:3943-3954.
40. Clarke DB, Bray GM, Aguayo AJ. Prolonged administration of NT-4/5 fails to rescue most axotomized retinal ganglion cells in adult rats. Vision Res 1998; 38:1517-1524.
41. Kermer P, Klocker N, Bahr M. Long-term effect of inhibition of ced 3-like caspases on the survival of axotomized retinal ganglion cells in vivo. Exp Neurol. 1999; 158:202-205.
42. Di Polo A, Aigner LJ, Dunn RJ et al. Prolonged delivery of brain-derived neurotrophic factor by adenovirus-infected Müller cells temporarily rescues injured retinal ganglion cells. Proc Natl Acad Sci USA 1998; 95:3978-3983.
43. Schmeer C, Straten G, Kugler S et al. Dose-dependent rescue of axotomized rat retinal ganglion cells by adenovirus-mediated expression of glial cell-line derived neurotrophic factor in vivo. Eur J Neurosci. 2002; 15:637-643.
44. Mo X, Yokoyama A, Oshitari T et al. Rescue of axotomized retinal ganglion cells by BDNF gene electroporation in adult rats. Invest Ophthalmol Vis Sci 2002; 43:2401-2405.
45. Cheng L, Sapicha P, Kittlerova P et al. TrkB gene transfer protects retinal ganglion cells from axotomy-induced death in vivo. J Neurosci 2002; 22:3977-3986.
46. Cui Q, Lu Q, So KF et al. CNTF, not other trophic factors, promotes axonal regeneration of axotomized retinal ganglion cells in adult hamsters. Invest Ophthalmol Vis Sci 1999; 40:760-766.
47. Cho KS, Chan PM, So KF et al. Ciliary neurotrophic factor promotes the regrowth capacity but not the survival of intraorbitally axotomized retinal ganglion cells in adult hamsters. Neuroscience 1999; 94:623-628.
48. Bonfanti L, Strettoi E, Chierzi S et al. Protection of retinal ganglion cells from natural and axotomy-induced cell death in neonatal transgenic mice overexpressing bcl-2. J Neurosci 1996; 16:4186-4194.
49. Inoue T, Hosokawa M, Morigiwa K et al. Bcl-2 overexpression does not enhance in vivo axonal regeneration of retinal ganglion cells after peripheral nerve transplantation in adult mice. J Neurosci 2002; 22:4468-4477.
50. Cenni MC, Bonfanti L, Martinou JC et al. Long-term survival of retinal ganglion cells following optic nerve section in adult bcl-2 transgenic mice. Eur J Neurosci 1996; 8:1735-1745.
51. Chierzi S, Strettoi E, Cenni MC et al. Optic nerve crush: axonal responses in wild-type and bcl-2 transgenic mice. J Neurosci 1999; 19:8367-8376.
52. Goldberg JL, Espinosa JS, Xu Y et al. Retinal ganglion cells do not extend axons by default: promotion by neurotrophic signaling and electrical activity. Neuron 2002; 33:689-702.
53. Fischer D, Heiduschka P, Thanos S. Lens-injury-stimulated axonal regeneration throughout the optic pathway of adult rats. Exp Neurol 2001; 172:257-272.
54. Zwimpfer TJ, Aguayo AJ, Bray GM. Synapse formation and preferential distribution in the granule cell layer by regenerating retinal ganglion cell axons guided to the cerebellum of adult hamsters. J Neurosci 1992; 12:1144-1159.
55. Avilés-Trigueros M, Sauvé Y, Lund RD et al. Selective innervation of retinorecipient brainstem nuclei by retinal ganglion cell axons regenerating through peripheral nerve grafts in adult rats. J Neurosci 2000; 20:361-374.

56. Knöll B, Isenmann S, Kilic E et al. Graded expression patterns of ephrin-As in the superior colliculus after lesion of the adult mouse optic nerve. Mech Dev 2001; 106:119-127.
57. Rodger J, Lindsey KA, Leaver SG et al. Expression of ephrin-A2 in the superior colliculus and EphA5 in the retina following optic nerve section in adult rat. Eur J Neurosci 2001; 14:1929-1936.
58. Stone J, Fukuda Y. Properties of cat retinal ganglion cells: a comparison of W-cells with X- and Y-cells. J Neurophysiol 1974; 37:722-748.
59. Fukuda Y, Stone J. Retinal distribution and central projections of Y-, X- and W-cells of the cat's retina. J Neurophysiol 1974; 37:749-772.
60. Cleland BG, Levick WR. Brisk and sluggish concentrically organized ganglion cells in the cat's retina. J Physiol 1974; 240:421-456.
61. Cleland BG, Levick WR. Properties of rarely encountered types of ganglion cells in the cat's retina and an overall classification. J Physiol 1974; 240:457-492.
62. Boycott BB, Wässle H. The morphological types of ganglion cells of the domestic cat's retina. J Physiol 1974; 240:397-419.
63. Dacey DM. Monoamine-accumulating ganglion cell type of the cat's retina. J Comp Neurol. 1989; 288:59-80.
64. Kawamura S, Fukushima N, Hattori S. Topographical origin and ganglion cell type of the retino-pulvinar projection in the cat. Brain Res 1979; 173:419-429.
65. O'Brien BJ, Isayama T, Richardson R et al. Intrinsic physiological properties of cat retinal ganglion cells. J Physiol 2001; 538:787-802.
66. Watanabe M, Sawai H, Fukuda Y. Axonal regeneration of retinal ganglion cells in the cat geniculocortical pathway. Brain Res 1991; 560:330-333.
67. Watanabe M, Sawai H, Fukuda Y. Number, distribution, and morphology of retinal ganglion cells with axons regenerated into peripheral nerve graft in adult cats. J Neurosci 1993; 13:2105-2117.
68. Watanabe M, Sawai H, Fukuda Y. Myelination of regenerated optic fibers in peripheral nerve graft of adult cats. Exp Brain Res 1994; 98:39-43.
69. Miyoshi T, Watanabe M, Sawai H et al. Receptive-field properties of adult cat's retinal ganglion cells with regenerated axons. Exp Brain Res 1999; 124:383-390.
70. Watanabe M, Sawai H, Fukuda Y. Number and dendritic morphology of retinal ganglion cells that survived after axotomy in adult cats. J Neurobiol 1995; 27:189-203.
71. Watanabe M, Inukai N, Fukuda Y. Survival of retinal ganglion cells after transection of the optic nerve in adult cats: a quantitative study within two weeks. Vis Neurosci 2001; 18:137-145.
72. Kurimoto T, Miyoshi T, Suzuki A et al. Apoptotic death of beta cells after optic nerve transection in adult cats. J Neurosci 2003; 23:4023-4028.
73. Takao M, Miyoshi T, Watanabe M et al. Changes in visual response properties of cat retinal ganglion cells within two weeks after axotomy. Exp Neurol 2002; 177:171-182.
74. Enroth-Cugell C, Robson JG. The contrast sensitivity of ganglion cells of the cat. J Physiol 1966; 187:517-552.
75. Watanabe M, Tokita Y, Kato M et al. Intravitreal injections of neurotrophic factors and forskolin enhance survival and axonal regeneration of axotomized beta ganglion cells in cat retina. Neuroscience 2003; 116:733-742.
76. Miyoshi T, Takao M, Watanabe M et al. Intraocular injections of neurotrophic factors rescue retinal ganglion cells from receptive field shrinkage after axotomy in adult cats. Inv Ophthalmol Vis Sci 2001; 42:S678.
77. Zhang LI, Poo M-M. Electrical activity and development of neural circuits. Nat Neurosci: 2001; 4 Suppl:1207-1214.
78. Penn AA, Riquelme PA, Feller MB et al. Competition in retinogeniculate patterning driven by spontaneous activity. Science. 1998; 279:2108-2112.
79. Weliky M, Katz LC. Correlational structure of spontaneous neuronal activity in the developing lateral geniculate nucleus in vivo. Science 1999; 285:599-604.
80. Yuste R, Bonhoeffer T. Morphological changes in dendritic spines associated with long-term synaptic plasticity. Annu Rev Neurosci 2001; 24:1071-1089.
81. Watanabe M, Inukai N, Fukuda Y. Environmental light enhances survival and axonal regeneration of axotomized retinal ganglion cells in adult cats. Exp Neurol 1999; 160:133-141.

82. Aigner LJ, Singel SA, Rasminsky M et al. Light exposure enhances the survival of adult axotomized retinal ganglion cells. Society for Neuroscience Abstr 1996; 305.
83. Morimoto T, Miyoshi T, Fujikado T et al. Electrical stimulation enhances the survival of axotomized retinal ganglion cells in vivo. Neuroreport 2002; 13:227-230.
84. Pham TA, Rubenstein JL, Silva AJ et al. The CRE/CREB pathway is transiently expressed in thalamic circuit development and contributes to refinement of retinogeniculate axons. Neuron 2001; 31:409-420.
85. Coffee PJ, Lund RD, Rawlins JN. Retinal transplant-mediated learning in the conditioned suppression task in rats. Proc Natl Acad Sci USA 1989; 86:7248-7249.
86. Brasted PJ, Watts C, Robbins TW et al. Associative plasticity in striatal transplants. Proc Natl Acad Sci USA 1999; 96:10524-10529.
87. Dobrossy MD, Dunnett SB. The influence of environment and experience on neural grafts. Nat Rev Neurosci. 2001; 2:871-9.
88. Yin Y, Cui Q, Li Y et al. Macrophage-derived factors stimulate optic nerve regeneration. J Neurosci 2003; 23:2284-2293.

Chapter 9

Brain Repair:

Experimental Treatment Strategies, Neuroprotective and Repair Strategies in the Lesioned Adult CNS

Mathias Bähr and Paul Lingor

Optic Nerve Transsection as Exemplary Model for CNS Lesions

The Retino-Tectal System

A number of functional and structural features make the retino-tectal projection an excellent model system for the study of neurodegeneration and protective strategies in the CNS. The knowledge about developmental procecces yields important information in order to better understand degenerative processes and the point of action of neuroprotective agents in this system: During ontogenesis ventro-lateral parts of the prosencephalic vesicle grow out to form the eye bud which stays attached to the forebrain by the optic stalk. Neuroblastic precursor cells located in the inner layer of the eye bud differentiate into retinal ganglion cells (RGCs) which project their axons to parts of the midbrain. Consequently they replace the optic stalk and create the optic nerve. It bears the only axons projecting from the retina and—as part of the CNS—is myelinated by oligodendrocytes.

In the human visual system, optic afferents project to the dorsal lateral geniculate nucleus (dLGN) and then to the primary visual cortex. In opposite, in the rat (one of the most common animal models studied) RGCs send the majority of their axons to the superior colliculus (SC) in the tectum, and only about 30% of them have collaterals to the lateral geniculate nucleus (LGN) in the thalamus.[1-4]

Thus, the retina and it's projections can be considered as an externalized central nucleus with all properties of the CNS. At the same time, due to it's anatomical localization the retino-tectal system has the advantage of being easily accessible for experimental procedures and evaluation.

Experimental Axotomy

The transection of the optic nerve (ON) in the rat is a classical lesion paradigm taking advantage of the characteristics of the retino-tectal system and is widely used in the study of neuronal cell death in the CNS. Since RGCs are the only neurons projecting via the optic nerve, it's lesioning leads to selective damage in this particular neuron population.

In comparison to other lesion paradigms, optic nerve transection can be performed reproducibly by a small surgical procedure and in a relatively short period of time: After

Brain Repair, edited by Mathias Bähr.

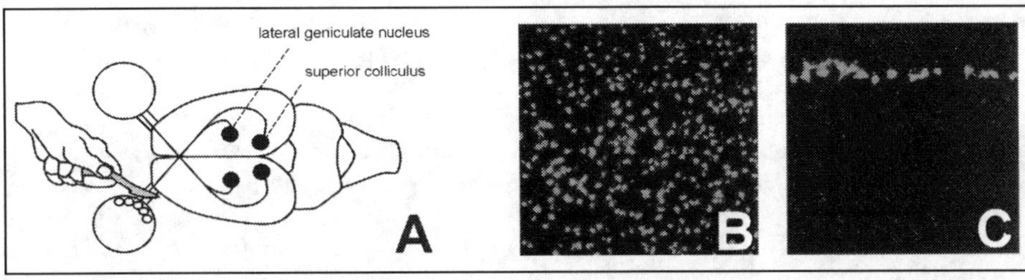

Figure 1. A) Model of the optic nerve transection approach. RGCs are labelled by injection of dye into the superior colliculus or at the nerve stump directly. B and C) Flatmount and section of retina displaying labelled RGCs.

narcotization of the animal and opening of the orbita, the optic nerve is transected close to the posterior eye pole. In order to later visualize the cell population of interest, different markers are used: fluorescent fast blue, fluorogold or Di-I can be applied directly at the ON stump after axotomy. Markers are then transported retrogradely and selectively label RGCs. In study setups requiring unlesioned animals as controls, RGCs alternatively can be labelled prior to lesioning by injection of Di-I into the superior colliculus.

Due to it's close spatial relationship to the inner retinal layers, the vitreous space of the posterior chamber serves as a convenient reservoir for the injection of drugs which shall reach the RGCs. Nevertheless, the use of this natural space sometimes poses problems in the determination of the biological half-life of applied substances and their concentration at the final destination. Factors like diffusion behaviour, interactions due to solvents and light-sensitivity of drugs may play important roles in the overall drug kinetics, and can be taken in account only approximatively.

Usually animals are sacrificed about 2h to 14 days after axotomy, because most of the cell death induced by the lesion will have taken place by then. Retinas then can easily be extracted and prepared for further examination. For RGC survival studies, the retina is flattened by radial incisions and mounted for further microscopical analysis. The selective retrograde prelabelling allows counting RGCs without additional histochemical stainings: with the exception of some endothelial or microglial cells, which can be distinguished by their morphology, only RGCs show characteristic fluorescence (see Fig. 1).

Alternatively the tissue can be processed for additional immunohistochemistry, protein expression analysis, RNA extraction or biochemical tests.[5]

Post-Traumatic and Ontogenetic Cell Death in the Retino-Tectal System

Cell Death after Axotomy

Following transection of the ON a selective degeneration of RGCs can be witnessed. The number of viable ganglion cells initially stays stable for about 4 days post-lesioning but then continuously starts to decline. Nearly 50% of RGCs are dead by day 7 and the majority (80-90%) of this population perish 14 days after axotomy. It is to note that the number of surviving RGCs increases with the distance of the lesion from the eye pole indicating a putative trophic support from the remaining axon. Studies looking at longer time frames revealed that the initial rapid phase of cell death is followed by a second one peaking 6 months after lesion.

Dying RGCs show characteristic morphological changes induced by axotomy with hallmarks such as a shrinkage of the cytoplasm, fragmentation of the nucleus and a degradation of cell organelles. Other than in necrotic cell death which is accompanied by swelling and hydrops of the cell soma, local inflammation and scar formation, post-axotomy cell death takes place in a well defined sequence and time range and does not lead to inflammation responses in surrounding tissue. In fact, there are numerous physiological and pathological conditions, when dying cells show similar morphological changes. While studying different types of cell death in cancer tissue, Kerr and colleagues already 1972 proposed to distinguish this morphologically very different cell death from necrosis and thus coined the term apoptosis.[6] As has been learned since, these morphologically visible changes are only the last steps in a complex process and are preceded by a cascade of protein interactions, which shall be reviewed later.

Cell Death during Ontogenesis

Curiously, post-traumatic apoptotic cell death in it's morphology is very similar to programmed cell death (PCD) during embryogenesis. In rats about 50% of RGCs die very soon after their axons reach their final targets in the superior colliculus (SC) and lateral geniculate nucleus (LGN), within the first postnatal week.[7] This, however, represents already the second wave of cell death during the development of the retinal projection, since many more of the RGC precursor cells die well before reaching the target region, in the stage of neurogenesis, migration and initial axon outgrowth. About 90% of all originally generated RGC precursor cells are eliminated during this first wave of developmental cell death.[8] The reason for this initial precursor overproduction followed by massive demise is not yet completely understood. Hypotheses about the function of PCD during ontogenesis include an adjustment of neuron numbers to neuronal and nonneuronal targets, the elimination of neurons with aberrant connections or the removal of cells which have only transitory functions during ontogenesis.[9]

More interestingly, these embryonic neurons are not killed by external stimuli, but internally launch a cascade of active enzymatic self-destruction, which, as we know now, is very similar in it's regulation to cell death in RGCs damaged by axotomy and employs similar pathways. The elucidation of mechanisms of neuronal cell death during embryogenesis thus yields important clues for the understanding of processes in experimental neuronal degeneration and finally in the course of neurodegenerative diseases, where apoptosis also seems to play an important role in cell destruction.

Key Players in Apoptotic Cell Death: Possible Targets for Therapeutic Strategies

Immediate-Early Genes (c-Fos/c-Jun)

One of the earliest events in RGCs determined to die following axotomy is an alteration in the expression pattern of so-called immediate-early genes (IEGs), such as *c-Fos* and *c-Jun* and it has thus been proposed, that IEGs act as important players in the apoptotic cascade. The products of these rapidly expressed genes have the ability to hetero- or homodimerize. C-Fos/c-Jun heterodimers lead to the formation of the transcription factor AP-1 (activator protein-1) which can bind to different promotor sites on the DNA and thus enhance the transcription of genes that encode growth promoting proteins. After phosphorylation by c-Jun N-terminal kinases (JNKs) the transcriptional activity of c-jun is enhanced.[10]

In RGCs Jun expression is not detectable under basal circumstances, but is rapidly induced following optic nerve transection. Expression of c-Jun is dependent on the lesion site: the shorter the remaining nerve stump the faster and longer an expression of c-Jun was observed and, as mentioned before, the number of surviving RGCs increased with the lenghth of the remaining axon.[11] In this paradigm, c-Jun levels thus seem either to indicate or to promote decreased survival in the RGC population. In sympathetic neurons microinjection of specific antibodies against c-Jun prevented apoptosis[12] and recent results from axotomy of the nigrostriatal pathway also favor a deleterious c-Jun function, since apoptotic neuronal death was attenuated by expression of dominant negative c-Jun.[13]

Results from optic nerve crush experiments, a lesion model where not all axons remain disconnected, however demonstrate the ambivalence of Jun signaling: strong Jun-immunoreactivity after the crush is seen in both disconnected as well as connected RGCs. Interestingly, only connected cells showed a strong coexpression of c-Jun and activating transcription factor 2 (ATF-2), whereas completely axotomized cells (commited to death) show a suppression of basal ATF-2 in addition to high c-Jun levels.[14] Combined c-Jun and ATF-2 expression therefore is associated with cell survival after lesioning, but c-Jun expression alone precedes apoptosis.

Experimental evidence from c-jun and c-fos null mutant mice further questions the role of these IEGs in neuronal apoptosis: other than could be expected, programmed cell death was observed in embryos lacking both c-fos and c-jun. It was thus postulated that, although frequently observed, IEG upregulation is only a secondary epiphenomenon in apoptotic cell death. Alternatively, the role of c-fos and c-jun in the knock-out animals could have been substituted by other proteins or even alternative pathways, which seems to be the more probable hypothesis.[15] The significance of c-Jun signaling in neuronal apoptosis therefore remains to be further clarified.

Caspases and Bcl-2 Family Members

The nematode *C.elegans* was one of the first species where programmed cell death (PCD) was the subject of intensive studies. The development of the worm follows a strict scheme leading to the production of 1090 somatic cells. Among those, 131 cells undergo apoptotic cell death, mainly representing cells of ectodermal origin, such as neurons and neuron-associated cells.[16] The study of mutations resulting in changes within the cell death pattern allowed the identification of several cell death related genes – the *Ced* gene family (cell death abnormal). *Ced-3* and *Ced-4* turned out to be essential for programmed cell death, whereas *Ced-9* acted in a protective manner enhancing cell survival. After the cloning of mammalian *Ced* homologues it became apparent that these genes are highly conserved between species and assume similar functions in the cascade of apoptotic cell death.

In mammals, the protease caspase 3 shows the closest homology to Ced-3. Caspase 3 and other caspases are members of a cysteinyl protease family, which share the common feature of cleaving substrates at aspartate residues. They reside as inactive forms in the cytoplasm and can be activated by proteolytic cleavage. Caspases 3, 6 and 7 make up the so-called "effector" caspases and stand at the end of the apoptotic pathway. After activation by "initiator" caspases 8, 9, 10 and 12, they are responsible for cleavage of cytoskeletal or nuclear proteins and thus degradation of cellular structures (see Fig. 2).

Knock-out animals for caspases underscore the significance of this protease family in developmental neuron death: Caspase 3 deficient mice show hyperplasias and disorganization in the developing brain, finally leading to premature death during the first postnatal weeks.[17] Very similarly, caspase 9 knockout mice show cerebral malformations due to brain enlargement caused by apoptosis inhibition. Both proteases are closely connected within the apoptotic cascade, since activated caspase 9 is able to cleave and thus activate pro-caspase 3, the inactive form of the zymogen, which explains the similarity of both knockout phenotypes. Supporting the hypothesis that optic nerve transection initiates apoptotic cell death, an activation of caspase 3 and caspase 9 could be demonstrated in axotomized RGCs.[18,19] Animals deficient for other members of the caspase family have less impressive neuronal phenotypes, but show dysfunctions in the execution of the cell death cascade in other cell types. For example, caspase $8^{-/-}$ mutants have an impaired development of the heart and die in utero. Caspase 8, functioning as an initiator caspase and standing at the very top of the apoptotic cascade, is activated by so-called

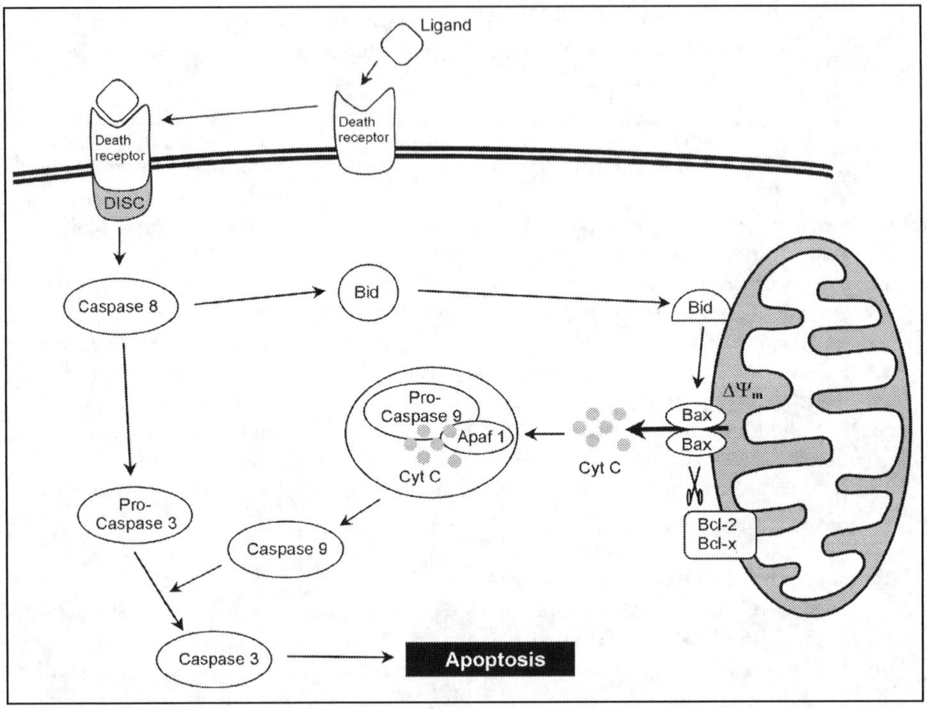

Figure 2. Apoptotic pathway diagram displaying the key players mentioned. After ligand binding to the death receptor, caspase 8 participates in initiating two main ways of cell death. The mitochondrium associated cascade is activated by caspase 8 induced Bid cleavage, followed Bax oligomerization and resulting in cytochrome C release. Cytochrome C plays an active role in apoptosome formation and activation of effector caspases. Caspase 8, however, is also able to directly activate caspase 3 omitting the mitochondrium associated cascade.

death receptors. At least six of these transmembrane receptors are known so far (e.g., TNF-receptor, CD95/Fas/Apo1, TRAIL-receptors). Their natural ligands are members of the TNF family, but also neurotrophic factors, such as NGF.[20] Because of their ability to induce cellular death upon ligand binding, they were attributed the name "death receptor". Fibroblasts derived from caspase $8^{-/-}$ mutants did not show any apoptotic response after death receptor stimulation while other signaling pathways were not affected.[21] Only recently an involvement of caspase 8 signaling in the course of neuronal cell death could be established in a lesion model of focal cerebral ischemia.[22]

Our current knowledge suggests that there are two main pathways which are capable to activate effector caspases: The so-called extrinsic pathway is activated by ligand binding to the death receptor family receptors (e.g., CD95). Upon binding signaling proteins associate with the receptor, which itself does not contain any enzymatic activity, and form the death inducing signaling complex (DISC). The DISC then converts cytosolic pro-caspase 8 into active caspase 8 subunits which then directly activate effector caspases by proteolysis. In fact, caspase 8 was first discovered when associated proteins of the CD95 DISC were analyzed and therefore has been known as FLICE (FADD-like ICE).[23]

Alternatively, caspase 8 activates the intrinsic apoptotic pathway by cleaving Bid. The intrinsic pathway or death receptor-independent pathway is regulated by proteins of the Bcl-2 family and is closely associated with mitochondrial function, thus also termed mitochondrial pathway.[24]

The highest resemblance to *Ced-4* in mammals is found in the apoptosis protease-activated factor 1 (Apaf 1) and it's knockout leads to a phenotype which mimicks those of caspases 3 and 9. Via it's caspase-recruitment domain (CARD) Apaf 1 binds to caspase 9 proenzyme (Apaf 3) and together with cytochrome c (Apaf 2) released from mitochondria forms the apoptosome. The apoptosome then allows proteolytic cleavage of pro-caspase 9 which in turn leads to activation of downstream effector caspases.

The third major member of the *Ced* gene family—*Ced-9*—relates to about 15 known mammalian homologues—the *Bcl-2* gene family.[25,26] Whereas *Ced-9* promotes cell survival in *C.elegans*, members of the Bcl-2 family in mammals have either pro- or anti-apoptotic properties. Structurally, Bcl-2 family members all share at least one Bcl-2 homology (BH) domain and some of them are able to homodimerize (i.e., Bcl-2/Bcl-2) or heterodimerize (i.e., Bcl-x_S/Bcl-2 or Bcl-x_S/Bcl-x_L).[27] They can translocate to intracellular membranes and are capable of pore formation.[28] It has been further shown that Bcl-2 family members regulate the opening of the permeability transition pore (PTP). Very likely, regulation of the PTP is crucial for the maintenance of mitochondrial integrity by regulation of the mitochondrial transmembrane potential ($\Delta\Psi_m$), as well as for downstream pro-apoptotic signaling by cytochrome c release from the mitochondrial intermembrane space.[29,30]

Amongst anti-apoptotic members we find Bcl-2, Bcl-w, Bcl-x_L, Mcl-1 and A1. The prototype of this group, Bcl-2 (which was first discovered as oncogene in human B-cell lymphomas), is membrane-bound and locates on the cytoplasmic side of the outer mitochondrial membrane, the endoplasmatic reticulum and the nuclear envelope. Members of this subfamily all share at least a BH1 and BH2 domain. Bcl-2 as well as Bcl-x_L prevent the release of cytochrome c from the mitochondrium thus inhibiting apoptosome formation.[31] Transgenic mice deficient for Bcl-2 survive the embryonic development, but later show an increased cell death in different tissues, including neurons. Therefore Bcl-2 seems to play a survival promoting role in adult neurons rather than in embryogenesis.[32] In opposite, Bcl-x_L knockouts exhibit a massive death of immature neurons and die in utero.[33]

Apoptosis promoting proteins can be subdivided into multidomain and BH3 domain-only proteins.

Bax is the prototype of the multidomain subfamily, which also includes Bak and Bok. It is able to oligomerize and forms pores in the outer mitochondrial membrane. This results in the release of mitochondrial proteins, such as cytochrome c, from the intermembrane space ultimatively leading to effector caspase activation and cell death.[34,35] Bax expression is regulated by p53, a sequence specific transcriptional activator, which in turn is activated by cellular stresses, such as growth factor deprivation or DNA damage. The role of Bax in developmental programmed cell death is demonstrated by knockout mice deficient for Bax: the numbers of surviving neuronal cells, including RGCs, as well as lymphoid cells is significantly increased. In adult RGCs Bcl-2, Bcl-x and Bax are expressed only at a basal level. Optic nerve transection, however, leads to loss of Bcl-2 and Bcl-x expression and an increase of pro-apoptotic Bax expression. Even more, application of Bax antisense oligonucleotides in this lesion paradigm resulted in decreased apoptosis.[36]

At least 10 BH3-only proteins are known in mammals, including Bad, Bik/Nbk, Bid, Bim/Bod, Bmf, Noxa and PUMA/Bbc3, while in *C. elegans* only one member of this group—EGL-1—could be identified. The nine amino acid BH3 domain is required for binding to Bcl-2-like survival promoting proteins. Bid plays a key role in the apoptotic cascade and unlike several other Bcl-2 family members does not act as a poreforming protein itself. Caspase 8 activates Bid in response to death receptor signaling and which is then translocated to the mitochondrial membrane were it is involved in the induction of Bax and Bak oligomerization.[37,38]

An activation of Bid preceding caspase 3 activation could be shown in neurons in vitro and in vivo, while cytochrome c release was attenuated in Bid$^{-/-}$ mice suggesting a role of Bid upstream of mitochondria.

Even though the above presented molecules and pathways play evident and established roles in apoptotic cell death, their intersection with survival promoting pathways may lead to unexpected results depending on the lesion paradigm. As would be expected, TNF-alpha should induce neuronal apoptosis via the death receptor signaling cascade. However, application of TNF-alpha to the vitreous chamber resulted rather in an increased survival of RGCs following axotomy.[39] An explication for this unexpected survival increase could be found in a TNF-mediated activation of IκBα and NF-κB followed by induction of IAP, an inhibitor of caspases 3,7 and 9. Thus, depending on the model used, identical cytokines can promote cell survival or cell death exploiting different signaling pathways.

Calcium and Excitatory Amino Acids

Calcium homeostasis is crucial in regulation of processes involved in the apoptotic cascade. Elevated levels of cytoplasmic calcium proved to be detrimental to cell survival and an influx of calcium can be observed in numerous lesion paradigms, such as trauma, ischemia or degenerative diseases and can be induced by exposure of cells to glutamate and other excitatory amino acids.[40] Binding of glutamate to it's ionotropic or metabotropic receptors (iGluRs or mGluRs) eventually leads to excessive influx of extracellular calcium and can further release calcium from intracellular stores. Excess cytoplasmic calcium enters cell organelles, e.g., mitochondria and nuclei, and leads to changes in phosphorylation states of proteins which in turn modulates signal transduction pathways or gene transcription and eventually may result in apoptotic cell death.[41]

For the retinal paradigm time-lapse studies with explanted retinas of new-born rat pups showed that a rise in intracellular calcium was closely associated with the death of the observed cell.[42]

Therapeutic Agents

Caspase Inhibitors

The enzymatic action of activated caspases can be blocked by synthetic tri- or tetrapeptide caspase inhibitors, like DEVD-fmk (specific for caspase 3), zLEHD-fmk (specific for caspase 9) and zIETD-fmk (specific for caspase 8). Certain viruses have developed evolutionary strategies to escape elimination by apoptosis of the infected cells by expression of anti-apoptotic proteins, so-called inhibitor of apoptosis proteins (IAPs). Neuronal apoptosis inhibiting protein (NAIP) was the first human IAP to be identified. Further members of the family are human IAP1 (HIAP1), human IAP2 (HIAP2), the X-chromosome linked IAP (XIAP) and survivin. CrmA (isolated from cowpox virus) and p35 (from baculovirus) are also able to potently inhibit caspases.

In the optic nerve transection model peptide inhibitors of apoptosis successfully were able to prevent RGC death after intraocular application.[43] This effect was especially pronounced when looking at rescued RGC numbers at early time points. However, this approach yields only a temporary delay of cell death rather than a definite protection as could be shown by studies looking at longer time frames (4 weeks) after axotomy. Direct caspase inhibition blocks the apoptotic cascade at a rather downstream level, therefore inhibiting the final execution step in the commited cell. Very likely, other—slower—pathways equally play a role in cell death induction thus eventually leading to cell death even after caspase inhibition.[44] Therefore, specific peptide caspase inhibitors are more likely to be of experimental than therapeutic value.

Neurotrophic Factors

Neurotrophic factors are proteins which belong to several structurally different molecule families and play an outstanding role in the regulation of ontogenetic PCD: In the peripheral nervous system, developing neurons generated in surplus compete for a limited amount of these molecules and neurons with deficient supply in trophic factors eventually die (so-called neurotrophin hypothesis). This effect was shown first in pioneering experiments by Levi-Montalcini, Hamburger and Cohen in the 1950s when it became clear that a component of snake venom was able to promote the neurite growth of chick sympathetic neurons - the molecule was the first neurotrophin isolated and became known as nerve growth factor (NGF). While certain neuron populations in the PNS are dependent on one specific growth factor, survival of CNS neurons requires a variety of growth factors (so-called multifactorial hypothesis).[45]

The group of neurotrophins consists of 5 molecules known to date (NGF, BDNF, NT-3, NT-4/5, NT-6), which play different roles in the development of the retino-tectal system controlling numbers of RGCs and having effects on their functional differentiation.

Neurotrophins use two kinds of receptor molecules for signaling – the high affinity receptors Trk A, B and C which are members of the tyrosin kinase family and the universal low affinity receptor $p75^{NGFR}$, which was the first neurotrophin receptor described.[46] Both receptor types can form homodimers Trk/Trk or p75/p75 as well as heterodimers Trk/p75.

Each neurotrophin preferrably binds and signals through one of the high affinity receptors, but is also able to bind to low affinity receptor dimers. Neurotrophin signaling is therefore dependent on the receptor type utilized, which is examplified best by NGF signaling: whereas NGF binding to TrkA/TrkA or TrkA/p75 is able to promote cell survival, p75 dimer signaling induces apoptosis in neuron-like cell lines.[47] In the embryonic retina, cell death is enhanced by NGF binding to p75, which could be effectively prevented by application of antibodies to NGF or p75.[48] Similarly, knock out mice lacking either NGF or p75 show a decreased embryonic PCD. One believes that apoptosis induction by NGF-p75 interaction follows the same intracellular pathway as the death receptor signaling, since p75 belongs the TNF receptor/Fas/CD40 superfamily.

Brain derived neurotrophic factor (BDNF), originally derived from pig brain, exerts trophic effects on RGCs.[49] Since it has been shown to be expressed during embryogenesis as well as in the SC of the adult, BDNF seems to have functions even beyond the developmental period. In vitro it is able to increase the survival of dissociated RGCs and in vivo exogenous BDNF applied to the developing SC reduces ontogenetic RGC death and promotes RGC axon arborization.[50] Thus, a survival promoting action of BDNF was presumed for models of CNS trauma. Mey and Thanos showed that axotomized RGCs are indeed rescued by intraocular BDNF administration in vivo.[51] Signaling of BDNF occurs in part through binding to it's high affinity receptor TrkB and consecutive activation of the phosphatidylinositide-3'-OH kinase (PI3K). PI3K in turn activates protein kinase B (PKB/Akt) which can exert numerous downstream effects, including survival promoting actions by it's interference with apoptotic pathways: PKB/Akt phosphorylates and inactivates caspase 9 thus blocking effector caspases. It is also able to phosphorylate pro-apoptotic Bad, which is then sequestered by 14-3-3 proteins in the cytosol thus counteracting cytochrome c release and cell death.

In the chick retina, application of antibodies to NT-3 dramatically reduces numbers of surviving RGCs at embryonic day 6 already. However, a diminished supply in NT-3 does not directly lead to PCD. It rather seems that the number of division cycles of RGC precursor cells is increased, while generated RGCs are dependent on NT-3 and thus subsequently die due to lack of trophic support. Interestingly, the knock out of NT-3 did not show any alterations in RGC numbers, which could be due to intrinsic compensatory mechanisms following gene ablation.

Very similarly, neurotrophin 4/5 (NT-4/5) decreases the PCD in developing RGCs and also is able to promote survival of RGCs following axotomy.[52]

As is true for members of the neurotrophin family IGF-I equally promotes survival of RGCs in the axotomy model apparently by decreasing caspase-3 activity. This effect can be blocked effectively by wortmannin (an inhibitor of PI3K). Thus, upon binding to it's receptor IGF-I also seems to utilize the PI3K/Akt pathway.[53] Several other factors (e.g., b-FGF, CNTF, GDNF) also proved efficient in prevention of degeneration of injured RGCs in the ONT model.[54]

Calcium Inhibiting Drugs

Based on the idea that excess calcium influx fosters cell death, pharmacological approaches focused on inhibiting glutamate release, antagonizing the NMDA receptors or blocking the calcium channels themselves. Several inhibitors of glutamate release, such as adenosine, BW619C89 (a lamotrigin derivative) and lubeluzole, have been tested experimentally and in clinical trials. But even lubeluzole, a benzothiazole derivative, which showed promising results among patients with mild to moderate ischaemic stroke was not able to keep up with it's expectations as has been shown by meta-analyses.[55]

Amongst NMDA receptor antagonists the noncompetitive antagonist dizlocipine (MK-801) showed positive effects on infarct size in rodent ischemia models, but lead to controversial results in retinal lesion paradigms.[56] Other clinical studies on NMDA receptor antagonists (such as Aptiganel/Cerestat/CNS 1102, selfotel, dextrorphan or eliprodil) had to be discontinued mostly due to side effects such as hypertension and psychotic episodes.

Nimodipine is the most widely evaluated calcium channel antagonist which acts on L-type calcium channels. It has been tested in different ischemic paradigms, but the latest data available does not support the hypothesis of beneficial effects in stroke patients. Hypotension has been mentioned in several study groups upon nimodipine administration and might have lead to unfavorable results compared to placebo.[57]

Ways to Enter the Brain

Blood-Brain Barrier

Access to the central nervous system is limited due to a histological boundary between the systemic circulation and the brain parenchyma. Tight junctions between adjacent endothelial cells of the capillary walls as well as astrocytes form a highly selective blood-brain barrier and effectively prevent the transfer of numerous substances into the CNS. Whether a molecule can pass beyond this limit depends on it's electric charge, conformation and most importantly it's molecular weight, with an observed bioavailability limit of about 600 Da. This particular anatomical situation of the CNS is both friend and foe when it comes to experimental or therapeutical approaches: Penetration of certain systemically administered drugs is prevented and can be used therapeutically, for example in the control of brain edema. However, the availability of substances specifically targeting the CNS is equally restricted.

Several approaches have been undertaken in order to overcome the blood-brain barrier by transient permeabilization:[58] osmotic agents (like mannitol) or mediators of inflammation (like bradykinin) can increase it's permeability. The selective bradykinin receptor 2 agonist Labradimil (Cereport; also formerly referred to as RMP-7), might play a role in the future treatment of gliomas or metastatic CNS tumors and is currently tested in preliminary clinical trials.[59] This approach might be usefull for the application of chemotherapeutical drugs, but is not able to induce long term permeabilization and additionally may induce unwanted side effects.

Besides their inability to overcome the blood-brain barrier, numerous molecules of interest have a short half-life, which would require a continous administration and thus not negligible technical effort. Finally, systemic drug application is only partially able to target specific cell populations, such as neurons.

Transplantation Approaches

In the search for local therapeutic strategies, transplantation of autologous or donor tissue to the lesion site assuming the function of a degenerated cell population seems to be a promising approach. Parkinson's disease was one of the first neurodegenerative conditions where tissue grafts were to be evaluated: it shows a characteristic degeneration of one neuronal subtype, dopaminergic neurons of the substantia nigra, thus calling for a targeted stereotactical restoration of dopamine producing cells.

Chromaffin cells from the patient's own adrenal medulla, also capable of dopamine synthesis, seemed to be a convenient graft source, but most studies showed only minor clinical improvements, possibly due to limited graft survival.[60] Fetal ventral mesencephalic grafts showed significantly better results in animal models. However, the use of embryonic tissue, even though promising concerning clinical outcome, raises ethical concerns about the therapeutic use of fetuses. Additionally it requires a substantial amount of tissue – about 3-10 foetuses per graft, depending on different authors. Here, embryonic stem cells could open a new direction towards a less limited supply of transplanted tissue.[61] Inspite of these perspectives, the development of vectors which are able to introduce proteins or genes into the living brain and ommiting transplantational approaches has as much of a priority as the search for therapeutic agents itself.

Gene Transfer by Viral Vectors

Viral vectors have proven to be an extremely efficient way of gene transfer in vitro and in experimental animal models in vivo. The virus particle enters the cell either by binding to cell surface receptors and subsequent endocytosis or via nonreceptor mediated endocytosis. Upon entrance the capsid is disassembled and the genetic material, either DNA or RNA, is transported to it's final destination in the nucleus or cytoplasm. Some viruses, e.g., retroviruses, introduce their genome into the host genome and may persist life-long in the host cell, whereas others, as the adenovirus, stay episomally. The viral genome is then transcribed and the protein of interest is expressed by the cell's own translational machinery (see Fig. 3).

In order to insert genes into the viral genome, certain viral genes have to be deleted. Thus, the cloning capacity in recombinant viruses—depending on their original genome size—is limited. Gene constructs therefore have to be designed as so-called mini-genes, which is usually cDNA reversely transcribed from mRNA lacking the intron sequences. Additionally, promoter sequences and regulatory sequences can be introduced, which is crucial in order to achieve a cell type specific expression pattern or for external regulation of gene expression. Promoters are able to influence kinetics, quantities and cell type specificity of expression: For example, the human cytomegalovirus (hCMV) promoter leads to fast and strong expression in most cell types, whereas the tubulin α1 (Tα1) promoter has a rather protracted expression start. Specific subpopulations of cells can be targeted by use of cell type specific promoters, such as the synapsin (SYN) promoter for neuronal expression.[62] By inserting a tetracyclin (tet)-operator-sequence in front of a promoter, expression of proteins can be regulated depending on the presence of the antibiotic, either as tet-on or tet-off system.[63] Finally, internal ribosome entry sites (IRES) —sequences where translation can be initiated outside a regular start codon—permit an expression of multiple gene products by one viral vector.[64]

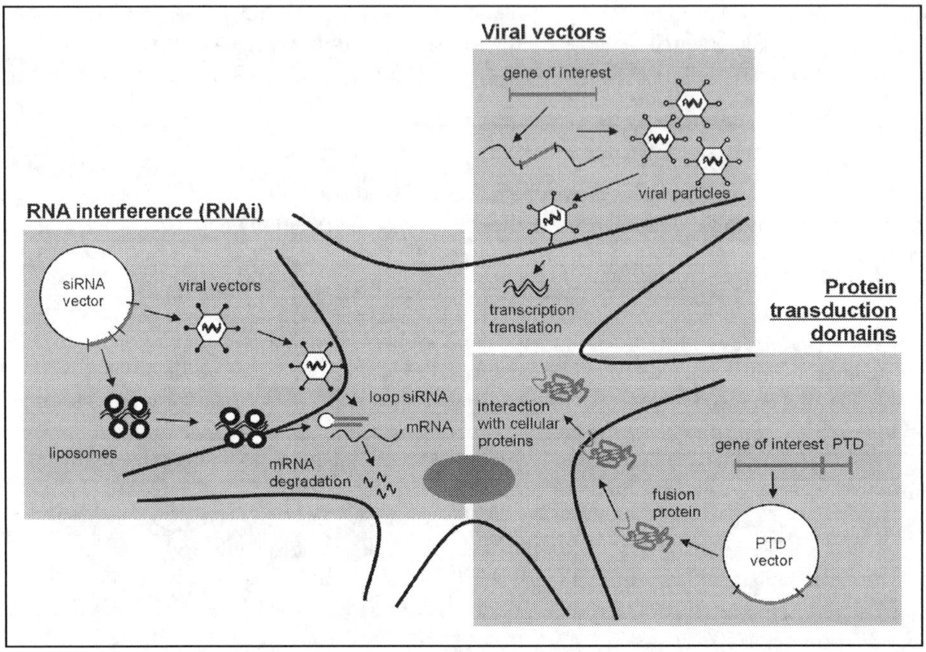

Figure 3. How to enter the brain: viral vectors are versatile tools for introduction of genetic material into neuronal cells in order to overexpress genes of interest. Protein transduction domains use their "shuttle" function and thus allow to enter desired proteins directly into the cell. By introduction of double stranded small interferring RNA mRNAs of interest are degraded thus producing a virtual knock out of a gene.

One of the most widely used viral vector systems is recombinant adenovirus, with most vectors based on the numan Ad5 genome. The viral particle is built of an icosaedric capsid and harbors a double-stranded DNA-genome with a size of 36kb. In the course of normal infection a battery of early genes (E1 through E4) are expressed prior to DNA replication. Deletion of the E1 gene, which is required for the subsequent expression of all other genes, eventuates in the generation of replication defective virions. Since safety considerations regarding the handling of viruses play a pivotal role, the E1 gene deletion is essential for the use of adenovirus as an experimental vector system. Replication and thus virus formation can thus be performed only in cell lines substituting the deleted E1 gene product, as do HEK 293 cells. Other genes, like E3, convey virus persistance in vivo and thus are not necessarily required for virus growth in vitro. Gene regions like E1 and E3 can therefore be employed as cloning regions for the insertion of mini-genes of interest.[65]

Adenoviral constructs have been extensively used for transduction of RGCs in the axotomy model in vivo and in cell culture studies in vitro. For example, overexpression of trophic factors after intravitreal injection of adenoviral particles coding for GDNF, CNTF or BDNF resulted in an increased survival of axotomized RGCs.[66] The antiapoptotic proteins p35 and XIAP could be overexpressed in RGCs after application of adenovirus on the optic nerve stump and similarly were able to effectively prevent neuronal apoptosis.[67,68]

Amongst other viruses that are used for gene transfer we find vector systems based on genomes from herpes simplex virus-1 (HSV-1), adeno-associated virus (AAV), Moloney murine leukemia virus (MoMLV) – a retrovirus, or the sindbis/semliki forest virus (SFV). The later represents a convenient system for the evaluation of cellular localization and distribution of desired transgenes in cell culture and in vivo.[69] In contrast to adenoviral vectors, even attenuated forms of the semliki forest virus have a considerable cytotoxicity in primary neuron cultures. However, a quick cloning procedure and several safety features make it a usefull tool when long-term transgene expression is not desired (own unpublished observations).

Introduction of Proteins Using Protein Transduction Domains (PTD)

Although previously described limitations set by the blood-brain barrier apply to most molecules, there seems to be a small group of proteins which seem to ignore this biological obstacle. Initially Green and Frankel independently showed that the HIV-1 TAT protein was able to enter cultured cells when simply added to the culture medium.[70,71] The protein not only entered the cells leaving the cellular membrane undisrupted, but also translocated to the nucleus (it's biological location) and was found to be functionally active. Some other proteins with similar capabilities have been discovered since, for example the *Drosophila* Antennapedia (Antp) homeotic transcription factor[72] and the Herpes-simplex-virus-1 DNA-binding protein VP22 (HSV VP22).[73]

The exact mechanism of cell entry employed by these peptides is not yet known, however it is clear that transduction of TAT, VP22 and Antp is not mediated by receptors, transporters or endocytosis. Most likely the presence of basic amino acids, as arginine and lysine, plays an important role in the interaction of PTD with negatively charged lipids of the cell membrane. TAT, for example, might directly penetrate the lipid bilayer when it's positive charge contacts the negatively charged outer membrane. It then translocates into the cytoplasm driven by it's momentum.

More importantly, not only TAT itself but also covalently attached "cargo"-proteins are equally transduced bypassing previous size limitations. It could be shown that fusion proteins with more than 1000 amino acids, like TAT-ß-gal, could be transduced in vitro and in vivo. After intra-peritoneal injection the active enzyme was found in all assayed tissue types, including the brain, without any indication for damage to the blood-brain barrier.[74]

There are numerous advantages in using protein transduction domains for protein delivery: All eukaryotic cell types tested so far with the exception of yeast (likely because of a less permeable cell wall) are susceptible to transduction. Since transduced proteins enter cells within a short time span (approximately 15 min), timing issues can be addressed unlike in conventional transfection approaches. Also, the concentration of intracellular protein can be precisely controlled with all cells containing approximately the same amount of protein.

Since the known PTDs do not show any selectivity for whatever cell type, one of the crucial questions remains the targeting of specific subpopulations. One possible approach could be the introduction of cleavage sites into the fusion protein which yields an active form of the fused protein only after proteolytic cleavage by cell type specific intracellular enzymes. TAT-caspase-3 with an HIV protease cleavage site, for example, was able to transduce into all cells, but only in cells infected by HIV and thus containing the HIV protease caspase-3 was processed into it's active form and induced apoptosis.[75]

Numerous other pathogens utilizing specific proteases, such as hepatitis C virus or *Plasmodium sp.*, could be targeted employing this so-called "Trojan horse" strategy.

Gene Silencing by RNA Interference (RNAi)

The introduction or overexpression of genes (by viral gene transfer) and functional proteins (by protein transduction domains) in order to interfere with apoptotic death cascades are already well established methods. Similarly, the knock out of certain genes of interest is a usefull tool for the study of gene function and can be accomplished by the creation of knock-out animals. However, the generation of knock-out animals is a laborious task and in some cases the phenotypes have only limited viability, forbidding the analysis of adult tissues. Furthermore knock-outs do not allow for direct regulation of the deleted gene and mechanisms can be induced compensating for the loss of function, which complicates the elucidation of the actual gene function.

Only recently it has been learned that gene function can be effectively silenced by introducing short double stranded RNA fragments into the cell – a process called RNA interference (RNAi). The phenomenon was first seen in plants and could be reproduced in a number of species including nematodes (*C. elegans*), insects (*Drosophila*) and recently even mammalian cells.[76]

When double stranded RNA (dsRNA) enters a cell in a natural context this is most likely caused by viral infection. In order to protect the organism, several pathways are activated in infected cells, such as the dimerization of PKR and phosphorylation of eIF2α resulting in a nonspecific shutdown of translation and the entire cellular metabolism.[77] However, long dsRNA also can be cleaved by the so-called DICER enzyme (an ortholog of ribonuclease III), which results in the generation of small dsRNA fragments measuring about 21-25 bp, so-called siRNAs.[78] Upon binding to the RISC complex (a multicomponent enzyme complex including endo- and exonucleases) siRNAs then target the homologous mRNA which is then cleaved becoming unfunctional.[79] Gene silencing by introduction of short interfering RNAs (siRNAs) virtually leads to knock-out phenotypes since it can silence up to 90% of the gene function at any time desired time point in transducible cell lines.

Currently, RNAi is becoming a more and more widespread instrument for the analysis of gene function. However, it should be only a matter of time that siRNAs will be employed for experimental therapeutic approaches. The construction of vectors coding for siRNAs (such as the pSUPER-Vector) will certainly simplify the use of RNAi making chemical synthesis of siRNAs obsolete.[80] Such vector systems will also allow to be introduced into viral genomes and thus open new possibilities for the transduction of cells which can not be reached by conventional transfection techniques.

RNAi together with gene transfer by viral vectors and protein transduction by PTD are thus tools which are already in wide current use when it comes to regulation of gene or protein function. They will possibly play important roles in therapeutic strategies, even though not all putative targets have yet been identified and unwanted side effects have to be considered when using these methods in the human beeing.

References

1. Lund RD, Land PW, Boles J. Normal and abnormal uncrossed retinotectal pathways in rats: An HRP study in adults. J Comp Neurol 1980; 189(4):711-20.
2. Beckstead RM, Frankfurter A. A direct projection from the retina to the intermediate gray layer of the superior colliculus demonstrated by anterograde transport of horseradish peroxidase in monkey, cat and rat. Exp Brain Res 1983; 52(2):261-8.
3. Dreher B, Sefton AJ, Ni SY et al. The morphology, number, distribution and central projections of Class I retinal ganglion cells in albino and hooded rats. Brain Behav Evol 1985; 26(1):10-48.
4. Cellerino A, Bahr M, Isenmann S. Apoptosis in the developing visual system. Cell Tissue Res Jul 2000; 301(1):53-69.
5. Kermer P, Klocker N, Weishaupt JH et al. Transection of the optic nerve in rats: Studying neuronal death and survival in vivo. Brain Res Brain Res Protoc 2001; 7(3):255-60.
6. Kerr FR, Wyllie AM, Currie AR. Apoptosis: A basic biological phenomenon with wider ranging implications in tissue kinetics. Br J Cancer 1972; 24:239-275.
7. Potts RA, Dreher B, Bennett MR. The loss of ganglion cells in the developing retina of the rat. Brain Res 1982; 255(3):481-6.
8. Galli-Resta L, Ensini M. An intrinsic time limit between genesis and death of individual neurons in the developing retinal ganglion cell layer. J Neurosci 1996; 16(7):2318-24.
9. Burek MJ, Oppenheim RW. Programmed cell death in the developing nervous system. Brain Pathol 1996; 6(4):427-46.

10. Gupta S, Barrett T, Whitmarsh AJ et al. Selective interaction of JNK protein kinase isoforms with transcription factors. EMBO J 1996; 15(11):2760-70.
11. Hüll M, Bähr M. Differential regulation of c-JUN expression in rat retinal ganglion cells after proximal and distal optic nerve transection. Neurosci Lett 1994; 178(1):39-42.
12. Estus S, Zaks WJ, Freeman RS et al. Altered gene expression in neurons during programmed cell death: Identification of c-jun as necessary for neuronal apoptosis. J Cell Biol 1994; 127:1717-27.
13. Crocker SJ, Lamba WR, Smith PD et al. c-Jun mediates axotomy-induced dopamine neuron death in vivo. Proc Natl Acad Sci USA 2001; 98(23):13385-90.
14. Martin-Villalba A, Winter C, Brecht S et al. Rapid and long-lasting suppression of the ATF-2 transcription factor is a common response to neuronal injury. Brain Res Mol Brain Res 1998; 62(2):158-66.
15. Roffler-Tarlov S, Brown JJ, Tarlov E et al. Programmed cell death in the absence of c-Fos and c-Jun. Development 1996; 122(1):1-9.
16. Sulston JE, Horvitz HR. Post-embryonic cell lineages of the nematode, *Caenorhabditis elegans*. Dev Biol 1977; 56(1):110-56.
17. Kuida K, Zheng TS, Na S et al. Decreased apoptosis in the brain and premature lethality in CPP32-deficient mice. Nature Nov 28 1996 ; 384(6607):368-72.
18. Kermer P, Klocker N, Labes M et al. Activation of caspase-3 in axotomized rat retinal ganglion cells in vivo. FEBS Lett 1999; 453(3):361-4.
19. Kermer P, Ankerhold R, Klocker N et al. Caspase-9: Involvement in secondary death of axotomized rat retinal ganglion cells in vivo. Brain Res Mol Brain Res 2000; 85(1-2):144-50.
20. Schulze-Osthoff K, Ferrari D, Los M et al. Apoptosis signaling by death receptors. Eur J Biochem 1998; 254(3):439-59.
21. Varfolomeev EE, Schuchmann M, Luria V et al. Targeted disruption of the mouse Caspase 8 gene ablates cell death induction by the TNF receptors, Fas/Apo1, and DR3 and is lethal prenatally. Immunity 1998; 9(2):267-76.
22. Plesnila N, Zinkel S, Le DA et al. BID mediates neuronal cell death after oxygen/ glucose deprivation and focal cerebral ischemia. Proc Natl Acad Sci USA 2001; 98(26):15318-23.
23. Alnemri ES, Livingston DJ, Nicholson DW et al. Human ICE/CED-3 protease nomenclature. Cell 1996; 87(2):171.
24. Schuler M, Green DR. Mechanisms of p53-dependent apoptosis. Biochem Soc Trans 2001; 29:684-8.
25. Reed JC. Bcl-2 family proteins. Oncogene 1998; 17(25):3225-36.
26. Green DR, Reed JC. Mitochondria and apoptosis. Science 1998; 281(5381):1309-12.
27. Lindenboim L, Borner C, Stein R. Bcl-x(S) can form homodimers and heterodimers and its apoptotic activity requires localization of Bcl-x(S) to the mitochondria and its BH3 and loop domains. Cell Death Differ 2001; 8(9):933-42.
28. Schendel SL, Montal M, Reed JC. Bcl-2 family proteins as ion-channels. Cell Death Differ 1998; 5(5):372-80.
29. Vander Heiden MG, Thompson CB. Bcl-2 proteins: Regulators of apoptosis or of mitochondrial homeostasis? Nat Cell Biol 1999; 1(8):E209-16.
30. Martinou JC. Apoptosis. Key to the mitochondrial gate. Nature 1999; 399(6735):411-2.
31. Newmeyer DD, Bossy-Wetzel E, Kluck RM et al. Bcl-xL does not inhibit the function of Apaf-1. Cell Death Differ 2000; 7(4):402-7.
32. Veis DJ, Sorenson CM, Shutter JR et al. Bcl-2-deficient mice demonstrate fulminant lymphoid apoptosis, polycystic kidneys, and hypopigmented hair. Cell 1993; 75(2):229-40.
33. Motoyama N, Wang F, Roth KA et al. Massive cell death of immature hematopoietic cells and neurons in Bcl-x-deficient mice. Science 1995; 267(5203):1506-10.
34. Brady HJ, Gil-Gomez G. Bax. The pro-apoptotic Bcl-2 family member, Bax. Int J Biochem Cell Biol 1998; 30(6):647-50.
35. Korsmeyer SJ, Wei MC, Saito M et al. Pro-apoptotic cascade activates BID, which oligomerizes BAK or BAX into pores that result in the release of cytochrome c. Cell Death Differ 2000; 7(12):1166-73.
36. Isenmann S, Wahl C, Krajewski S et al. Up-regulation of Bax protein in degenerating retinal ganglion cells precedes apoptotic cell death after optic nerve lesion in the rat. Eur J Neurosci 1997; 9(8):1763-72.

37. Yin XM. Signal transduction mediated by Bid, a pro-death Bcl-2 family proteins, connects the death receptor and mitochondria apoptosis pathways. Cell Res 2000; 10(3):161-7.
38. Antonsson B. Bax and other pro-apoptotic Bcl-2 family "killer-proteins" and their victim the mitochondrion. Cell Tissue Res 2001 Dec; 306(3):347-61.
39. Diem R, Meyer R, Weishaupt JH et al. Reduction of potassium currents and phosphatidylinositol 3-kinase-dependent AKT phosphorylation by tumor necrosis factor-(alpha) rescues axotomized retinal ganglion cells from retrograde cell death in vivo. J Neurosci 2001; 21(6):2058-66.
40. Choi DW. Glutamate neurotoxicity and diseases of the nervous system. Neuron 1988; 1(8):623-634.
41. Ermak G, Davies KJ. Calcium and oxidative stress: From cell signaling to cell death. Mol Immunol 2002; 38(10):713-21.
42. Cellerino A, Galli-Resta L, Colombaioni L. The dynamics of neuronal death: A time-lapse study in the retina. J Neurosci 2000; 20(16):RC92.
43. Chaudhary P, Ahmed F, Quebada P et al. Caspase inhibitors block the retinal ganglion cell death following optic nerve transection. Brain Res Mol Brain Res 1999; 67(1):36-45.
44. Kermer P, Klocker N, Bahr M. Long-term effect of inhibition of ced 3-like caspases on the survival ofaxotomized retinal ganglion cells in vivo. Exp Neurol 1999 Jul; 158(1):202-5.
45. Oppenheim RW. The concept of uptake and retrograde transport of neurotrophic molecules during development: History and present status. Neurochem Res 1996; 21(7):769-77.
46. Johnson D, Lanahan A, Buck CR et al. Expression and structure of the human NGF receptor. Cell 1986; 47(4):545-54.
47. Niederhauser O, Mangold M, Schubenel R et al. NGF ligand alters NGF signaling via p75(NTR) and trkA. J Neurosci Res 2000; 61(3):263-72.
48. Frade JM, Rodriguez-Tebar A, Barde YA. Induction of cell death by endogenous nerve growth factor through its p75 receptor. Nature 1996; 383(6596):166-8.
49. Barde YA, Edgar D, Thoenen H. Purification of a new neurotrophic factor from mammalian brain. EMBO J 1982; 1(5):549-53.
50. Cohen-Cory S, Fraser SE. Effects of brain-derived neurotrophic factor on optic axon branching and remodelling in vivo. Nature 1995; 378(6553):192-6.
51. Mey J, Thanos S. Intravitreal injections of neurotrophic factors support the survival of axotomized retinal ganglion cells in adult rats in vivo. Brain Res 1993; 602(2):304-17.
52. Peinado-Ramon P, Salvador M, Villegas-Perez MP et al. Effects of axotomy and intraocular administration of NT-4, NT-3, and brain-derived neurotrophic factor on the survival of adult rat retinal ganglion cells. A quantitative in vivo study. Invest Ophthalmol Vis Sci 1996; 37(4):489-500.
53. Kermer P, Klöcker N, Labes M et al. Insulin-like growth factor-I protects axotomized rat retinal ganglion cells from secondary death via PI3-K-dependent Akt phosphorylation and inhibition of caspase-3 In vivo. J Neurosci 2000; 20:2-8.
54. Cellerino A, Bahr M, Isenmann S. Apoptosis in the developing visual system. Cell Tissue Res 2000; 301(1):53-69.
55. Gandolfo C, Sandercock P, Conti M. Lubeluzole for acute ischaemic stroke. Cochrane Database Syst Rev 2002; (1):CD001924.
56. Schmitt U, Sabel BA. MK-801 reduces retinal ganglion cell survival but improves visual performance after controlled optic nerve crush. J Neurotrauma 1996; 13(12):791-800.
57. Horn J, de Haan RJ, Vermeulen M et al. Very early nimodipine use in stroke (VENUS): A randomized, double-blind, placebo-controlled trial. Stroke 2001; 32(2):461-5.
58. Mayhan WG. Regulation of blood-brain barrier permeability. Microcirculation 2001; 8(2):89-104.
59. Emerich DF, Dean RL, Osborn C et al. The development of the bradykinin agonist labradimil as a means to increase the permeability of the blood-brain barrier: From concept to clinical evaluation. Clin Pharmacokinet 2001; 40(2):105-23.
60. Kordower JH, Goetz CG, Freeman TB et al. Dopaminergic transplants in patients with Parkinson's disease: Neuroanatomical correlates of clinical recovery. Exp Neurol 1997; 144(1):41-6.
61. Freed CR. Will embryonic stem cells be a useful source of dopamine neurons for transplant into patients with Parkinson's disease? Proc Natl Acad Sci USA 2002; 99(4):1755-7.
62. Kugler S, Meyn L, Holzmuller H et al. Neuron-specific expression of therapeutic proteins: Evaluation of different cellular promoters in recombinant adenoviral vectors. Mol Cell Neurosci 2001; 17(1):78-96.

63. Gossen M, Bujard H. Tight control of gene expression in mammalian cells by tetracycline-responsive promoters. Proc Natl Acad Sci USA 1992; 89(12):5547-51.
64. Hellen CU, Sarnow P. Internal ribosome entry sites in eukaryotic mRNA molecules. Genes Dev 2001; 15(13):1593-612.
65. Hitt MM, Addison CL, Graham FL. Human adenovirus vectors for gene transfer into mammalian cells. Adv Pharmacol 1997; 40:137-206.
66. Schmeer C, Straten G, Kugler S et al. Dose-dependent rescue of axotomized rat retinal ganglion cells by adenovirus-mediated expression of glial cell-line derived neurotrophic factor in vivo. Eur J Neurosci 2002; 15(4):637-43.
67. Kugler S, Klocker N, Kermer P et al. Transduction of axotomized retinal ganglion cells by adenoviral vector administration at the optic nerve stump: An in vivo model system for the inhibition of neuronal apoptotic cell death. Gene Ther 1999; 6(10):1759-67.
68. Kugler S, Straten G, Kreppel F et al. The X-linked inhibitor of apoptosis (XIAP) prevents cell death in axotomized CNS neurons in vivo. Cell Death Differ 2000; 7(9):815-24.
69. Lundstrom K, Schweitzer C, Rotmann D et al. Semliki Forest virus vectors: Efficient vehicles for in vitro and in vivo gene delivery. FEBS Lett 2001; 504(3):99-103.
70. Green M, Loewenstein PM. Autonomous functional domains of chemically synthesized human immunodeficiency virus tat trans-activator protein. Cell 1988; 55(6):1179-88.
71. Frankel AD, Pabo CO. Cellular uptake of the tat protein from human immunodeficiency virus. Cell 1988; 55(6):1189-93.
72. Joliot A, Pernelle C, Deagostini-Bazin H et al. Antennapedia homeobox peptide regulates neural morphogenesis. Proc Natl Acad Sci USA 1991; 88(5):1864-8.
73. Elliott G, O'Hare P. Intercellular trafficking and protein delivery by a herpesvirus structural protein. Cell 1997; 88(2):223-33.
74. Schwarze SR, Ho A, Vocero-Akbani A et al. In vivo protein transduction: Delivery of a biologically active protein into the mouse. Science 1999; 285(5433):1569-72.
75. Vocero-Akbani AM, Heyden NV, Lissy NA et al. Killing HIV-infected cells by transduction with an HIV protease-activated caspase-3 protein. Nat Med 1999; 5(1):29-33.
76. Caplen NJ, Parrish S, Imani F et al. Specific inhibition of gene expression by small double-stranded RNAs in invertebrate and vertebrate systems. Proc Natl Acad Sci USA. 2001; 98(17):9742-7.
77. Clemens MJ, Elia A. The double-stranded RNA-dependent protein kinase PKR: Structure and function. J Interferon Cytokine Res 1997; 17(9):503-24.
78. Nicholson RH, Nicholson AW. Molecular characterization of a mouse cDNA encoding Dicer, a ribonuclease III ortholog involved in RNA interference. Mamm Genome 2002; 13(2):67-73.
79. Caplen NJ. A new approach to the inhibition of gene expression. Trends Biotechnol 2002; 20(2):49-51.
80. Brummelkamp TR, Bernards R, Agami R. A system for stable expression of short interfering RNAs in mammalian cells. Science 2002; 296(5567):550-3.

CHAPTER 10

Neuroprotection by cAMP:

Another Brick in the Wall

Mariana S. Silveira and Rafael Linden

Abstract

Programmed cell death occurs in the nervous system both in normal development as well as in pathologic conditions, and is a key issue related to both brain repair and neurodegenerative diseases. Modulation of cell death in the nervous system may involve neurotrophic factors and other peptides, neurotransmitters and neuromodulators, that activate various signal transduction pathways, which in turn interact with the cell death execution machinery. Here we discuss the role of the second messenger cyclic adenosine 3'5'-monophosphate (cAMP) in cell death, and summarize current evidence that cAMP is a nodal point of neuroprotective signaling pathways.

Programmed Cell Death and Signal Transduction Pathways

Physiological mechanisms of cell death are effective in multicellular organisms both to control the number of cells and to remove seriously damaged, infected or mutated cells.[1] Deregulation of programmed cell death may lead to several pathologies, such as cancer, autoimmune diseases and neurodegenerative disorders.[2-5]

The sensitivity to cell death is also a key issue related to repair of damaged tissues. In the nervous system, this affects the ability to maintain neurons alive until the re-establishment of connections that may provide neurotrophic support.[6] It is also germane to the efforts to use stem cells in order to repopulate depleted areas.[7,8]

Cells in all tissues are constantly exposed both to a wealth of signaling molecules secreted by their neighbors, as well as to components of the extracellular matrix. These signals interact in the multivariate metabolic network that controls the decision to enter the cell death execution pathways.[9] Within the heterogeneous and highly organized cell populations that compose the nervous system, both neurons and glia constantly secrete neuroactive molecules that modulate various signal transduction pathways, many of which directly affect the cell death machinery. Distinct cell populations and specific repertoires of neuroactive molecules in the various areas of the nervous system produce complex scenarios with regard to the control of cell death execution pathways.

Cyclic adenosine 3',5' monophosphate, or cyclic AMP (cAMP) was the first second messenger described that mediates responses to extracellular hormones and other ligands of a variety of cell membrane receptors.[10-12] This cyclic nucleotide is generated through the activity of adenylyl cyclases, which are encoded by a large number of genes and show differences in both

Brain Repair, edited by Mathias Bähr. ©2006 Eurekah.com
and Kluwer Academic / Plenum Publishers.

distribution and regulation.[13] Degradation of cAMP is achieved by phosphodiesterases, of which at least 11 distinct families have been identified so far.[14] Cyclic AMP binds to the regulatory subunits of cAMP-dependent protein kinases (PKA), thus releasing the catalytic subunits from inhibition, and allowing the phosphorylation of various downstream substrates.[15]

The purpose of this article is to review evidence that signal transduction pathways mediated by the ubiquitous second messenger cAMP are critically involved in the modulation of cell death, and particularly that several neuroactive molecules may have a cAMP-dependent neuroprotective role.

Cell Death by Apoptosis and the Role of cAMP-Dependent Protein Kinase in Apoptotic Execution Pathways

Notwithstanding the increasing evidence for a variety of cell death programs,[16-20] the best known form of physiological cell death is apoptosis.[21] This form of programmed cell death was initially described on morphological grounds, by the blebbing of the plasma membrane and condensation of both the cytoplasm and the chromatin, followed by fragmentation of the cell into multiple membrane-enclosed corpses dubbed apoptotic bodies.[22] These morphological hallmarks were highlighted in contrast with necrotic cell death, which is characterized by cell swelling and rupture of the plasma membrane.

In recent years, a detailed account of biochemical pathways of apoptosis has emerged from experimental studies.[23] Key components of the apoptotic cell death pathways are the aspartate-directed cysteine proteases called caspases.[24] These enzymes selectively cleave a plethora of substrates,[25] and their activation is precisely regulated.[26] Caspases may be classified as either initiator or effector (Fig. 1). For example, caspases 2, 8, 9 and 10 are classified as initiator caspases, and are responsible for the transduction of various signals into the proteolytic activation of downstream effector caspases 3, 6 and 7. The latter enzymes target most of the apoptotic substrates described to date.[25]

Biochemical and genetic analyses showed that mitochondria-based pathways of activation of caspases are regulated by the Bcl-2 protein family.[27] This family is composed of proteins characterized by the presence of one or more Bcl-2 homology (BH) domains, which play either pro-survival or pro-apoptotic roles.[28,29] The mechanisms by which Bcl-2 family proteins function are still a matter of controversy. However, their ability to either homo- or heterodimerize has led to the hypothesis that the balance between pro-survival and pro-apoptotic proteins determines whether a cell will live or die.[30]

According to this hypothesis, survival is favored when pro-apoptotic Bcl-2 family members are prevented from interacting with, and therefore fail to inactivate pro-survival members. Among the pro-apoptotic Bcl-2 family members a major component is the BH3-only protein BAD.[31] Inactivation of BAD is considered a key step in the transduction of various survival signals. In the absence of survival factors, endogenous BAD is dephosphorylated and this favors its binding, through the BH3 domain, to anti-apoptotic Bcl-2 family members, such as Bcl-X_L, whereupon BAD is targeted to the outer mitochondrial membrane. In contrast, the activation of certain kinases in response to survival factors leads to the phosphorylation of one or more sites in BAD. This results in translocation of BAD from the mitochondria to the cytoplasm bound to 14-3-3 adapter proteins, thus blocking its pro-apoptotic action (Fig. 1).[32]

Four distinct phosphorylation sites in BAD have been associated with the control of apoptosis: Ser112, Ser136, Ser155, and Ser170. The contribution of each of these sites to the prevention of the pro-apoptotic function of BAD is still in debate. Initially, inhibition of BAD-dependent cell death was ascribed to phosphorylation of either Ser136 by Akt/PKB,[33] or of Ser112 by a cyclic AMP-dependent protein kinase (PKA) anchored to mitochondria.[34] However, other enzymes were also defined as BAD kinases, such as MAPK-activated p90 ribosomal S6 kinase (Rsk)[35,36] for Ser112, and the 70-kDa ribosomal protein S6 kinase (p70S6K)

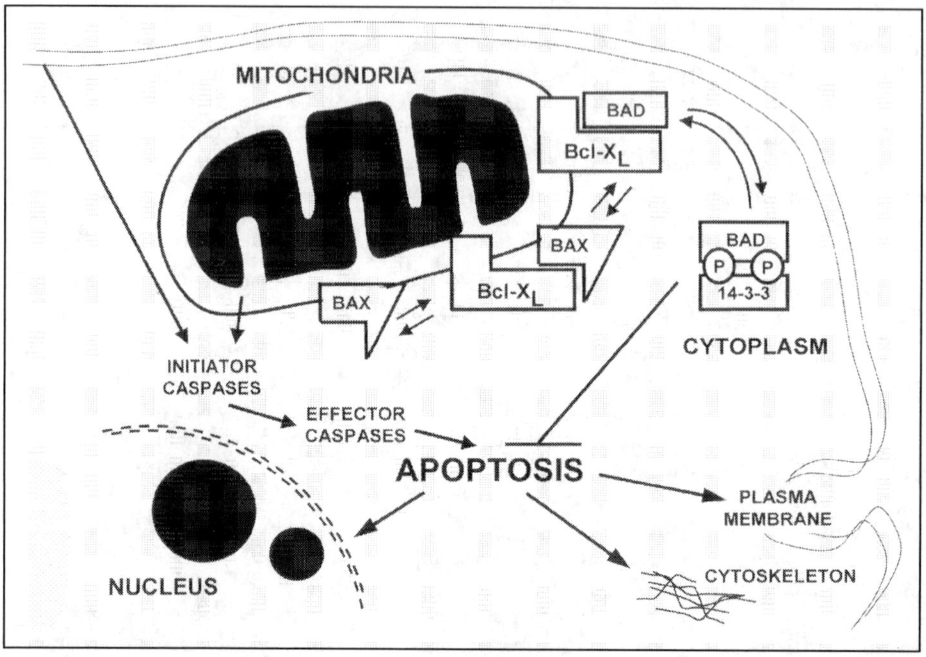

Figure 1. Schematic diagram of the major components of the apoptosis execution pathways, with emphasis on the Bcl-2 family members subject to regulation by cAMP-dependent protein kinase. Signals from either the plasma membrane or the mitochondria activate initiator caspases that, in turn, activate effector caspases, such as caspase-3. The latter cleave a variety of nuclear, cytoskeletal and membrane substrates to produce the morphological features of apoptosis. The effects of Bcl-2 family members are based on a balance of pro- and anti-apoptotic members. A key component of this balance is the protein BAD, that can be phosphorylated by various protein kinases, including PKA. Phosphorylation of BAD leads to its binding to 14-3-3 and cytoplasmic retention. In its dephosphorylated form, BAD can interact with anti-apoptotic proteins such as Bcl-Xl, and prevent the latter to heterodimerize with the pro-apoptotic Bax protein. Free Bax, in turn, can induce mitochondrial-mediated apoptosis. Phosphorylation of BAD, therefore, prevents apoptosis.

for Ser136.[37] Ser155 was described by several groups as the most important phosphorylation site,[38-40] and a cooperative model was proposed, according to which phosphorylation at Ser155 is needed for the dissociation of BAD from Bcl-X_L, but the availability of Ser155 to kinases depends in turn on the phosphorylation of either Ser112 and/or Ser136, which leads to interacton of BAD with 14-3-3.[40] More recently, an additional phosphorylation site has been found in Ser170, which seems to be important for the modulation of BAD pro-apoptotic activity by cytokines, but the kinase involved was not identified.[41]

Within the context of the role of cyclic AMP-activated pathways, it is particularly relevant that both Ser112 and Ser155 in the BAD protein were described as sites phosphorylated by PKA, therefore resulting in protection from apoptosis. These findings raise the hypothesis that modulation of the intracellular concentration of cAMP may play a major role in the control of sensitivity to programmed cell death.

Cyclic AMP and Cell Death

Experiments with various cell types have implicated cAMP in the control of cell death. In early studies, it was shown that a variety of cAMP analogues, as well as cholera toxin and prostaglandins, both of which increased intracellular cAMP, induce apoptosis in a myeloid leukemia cell line,[42] depending on activation of PKA.[43] Both forskolin, an activator of adenylyl cyclases that induces a large increase in intracellular cAMP, and cell-permeant cAMP analogues also induced apoptotic cell death in resting human B lymphocytes,[44] and evidence was presented for teophylline-induced cell death in chronic lymphocytic leukemia cells through activation of PKA due to increased cAMP.[45]

In contrast, elevated cAMP was shown to protect certain cells from induced cell death. For example, cAMP analogues prevented apoptosis of osteoclasts in vitro,[46] and nitric oxide-induced apoptosis of macrophages.[47] Apoptosis of pancreatic cancer cells induced by inhibition of the Erk pathway was prevented by either a combination of the adenylyl cyclase activator forskolin plus the wide-spectrum phosphodiesterase inhibitor 3-isobutyl-1-methylxanthine (IBMX), or with the cell-permeant analogue 8-bromo-cAMP (8-Br-cAMP).[48] In addition, di-butyryl cyclic AMP (db-cAMP) also protected the livers of mice from apoptosis induced in vivo by TNF-α.[49]

Db-cAMP protected NGF-pretreated PC12 cells from cell death induced by withdrawal of both NGF and serum, an effect that was also produced by the Pituitary Adenylyl Cyclase Activating Polypeptide (PACAP). The protective effect of PACAP was abrogated by the PKA inhibitor Adenosine-3'-5'-cyclic Monophosphorothioate-Rp isomer (Rp-cAMPS).[50] These data indicated that PACAP-induced elevation of intracellular cAMP may protect neuron-like cells from degeneration induced by trophic factor withdrawal.

Neuroprotection by Cyclic AMP

Neurotrophins of the Nerve Growth Factor (NGF) family are the best known extracellular neuroprotective agents.[51,52] However, an increasing number of studies have addressed the effects upon cell death of other growth factors as well as neurotransmitters and neuromodulators.[53-56] Many of the latter bind to Gs-protein coupled receptors that stimulate adenylyl cyclase to increase the concentration of intracellular cAMP.[57]

Early evidence for a role of cAMP as a neuroprotective second messenger was gathered with the use of either forskolin or with cell permeant cAMP analogues, such as db-cAMP, 8-bromo-cAMP (8-Br-cAMP), and chlorophenylthio-cAMP (CPT-cAMP). In dissociated cell cultures from the spinal cord and dorsal root ganglion, cell death induced by blockade of electrical activity with tetrodotoxin (TTX) was attenuated with 8-Br-cAMP, whereas treatment with TTX decreased intracellular cAMP.[58] A similar protective role of either forskolin or cAMP analogues was described for sympathetic and sensory neurons deprived of NGF,[59,60] developing septal cholinergic neurons deprived of NGF in low-potassium medium,[61] and mesencephalic dopaminergic neurons exposed to the toxin 1-methyl-4-phenyl-pyridinium ion (MPP+).[62]

The search for physiological modulators of cAMP involved in the control of sensitivity to cell death took advantage of various experimental models. An early study[63] showed that either Vasoactive Intestinal Peptide (VIP) or 8-Br-cAMP prevented the death of retinal ganglion cells induced by TTX in dissociated cell cultures of postnatal rat retinae. TTX reduced the concentration of cAMP in the culture, an effect that was antagonized by VIP, whereas an antagonist of VIP that prevents the peptide-induced increase in cAMP also reduced the number of surviving ganglion cells. These data are consistent with a cAMP-dependent neuroprotective role of VIP.

A similar protective effect of VIP was observed in spinal cord-dorsal root ganglion cultures, in which 8-bromo-cAMP had a neuroprotective effect. Nonetheless, in the latter, the effect of VIP upon spinal cord neurons was indirectly mediated by non-neuronal cells.[64]

CPT-cAMP as well as the neuropeptide PACAP protected cerebellar granule neurons in dissociated cell cultures from cell death induced by low potassium.[65] Distinct from the previous work in retinal cell cultures,[63] it was concluded that the drop in cAMP content that occurred during programmed cell death of cerebellar granule cells following potassium withdrawal was a passive event. However, the increased levels of cAMP associated with treatment with PACAP are consistent with a cAMP-mediated trophic effect of the peptide.[64] Indeed, it was shown that inhibition of cAMP-dependent protein kinase prevented the protection of granule cells by PACAP in a low potassium medium.[66]

Both PACAP and CPT-cAMP were also able to protect NGF-deprived sympathetic neurons from cell death.[67] The effect of PACAP was dose-dependent, and neuroprotection with low doses of PACAP was potentiated by concurrent incubation with IBMX, that blocks the degradation of cAMP. Although these data seem to imply the PACAP-induced increase of cAMP as the neuroprotective signal, the authors failed to examine the effects of IBMX alone, which by itself was shown to have neuroprotective effects in other cells.[68]

Both PACAP and db-cAMP protected dissociated cerebral cortical neurons from glutamate cytotoxicity, and it was suggested that a PACAP-induced elevation of cAMP caused the neuroprotective effect.[69] Protection against excitotoxic insults was also investigated in dissociated cell cultures from the retina of neonatal rats,[70] and it was found that VIP, forskolin and 8-Br-cAMP reduced cell death induced by high concentrations of glutamate. The protective effect of the neuropeptide was prevented either by a VIP receptor antagonist or by the PKA inhibitor N-[2-(p-Bromocinnarnylamino)ethyl]-5-isoquinolinesulfonamide (H89), showing that an increase in cAMP produced by VIP receptor activation results in PKA-mediated protection against excitotoxicity. Interestingly, the VIP-induced increase in cAMP in the retinal cell cultures was potentiated by concurrent glutamate. In another study, adenosine, as well as agonists of its A2a receptors protected dissociated chick retinal cells from glutamate excitotoxicity.[71] The A2a adenosine receptors are coupled to Gs proteins and stimulate adenylyl cyclase, and accordingly, cAMP analogues mimicked the effect of the A2a agonists. These data suggest that adenosine inhibits glutamate neurotoxicity through elevation of cAMP levels.

Recently, it was shown that the stress-related neuropeptides corticotropin-releasing hormone (CRH) and urocortin (URC), protect hippocampal neurons in dissociated cell cultures from cell death induced by various insults. Among the various signal transduction pathways activated by these peptides, the protective effects of both CRH and URC were blocked by H89, showing the involvement of cAMP-dependent protein kinase in the neuroprotection.[72,73]

In cultures of purified retinal ganglion cells, an increase in intracellular cAMP was required for responsiveness to various trophic factors.[74] One mechanism for the enhanced responsiveness of these neurons to the neurotrophins Brain-Derived Neurotrophic Factor (BDNF) and Neurotrophin-4 (NT4) is the recruitment of their high-affinity TrkB neurotrophin receptors to the plasma membrane by translocation from intracellular stores.[75] In contrast, forskolin plus IBMX alone were sufficient to protect cultures of highly purified spinal motor neurons in vitro, with a potency that could only be matched by combinations of more than 5 trophic factors.[76] These data highlight the distinct requirements of intracellular cAMP for the survival of various neuron types, but they may be determined by conditions of the purified neuron cultures.

Studies in our laboratory have been directed at the effects of cAMP upon the sensitivity to cell death within the organized retinal tissue. Using histotypic retinal explants in vitro,[77] we have shown that the protein synthesis inhibitor anisomycin induces apoptosis in undifferentiated cells that recently left the cell cycle, within the neuroblastic (ventricular) layer (NBL) of the retina of neonatal rodents.[78,79] Cell death in the NBL was blocked by concurrent incubation of the explants with forskolin, suggesting that cAMP might have a protective effect in retinal tissue. Indeed, both 8-Br-cAMP, as well as IBMX and the cAMP-phosphodiesterase specific inhibitor 4-(3-Butoxy-4-methoxybenzyl)-2-imidazolidinone (Ro-201724) prevented cell death in the NBL, whereas the cGMP-phosphodiesterase inhibitor 1,4-Dihydro-5-(2-propoxyphenyl)-7H-1,2,3-triazolo[4,5-d]pyrimidine-7-one (Zaprinast) had no effect. These data strongly supported the hypothesis that cAMP has a neuroprotective role within the developing retina.[68] It was also shown that the PKA inhibitor H89 abrogated the protective effect of forskolin, thus implying PKA in the protective pathway.[68] Further studies showed that, while cell death induced in proliferating retinal cells by the topoisomerase II inhibitor etoposide was insensitive to forskolin, the activation of adenylyl-cyclase prevented cell death of photoreceptors induced by thapsigargin.[80] These data suggest that the neurotrophic response of retinal cells to cAMP is developmentally regulated.

In contrast with either TTX-induced cell death in clusters of retinal cells following dissociation,[63] or with the death of purified retinal ganglion cells,[74] neither forskolin nor neuroactive molecules such as dopamine or PACAP that increase intracellular cAMP, prevented the degeneration of axotomized ganglion cells in retinal explants. The reason for this discrepancy is not known. However, the data showing that electrical blockade of neural activity with TTX leads to a decrease in intracellular cAMP[58,63] raises the hypothesis that in the histotypical retinal explants, ganglion cells may sustain higher concentrations of cAMP than in dissociated cell cultures due to synaptic action of growing afferent inputs,[81] even though synaptic profiles are still rudimentary at early postnatal stages.[82] This, in turn, could obviate the need for either depolarization or cAMP as ancillary survival factors.[75]

The protective effect of the phosphodiesterase inhibitors alone[68] indicated that endogenous neuroactive molecules are involved in cAMP-mediated neuroprotection within the retina. We investigated the effects of dopamine, a retinal neurotransmitter present in the retina of neonatal rats together with D1-like functional receptors,[83-86] which stimulate adenylyl cyclase and mediate the functions of many dopaminergic inputs.[87] Indeed, dopamine also prevented anisomycin-induced cell death in this model.[88] The protection by dopamine was mimicked by the D1-like receptor agonists SKF 38393, 6-Chloro-PB and ADTN, but the D1-like receptor antagonist SCH 23390 did not prevent protection by the agonists. Distinct from the complete antagonism revealed in chick retinal tissue, even high concentrations of SCH 23390 were unable to completely block the increase of intracellular cAMP induced by either dopamine or 6-Cl-PB in neonatal rat retinas. In addition, H89 again blocked the protective effect of D1-like receptor agonists. Therefore, the data indicated that activation of a D1-like receptor by dopamine in retinal tissue induced an increase in intracellular cAMP with a consequent neuroprotective effect mediated by PKA.[88]

We also showed that PACAP has a strong neuroprotective effect in retinal tissue. Low concentrations of the peptide prevented both the anisomycin-induced degeneration of undifferentiated cells within the NBL, as well as photoreceptor cell death induced by thapsigargin in the outer nuclear layer.[89] In both cases, forskolin had a similar neuroprotective effect.[80] The neuroprotective effect of PACAP is mediated at least in part through the PAC1 receptor (PAC1R), which is expressed in the neonatal retina. Accordingly, PAC1R antagonists prevented the neuroprotection by the peptide. We also demonstrated that PACAP induces an increase in the intracellular concentration of cAMP in the retina, and that the PKA inhibitor Rp-cAMPS prevented neuroprotection by PACAP. The data show, therefore, that an increased concentration of cAMP induced by PACAP through PAC1R, lowers the sensitivity to cell death in retinal tissue through activation of PKA.[89]

A further example of cAMP-dependent neuroprotection was obtained with engagement of the cellular prion protein—PrP^c.[90,91] We found that treatment of retinal explants with several PrP^c-binding peptides or with the PrP^c-binding protein STI-1 induced an increase in intracellular cAMP, and protected cells within the NBL from anisomycin-induced cell death in wild-type rodents but not in PrP^c-knockout mice. Similar cytoprotection was also observed upon other cell types under distinct inducers of cell death. The neuroprotection mediated by PrP^c was blocked by Rp-cAMPS, thus suggesting a neuroprotective cascade composed in succession by STI-1, PrP^c, cAMP and PKA.[90,91]

The overall data reviewed above support the hypothesis that an increase in the concentration of cAMP has, in general, a protective effect upon cells within the nervous system. Among

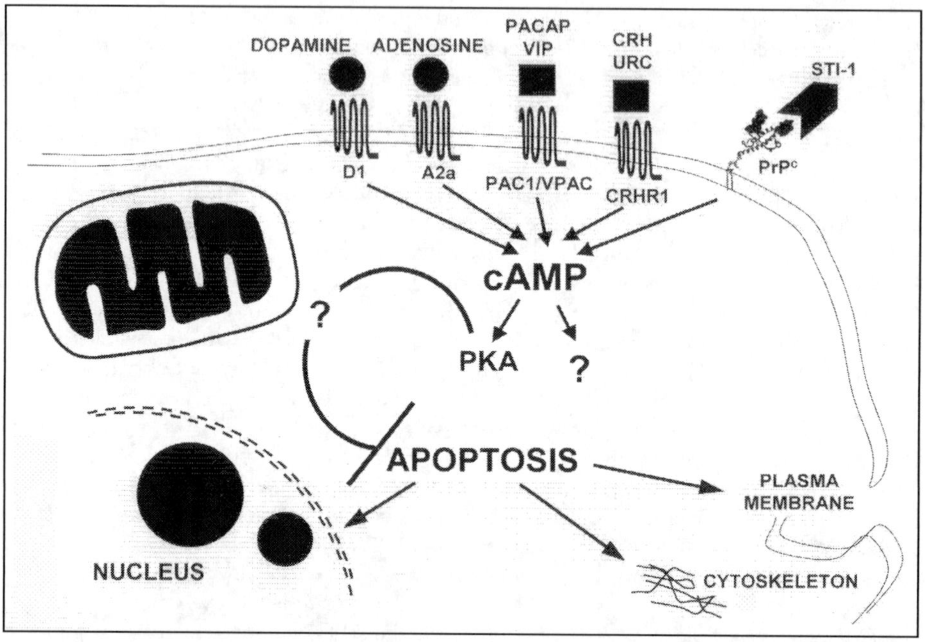

Figure 2. Cyclic AMP is a nodal point for neuroprotective signaling pathways. Various neurotransmitters (circles), neuropeptides (squares) or even proteins bind to their receptors or membrane ligands, and induce an increase in the intracellular concentration of cAMP. One major consequence of this increase is the activation of PKA, which leads to blockade of apoptosis. The exact level at which PKA blocks the apoptotic execution pathways is not known, but may be related to the interplay of Bcl-2 family members shown in Figure 1. Other pathways are also activated by cAMP, but their relationship with the apoptotic execution pathways is not known.

the neuroactive molecules that modulate the production of cAMP, neurotransmitters, neuropeptides and at least one protein may be involved in physiological neuroprotection mediated by cAMP-PKA signal transduction pathways (Fig. 2).

Signal Transduction Pathways Regulated by cAMP

The classical target of cAMP is the cyclic AMP-dependent protein kinase, or PKA.[92] The data showing PKA-dependent neuroprotection by various neuroactive molecules that increase the concentration of cAMP, raise the hypothesis that this neuroprotection may be mediated by direct phosphorylation of either Ser112 or Ser155 in BAD.

An additional substrate of PKA that may be involved in neuroprotection is the cAMP-reponsive element binding protein (CREB), a transcription factor that is phosphorylated by PKA, among other kinases. CREB-dependent induction of Bcl-2 has been found associated with the NGF-dependent survival of sympathetic neurons,[93] and CREB-mediated transcription has been implicated in BDNF-induced cerebellar neuron survival,[35] although in the latter case, the survival promoting activity was associated with phosphorylation of both CREB as well as BAD by the pp90 ribosomal S6 kinase Rsk2, which is activated by mitogen activated protein (MAP) kinases independent of PKA.[94] It is also noteworthy that deletion of CREB has been reported to impair the survival of peripheral, but not of central neurons.[95]

Cross talk between cAMP and MAP kinase signaling pathways has been examined in various biological responses.[96] Cyclic AMP may lead to either activation or inhibition of extracellular signal-regulated kinase (Erk), depending on cell type. In the context of cell death, it has been reported that activation of PKA by cAMP analogues protected a hypothalamic neuronal cell line transfected with TrkA from NGF withdrawal, through activation of Erk mediated by B-Raf.[97] Consistent with these data, the protection of granule cells by PACAP in a low potassium medium was blocked by inhibition of the Erk pathway, whereas inhibition of PKA blocked the activation of Erk.[66] This pathway, however, is unlikely to apply to every cell type in the nervous system, because, for example, the anisomycin-induced cell death that is blocked by cAMP is also prevented by inhibition, rather than activation of the Erk pathway, and the neuroprotective effect of engagement of PrP^c was potentiated by concurrent inhibition of the Erk pathway.[90]

Cyclic AMP also binds to and modulates the activity of cyclic nucleotide-gated membrane channels.[98-100] These are non-selective cation channels that conduct both Ca^{2+}, which in turn may activate intracellular pathways associated with cell death,[101] as well as Na^+ and K^+ which may alter the membrane potential and disturb ionic balance, also affecting cell death.[102] Both depolarization and calcium fluxes have been associated with neuroprotection in certain systems,[103] but there is so far no evidence that cAMP-gated membrane channels are involved in neuroprotection.

More recently, it was described that cAMP directly modulates GTP-exchange factors (GEFs) for the small GTPases Rap1 and Ras. The protein Epac (Exchange Protein Activated by cAMP) binds cAMP in vitro and has a strong GEF activity towards Rap1.[104,105] Also, a Ras GEF that responds to both cAMP and cGMP was described to activate Rap1 constitutively, and Ras in response to cyclic nucleotide binding.[106] These small GTPases are known to modulate MAP Kinase signal transduction pathways. Thus, GEFs directly responsive to cAMP may be linked to cell death pathways through their interaction with MAP kinases independently of PKA. However, to date the only study of Epac in the context of cytoprotective pathways was done following treatment of HEK293 cell lines with forskolin, and suggested opposing effects of PKA and Epac upon the activity of PKB/Akt.[107]

Thus, current data show that some cytoprotective effects of cAMP may be mediated by Erk pathways, and that kinases involved in cytoprotection may be modulated independent of PKA. Still, the hypothesis of a direct link from cAMP to apoptosis execution pathways through phosphorylation of BAD by PKA remains a major possibility.

Conclusion and Perspectives

The overall data reviewed in this chapter indicates that an increase in the intracellular concentration of cAMP is generally associated with cytoprotection in the nervous system, essentially through the activation of PKA. Elevated levels of cAMP may be the consequence of activation of membrane receptors by certain classical neurotransmitters, neuropeptides or by other protein-ligand interactions. The pathways that link the extracellular neuroactive molecules with the cAMP/PKA-mediated neuroprotection are likely an integral part of the control of sensitivity to cell death within the nervous system. In turn, the neuroprotective events downstream of PKA remain to be clarified.

Recent studies of the kinetics of cell loss in models of neurodegenerative disease suggest that cells constantly integrate the activity of various metabolic pathways that converge upon the cell death execution machinery. This leads each individual cell to a homeostatic state upon which single catastrophic events may induce cell death depending on the balance between pro-apoptotic and pro-survival components.[108] The cAMP/PKA pathway feeds into this metabolic network as a neuroprotective component, and therefore the state of the various extracellular mediators that signal through cAMP are of major concern in issues related to brain repair as well as to neural cell survival in general.

Most of the available data on the intracellular pathways were, however, obtained either from gene overexpression in cell lines or from studies of primary cultures of dissociated cells, and must be confirmed both in tissue models and in situ. The growing evidence for alternative cell signaling pathways mediated by cAMP[109] warrants further examination of the mechanisms of control of cell death by cyclic nucleotides. In addition, the cross-talk between cAMP-mediated pathways and the classical neurotrophic factor-induced signaling pathways deserves particular attention.

Current tools available for experimentation should allow progress in studies of the metabolic pathways that control cell death related both with neural development and repair, as well as with neurodegeneration. Further work shall lead to the understanding of the integration of individual signaling cascades into the network of cell survival control, both in health and disease, much like the metaphoric construction and eventual deconstruction of Pink Floyd's wall.

Acknowledgements

Work in the Laboratory of Neurogenesis is supported by grants and fellowships from CNPq, FAPERJ, PRONEX-MCT, FAPESP, CAPES and a fellowship from the John Simon Guggenheim Memorial Foundation to R.L. We are indebted to all our students and colleagues who participated in the various experimental studies cited in the text, and to J. Nilson dos Santos and José Francisco Tiburcio for technical support. The title of this article quotes lyrics from the rock opera 'The Wall', by Pink Floyd.

References

1. Lockshin RA, Zakeri Z. Programmed cell death and apoptosis: origins of the theory. Nat Rev Mol Cell Biol 2001: 2(7):545-550.
2. Uren AG, Vaux DL. Molecular and clinical aspects of apoptosis. Pharmacol Ther 1996; 72(1):37-50.
3. Gibson RM. Does apoptosis have a role in neurodegeneration? Brit Med J 2001; 322(7301):1539-1540.
4. Mattson MP. Apoptosis in neurodegenerative disorders. Nat Rev Mol Cell Biol 2000; 1(2):120-129.
5. Evan GI, Vousden KH. Proliferation, cell cycle and apoptosis in cancer. Nature 2001; 411(6835):342-348.
6. Aguayo AJ, Clarke DB, Jelsma TN et al. Effects of neurotrophins on the survival and regrowth of injured retinal neurons. Ciba Found Symp 1996; 196:135-144.
7. Bjorklund A. Cell replacement strategies for neurodegenerative disorders. Novartis Found Symp 2000; 231:7-15.
8. Rossi F, Cattaneo E. Opinion: neural stem cell therapy for neurological diseases: dreams and reality. Nat Rev Neurosci 2002; 3(5):401-409.
9. Raff, MC. Social controls on cell survival and cell death. Nature 1992; 356(6368):397-400.
10. Robinson GA, Butcher RW, Sutherland EW. Cyclic AMP. Annu Rev Biochem 1968; 37:149-174.
11. McKnight GS. Cyclic AMP second messenger systems. Curr Opin Cell Biol 1991; 3(2):213-217.
12. Selbie LA, Hill SJ. G protein-coupled-receptor cross-talk: the fine-tuning of multiple receptor-signalling pathways. Trends Pharmacol Sci 1998; 19(3):87-93.
13. Hanoune J, Defer N. Regulation and role of adenylyl cyclase isoforms. Annu Rev Pharmacol Toxicol 2001; 41:145-174.
14. Mehats C, Andersen CB, Filopanti M et al. Cyclic nucleotide phosphodiesterases and their role in endocrine cell signaling. Trends Endocrinol Metab 2002; 13(1):29-35.
15. Walsh DA, Van Patten SM. Multiple pathway signal transduction by the cAMP-dependent protein kinase. FASEB J 1994; 8(15):1227-136.
16. Clarke PGH. Developmental cell death: morphological diversity and multiple mechanisms. Anat Embryol (Berl) 1990; 181(3):195-213.
17. Dunn WA, Jr. Studies on the mechanisms of autophagy: formation of the autophagic vacuole. J Cell Biol 1990; 110: 1923-1933.

18. Dunn WA, Jr. Studies on the mechanisms of autophagy: maturation of the autophagic vacuole. J Cell Biol 1990; 110:1935-1945.
19. Bursch W. The autophagosomal-lysosomal compartment in programmed cell death. Cell Death Differ 2001; 8(6):569-581.
20. Tolkovsky AM, Xue L, Fletcher GC et al. Mitochondrial disappearance from cells: a clue to the role of autophagy in programmed cell death and disease? Biochimie 2002; 84(2-3):233-240.
21. Wyllie AH. Apoptosis: an overview. Br Med Bull 1997; 53(3):451-465.
22. Kerr JFR, Wyllie AH, Currie AR. Apoptosis: A basic biological phenomenon with wide-ranging implications in tissue kinetics. Br J Cancer 1972; 26:239-257.
23. Hengartner MO. The biochemistry of apoptosis. Nature 2000; 407(6805):770-776.
24. Salvesen GS, Dixit VM. Caspases: intracellular signaling by proteolysis. Cell 1997; 91(4):443-446.
25. Wolf BB, Green DR. Suicidal tendencies: apoptotic cell death by caspase family proteinases. J Biol Chem 1999; 274(29):20049-20052.
26. Stennicke HR, Ryan CA, Salvesen GS. Reprieval from execution: the molecular basis of caspase inhibition. Trends Biochem Sci 2002; 27(2):94-101.
27. Reed J. Bcl-2 family proteins. Oncogene 1998; 17(25):3225-36.
28. Gross A, McDonnell JM, Korsmeyer SJ. BCL-2 family members and the mitochondria in apoptosis. Genes Dev 1999;13(15):1899-9111.
29. Strasser A, O'Connor L, Dixit VM. Apoptosis signaling. Annu Rev Biochem 2000; 69:217-245.
30. Oltvai ZN, Korsmeyer SJ. Checkpoints of dueling dimers foil death wishes. Cell 1994; 79:189-192.
31. Yang E, Zha J, Jockel J et al. Bad, a heterodimeric partner for Bcl-XL and Bcl-2, displaces Bax and promotes cell death. Cell 1995; 80(2):285-291.
32. Datta SR, Dudek H, Tao X et al. Akt phosphorylation of BAD couples survival signals to the cell-intrinsic death machinery. Cell 1997; 91(2):231-241.
33. Datta SR, Brunet A, Greenberg ME. Cellular survival: a play in three Akts. Genes Dev 1999; 13(22):2905-2927.
34. Harada H, Becknell B, Wilm M et al. Phosphorylation and inactivation of BAD by mitochondria-anchored protein kinase A. Mol Cell 1999; 3(4):413-422.
35. Bonni A, Brunet A, West AE et al. Cell survival promotrd by the Ras-MAPK signaling pathway by transcription-dependent and –independent mechanisms. Science 1999; 286:1358-1362.
36. Bertolotto C, Maulon L, Filippa N et al. Protein kinase C q and e promote T-cell survival by a Rsk-dependent phosphorylation and inactivation of BAD. J Biol Chem 2000; 275:37246-37250.
37. Harada H, Andersen JS, Mann M et al. P70S6 kinase signals cell survival as well as growth, inactivating the pro-apoptotic molecule BAD. Proc Nat Acad Sci USA 2001; 98:9666-9670.
38. Tan Y, Demeter MR, Ruan H et al. BAD Ser-155 phosphorylation regulates BAD/Bcl-XL interaction and cell survival. J Biol Chem 2000; 275:25865-25869.
39. Zhou X, Liu Y, Payne G et al. Growth factors inactivate the cell death promoter BAD by phosphorylation of its BH3 domain on Ser155. J Biol Chem 2000; 275:25046-25051.
40. Datta SR, Katsov A, Hu L et al. 14-3-3 proteins and survival kinases cooperate to inactivate BAD by BH3 domain phosphorylation. Mol Cell 2000; 6:41-51.
41. Dramsi S, Scheid MP, Maiti A et al. Identification of a novel phosphorylation site, Ser170, as a regulator of BAD pro-apoptotic activity. J Biol Chem 2002; 277:6399-6405.
42. Lanotte M, Riviere JB, Hermouet S et al. Programmed cell death (apoptosis) is induced rapidly and with positive cooperativity by activation of cyclic adenosine monophosphate-kinase I in a myeloid leukemia cell line. J Cell Physiol 1991;146(1):73-80.
43. Duprez E, Gjertsen BT, Bernard O et al. Antiapoptotic effect of heterozygously expressed mutant RI (Ala336—>Asp) subunit of cAMP kinase I in a rat leukemia cell line. J Biol Chem 1993; 268(11):8332-8340.
44. Lomo J, Blomhoff HK, Beiske K et al. TGF-beta 1 and cyclic AMP promote apoptosis in resting human B lymphocytes. J Immunol 1995; 154(4):1634-1643.
45. Mentz F, Merle-Beral H, Ouaaz F et al. Theophylline, a new inducer of apoptosis in B-CLL: role of cyclic nucleotides. Br J Haematol 1995; 90(4):957-959.
46. Selander KS, Harkonen PL, Valve E et al. Calcitonin promotes osteoclast survival in vitro. Mol Cell Endocrinol 1996; 122(2):119-129.

47. von Knethen A, Brune B. Attenuation of macrophage apoptosis by the cAMP-signaling system. Mol Cell Biochem 2000; 212(1-2):35-43.
48. Boucher MJ, Duchesne C, Laine J et al. cAMP protection of pancreatic cancer cells against apoptosis induced by ERK inhibition. Biochem Biophys Res Commun 2001; 285(2):207-216.
49. Takano M, Arai T, Mokuno Y et al. Dibutyryl cyclic adenosine monophosphate protects mice against tumor necrosis factor-alpha-induced hepatocyte apoptosis accompanied by increased heat shock protein 70 expression. Cell Stress Chaperones 1998; 3(2):109-117.
50. Tanaka J, Koshimura K, Murakami Y et al. Neuronal protection from apoptosis by pituitary adenylate cyclase-activating polypeptide. Regul Pept 1997; 72(1):1-8.
51. Barbacid M. Neurotrophic factors and their receptors. Curr Opin Cell Biol 1995; 7(2):148-155.
52. Heerssen HM, Segal RA. Location, location, location: a spatial view of neurotrophin signal transduction. Trends Neurosci 2002; 25(3):160-165.
53. Cardona-Gomez GP, Mendez P, DonCarlos LL et al. Interactions of estrogens and insulin-like growth factor-I in the brain: implications for neuroprotection. Brain Res Brain Res Rev 2001; 37(1-3):320-334.
54. Gozes I, Brenneman DE. A new concept in the pharmacology of neuroprotection. J Mol Neurosci 2000; 14(1-2):61-68.
55. Airaksinen MS, Saarma M. The GDNF family: signalling, biological functions and therapeutic value. Nat Rev Neurosci 2002; 3(5):383-394.
56. Linden R. The anti-death league: Associative control of apoptosis in developing retinal tissue. Brain Res Brain Res Rev 2000; 32(1):146-158.
57. Rodbell M. The complex regulation of receptor-coupled G-proteins. Adv Enzyme Regul 1997; 37:427-435.
58. Brenneman DE, Fitzgerald S, Litzinger MJ. Neuronal survival during electrical blockade is increased by 8-bromo cyclic adenosine 3',5' monophosphate. J Pharm Exp Ther 1985; 233:402-408.
59. Rydel RE, Greene LA. cAMP analogs promote survival and neurite outgrowth in cultures of rat sympathetic and sensory neurons independently of nerve growth factor. Proc Natl Acad Sci USA 1988; 85:1257-1261.
60. Edwards SN, Buckmaster AE, Tolkovsky AM. The death programme in cultured sympathetic neurones can be supressed at the posttranslational level by nerve growth factor, cyclic AMP, and depolarization. J Neurochem 1991; 57:2140-2143.
61. Kew JNC, Smith DW, Sofroniew MV. Nerve growth factor withdrawal induces the apoptotic death of developing septal cholinergic neurons in vitro: protection by cyclic AMP analogues and high potassium. Neuroscience 1996; 70:329-339.
62. Hartikka J, Staufenbiel M, Lübbert H. Cyclic AMP, but not basic FGF, increases the in vitro survival of mesencephalic dopaminergic neurons and protects them from MPP+-induced degeneration. J. Neurosci Res 1992; 32:190-201.
63. Kaiser PK, Lipton SA. VIP-mediated increase in cAMP prevents tetrodotoxin-induced retinal ganglion cell death in vitro. Neuron 1990; 5:373-381.
64. Brenneman DE, Neale EA, Foster GA et al. Nonneuronal Cells Mediate Neurotrophic Action of Vasoactive Intestinal Peptide. J Cell Biology 1987; 104:1603-1610.
65. Chang JY, Korolev VV, Wang JZ. Cyclic AMP and pituitary adenylate cyclase-activating polypeptide (PACAP) prevent programmed cell death of cultured rat cerebellar granule cells. Neurosci Lett 1996; 206(2-3):181-184.
66. Villalba M, Biackaert J, Journot L. Pituitary adenylate cyclase-activating polypeptide (PACAP38) protects cerebellar granule neurons from apoptosis by activating the mitogen-activated protein kinase (MAP Kinase) pathway. J Neurosci 1997; 17(1):83-90.
67. Chang JY, Korolev VV. Cyclic AMP and sympathetic neuronal programmed cell death. Neurochem Int 1997; 31(2):161-167.
68. Varella MH, Correa DF, Campos CBL et al. Protein kinases selectively modulate apoptosis in the developing retina in vitro. Neurochem Int 1997; 31:217-227.
69. Morio H, Tatsuno I, Hirai A et al. Pituitary adenylate cyclase-activating polypeptide protects rat-cultured cortical neurons from glutamate-induced cytotoxicity. Brain Res 1996; 741(1-2):82-88.
70. Shoge KS, Mishima HK, Saitoh T et al. Protective effects of vasoactive intestinal peptide against delayed glutamate neurotoxicity in cultured retina. Brain Res 1998; 809:127-136.

71. Ferreira JM, Paes-de-Carvalho R. Long-term activation of adenosine A(2a) receptors blocks glutamate excitotoxicity in cultures of avian retinal neurons. Brain Res 2001; 900(2):169-176.
72. Pedersen WA, McCullers D, Culmsee C et al. Corticotropin-releasing hormone protects neurons against insults relevant to the pathogenesis of Alzheimer's disease. Neurobiol Dis 2001; 8(3):492-503.
73. Pedersen WA, Wan R, Zhang P et al. Urocortin, but not urocortin II, protects cultured hippocampal neurons from oxidative and excitotoxic cell death via corticotropin-releasing hormone receptor type I. J Neurosci 2002; 22(2):404-412.
74. Meyer-Franke A, Kaplan MR, Pfrieger FW et al. Characterization of the signaling interactions that promote the survival and growth of developing retinal ganglion cells in culture. Neuron 1995; 15(4):805-19.
75. Meyer-Franke A, Wilkinson GA, Kruttgen A et al. Depolarization and cAMP elevation rapidly recruit TrkB to the plasma membrane of CNS neurons. Neuron 1998; 21(4):681-693.
76. Hanson MG Jr, Shen S, Wiemelt AP et al. Cyclic AMP elevation is sufficient to promote the survival of spinal motor neurons in vitro. J Neurosci 1998; 18(18):7361-7371.
77. Linden R, Rehen SK, Chiarini LB. Apoptosis in developing retinal tissue. Progr Retinal Eye Res 1999; 18:133-165.
78. Rehen SK, Varella MH, Freitas FG et al. Contrasting effects of protein synthesis and of cyclic AMP on apoptosis in the developing retina. Development 1996; 122:1439-1448.
79. Rehen SK, Neves DDC, Fragel-Madeira L et al. Selective sensitivity of early post-mitotic retinal cells to apoptosis induced by inhibition of protein synthesis. Eur J Neurosci 1999; 11:349-356.
80. Chiarini LB, Leal-Ferreira ML, de Freitas FG et al. Changing sensitivity to cell death during development of retinal photoreceptors. J Neurosci Res 2003; 74(6):875-83.
81. Young SH, Poo MM. Spontaneous release of transmitter from growth cones of embryonic neurones. Nature 1983; 305(5935):634-637.
82. Fernandes AJ, Martinez AMB, Linden R. Estruturas juncionais precoces na camada plexiforme interna da retina do rato em desenvolvimento. Res XII Col Soc Bras Micr Eletr 1988; 225-226.
83. Shearman LP, Zeitzer J, Weaver DR. Widespread expression of functional D1-dopamine receptors in fetal rat brain. Brain Res Dev Brain Res 1997; 102(1):105-115.
84. Wu DK, Cepko CL. Development of dopaminergic neurons is insensitive to optic nerve section in the neonatal rat retina. Brain Res Dev Brain Res 1993; 20;74(2):253-260.
85. Schambra UB, Duncan GE, Breese GR et al. Ontogeny of D1A and D2 dopamine receptor subtypes in rat brain using in situ hybridization and receptor binding. Neuroscience 1994; 62(1):65-85.
86. Shelke RR, Lakshmana MK, Ramamohan Y et al. Levels of dopamine and noradrenaline in the developing of retina—Effect of light deprivation. Int J Dev Neurosci 1997; 15(1):139-143.
87. Witkovsky P, Schutte M. The organization of dopaminergic neurons in vertebrate retinas. Vis Neurosci 1991; 7(1-2):113-124.
88. Varella MH, Mello FG, Linden R. Evidence for an antiapoptotic role of dopamine in developing retinal tissue. J Neurochem 1999; 73: 485-492.
89. Silveira MS, Costa MR, Bozza M et al. Pituitary Adenylyl Cyclase–Activating Polypeptide Prevents Induced Cell Death In Retinal Tissue Through Activation Of Cyclic AMP-Dependent Protein Kinase. J Biol. Chem 2002; 27(18): 16075-16080.
90. Chiarini LB, Freitas ARO, Martins VM et al. Cellular prion protein transduces neuroprotective signals. EMBO J 2002; 21: 3317-3326.
91. Zanata SM, Lopes MH, Mercadante AF et al. The stress- inducible protein 1 is a cell surface ligand for cellular prion that triggers neuroprotection. EMBO J 2002; 21:3307-3316.
92. Taylor SS. cAMP-dependent protein kinase. Model for an enzyme family. J Biol Chem 1989; 264(15):8443-8446.
93. Riccio A, Ahn S, Davenport CM et al. Mediation by a CREB family transcription factor of NGF-dependent survival of sympathetic neurons. Science 1999; 286(5448):2358-2361.
94. Frodin M, Gammeltoft S. Role and regulation of 90 kDa ribosomal S6 kinase (RSK) in signal transduction. Mol Cell Endocrinol 1999; 151(1-2):65-77.
95. Lonze BE, Riccio A, Cohen S et al. Apoptosis, axonal growth defects, and degeneration of peripheral neurons in mice lacking CREB Neuron 2002; 34(3):371-385.
96. Stork PJ, Schmitt JM. Crosstalk between cAMP and MAP kinase signaling in the regulation of cell proliferation Trends Cell Biol 2002; 12(6):258-266.

97. Dugan LL, Kim JS, Zhang Y et al. Differential effects of cAMP in neurons and astrocytes. Role of B-raf. J Biol Chem 1999; 274(36):25842-25848.
98. Nakamura T, Gold GH. A cyclic nucleotide-gated condutance in olfactory cilia. Nature 1987; 325: 342-344.
99. Dhallan RS, Yau K, Scrader K et al. Primary structure and functional expression of a cyclic nucleotide-activated channel from olfactory neurons. Nature 1990; 347:184-187.
100. Goulding EH, Ngai J, Kramer RH et al. Molecular cloning and single-channel properties of the cyclic nucleotide-gated channel from catfish olfactory. Neuron 1992; 8(1):45-58.
101. McConkey DJ, Orrenius S. The role of calcium in the regulation of apoptosis. Biochem Biophys Res Commun. 1997; 239(2):357-366.
102. Xiao AY, Wei L, Xia S et al. Ionic mechanism of ouabain-induced concurrent apoptosis and necrosis in individual cultured cortical neurons. J Neurosci 2002; 22(4):1350-1362.
103. Gallo V, Kingsbury A, Balazs R et al. The role of depolarization in the survival and differentiation of cerebellar granule cells in culture. J Neurosci 1987; 7(7):2203-2213.
104. Rooji J, Zwartkruis FJT, Verheijen MHG et al. Epac is a Rap1 guanine nucleotide-exchange factor directly activated by cyclic AMP. Nature 1998; 396:474-477.
105. Kawasaki H, Springett GM, Mochizuki N et al. A family of cAMP-binding proteins that directly activate Rap1. Science 1998; 282:2275-2279.
106. Pham N, Cheglakov I, Koch CA et al. The guanine nucleotide exchange factor CNrasGEF activates Ras in response to cAMP and cGMP. Curr Biol 2000; 10:555-558.
107. Mei FC, Qiao J, Tsygankova OM et al. Differential signaling of cyclic AMP: opposing effects of exchange protein directly activated by cyclic AMP and cAMP-dependent protein kinase on protein kinase B activation. J Biol Chem 2002; 277(13):11497-11504.
108. Clarke G, Collins RA, Leavitt BR et al. A one-hit model of cell death in inherited neuronal degenerations. Nature 2000; 406(6792):195-199.
109. Richards JS. New signaling pathways for hormones and cyclic adenosine 3',5'-monophosphate action in endocrine cells. Mol Endocrinol 2001; 15(2):209-218.

Chapter 11

The Collagenous Wound Healing Scar in the Injured Central Nervous System Inhibits Axonal Regeneration

Susanne Hermanns, Nicole Klapka, Marcia Gasis and Hans Werner Müller

Abstract

Following traumatic injuries of the central nervous system (CNS) a wound healing scar, resembling the molecular structure of a basement membrane and mainly composed of Collagen type IV and associated glycoproteins and proteoglycans, is formed. It is well known that CNS fibers poorly regenerate after traumatic injuries. In this article we summarize data showing that prevention of collagen scar formation enables severed axons in brain and spinal cord to regrow across the lesion site and to elongate in uninjured CNS tissue. We observed that regenerating fibers grow back to their former target where they develop chemical synapses, become remyelinated by resident oligodendrocytes and conduct action potentials.

Peripheral and Central Nervous System Responses to Axotomy

The peripheral nervous system (PNS) and the CNS respond differently to traumatic injuries (Fig. 1). Following axotomy Wallerian degeneration of the distal stump of the lesioned nerve occurs in both, PNS and CNS, but PNS axons initiate regenerative sprouting and elongate through the Bands of Büngner towards the peripheral target.

In CNS tissue neurons either die, if the lesion is close to the cell bodies, or the proximal axon stumps retract for a short distance (approx. 2-500 μm) from the lesion zone. Subsequently, the injured axons grow back but fail to cross the lesion site. These observations lead to the hypothesis that a molecular barrier at the lesion site may impede axonal regeneration.[1]

Factors in the Lesion Area Impede Axonal Regeneration

In recent literature many factors impeding axonal regeneration at the CNS lesion site could be identified. Myelin constituents[2,3] as well as the presence of a glial scar[4-6] were shown to contribute to the regenerative failure of adult CNS axons. Guidance molecules like, e.g., semaphorins,[7] and ephrins[8] are present at the lesion site and are considered to repel regenerating axons in the CNS.

Other putative neurite outgrowth inhibitors, especially extracellular matrix (ECM) components become increasingly important in the field of CNS repair. Recent in vitro and in vivo studies identified a variety of ECM molecules that show inhibitory properties for axonal

Brain Repair, edited by Mathias Bähr. ©2006 Eurekah.com and Kluwer Academic / Plenum Publishers.

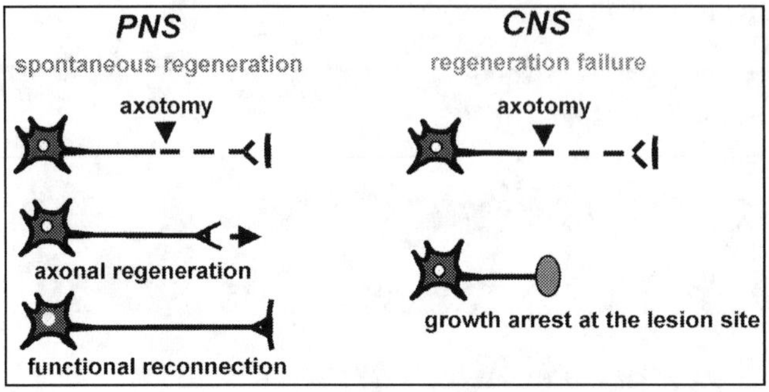

Figure 1. Different responses following axotomy in the peripheral and the central nervous system of adult mammals.

regeneration.[9-14] Some of these proteins are associated with the collagenous basement membrane (BM) that develops at the lesion site in the injured CNS.[1,15] The role of the lesion-induced collagenous BM as regeneration barrier will be described in this article.

Basement Membrane Formation in the Injured CNS Coincides with Axonal Growth Arrest

Regeneration studies were performed in the transected postcommissural fornix of the adult rat. This unidirectional fiber tract projects from the subiculum to the mammillary body. The lesion was performed with a Scouten wire knife as illustrated in Figure 2. A preshaped tungsten wire is located in a guidance cannula and the cannula is stereotactically placed into the rat brain below the transection level, then the wire is pushed out and the knife is lifted up to transect the fornix approximately 1,2 mm rostral to the target area.

Four days after transection some basement membrane (BM) sheets appeared in the lesion zone. After 6 days the BM became more dense and at 7 to 14 days an extensive BM was deposited at the lesion site (Fig. 3).

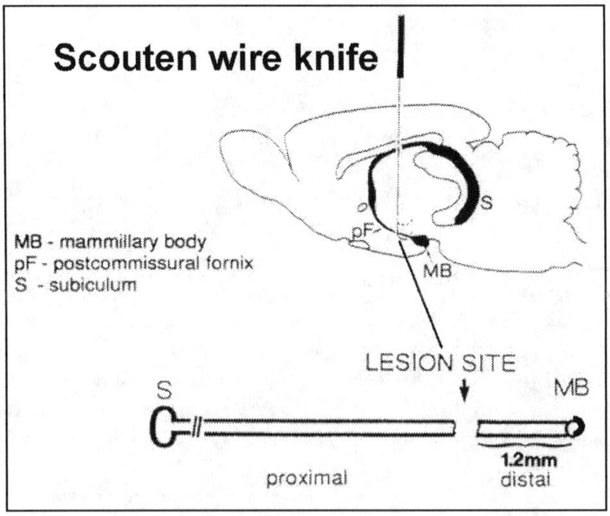

Figure 2. Transection of postcommissural fornix with a Scouten wire knife.

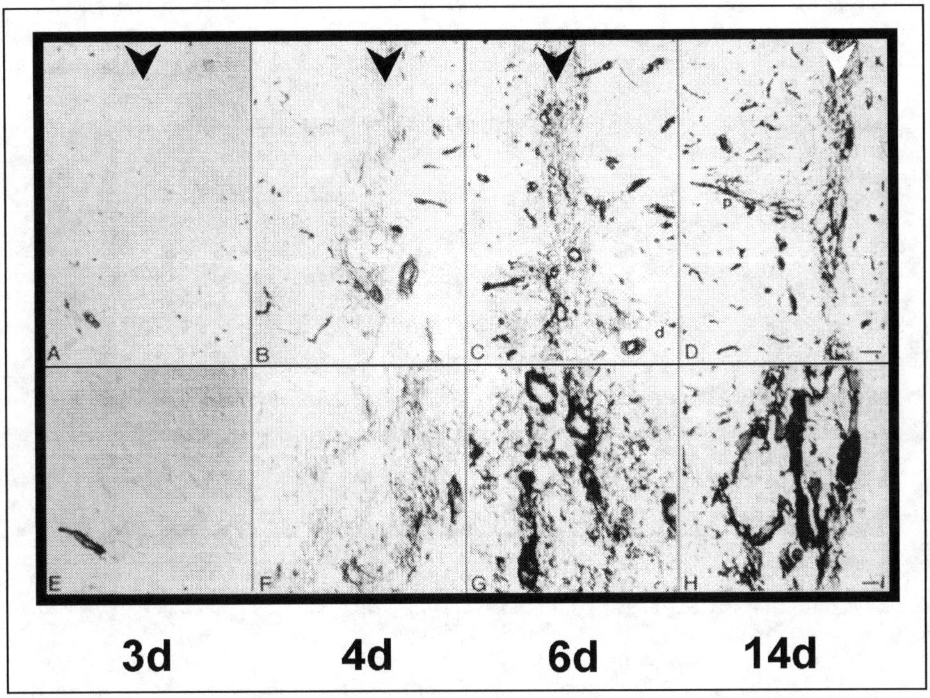

Figure 3. Time course of BM formation in the transected postcommissural fornix. First collagen type IV immunopositive structures are detectable at 4 days post lesion. Arrows mark the lesion site. Magnification bar in D for A-D: 50 μm, in H for E-H: 10 μm.

The extracellular localization of lesion-induced collagenous BM coincided exactly with the region of growth arrest of sprouting fibers, thus suggesting a role as barrier for axonal regeneration (Fig. 4).

The lesion scar was previously considered an obstacle for axonal regeneration by other researchers. Ramon y Cajal noted at the beginning of the last century that scar formation in the injured CNS contributes somehow to the regenerative failure of severed axons.[16] Thus it was not surprising that the lesion-induced BM has been targeted in CNS lesions before. In the early seventies of the last century Russian scientists reported a functional recovery in spinal cord injured animals treated with proteolytic enzymes.[17] Attempts of other groups to reproduce these experiments failed due to extensive bleeding following degradation of the intact BM of blood vessels by the enzymes used, thus enhancing secondary damage markedly.[18-20]

Inhibition of Collagen Biosynthesis in the Injured CNS

Our approach to suppress BM formation was based on two alternative treatments. The pharmacological approach comprised the inhibition of a key enzyme of collagen biosynthesis, prolyl 4-hydroxylase (PH). This enzyme catalizes hydroxylation of prolyl residues, a step in collagen biosynthesis that is crucial for thermal stability of procollagen triplehelices.[21] Since PH requires bivalent iron as a cofactor we injected the iron chelator 2,2'-bipyridine (BPY) stereotactically into the injury site. An alternative treatment to prevent BM formation was based on an immunochemical approach including injection of anti-collagen IV antibodies to neutralize secreted tropocollagen (Fig. 5).

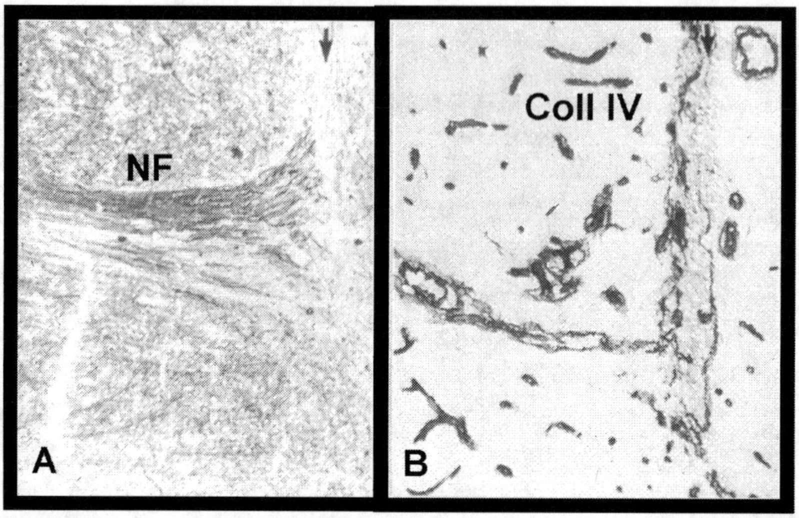

Figure 4. Neurofilament stained regrowing fornix fibers (A) stop elongating at the lesion-induced Collagen type IV immunopositive BM (B). Arrows mark the lesion site.

Both approaches were successful in preventing BM formation after fornix transection without affecting pre-existing blood vessels (Fig. 6).

In contrast to earlier attempts to remove the lesion-induced fibrous scar proteolytically the present approach allows a transient suppression of BM formation at the lesion site for approximately 14 days without destroying already existing functional BM in the brain parenchyma.

Suppression of BM Formation Leads to Extensive Axonal Regeneration of Severed Fibers

Suppression of BM formation after transection of the postcommissural fornix had an enormous effect on outgrowing fibers. The proximal stump of fibers in control animals retracted for a short distance after lesion, grew out again and stopped elongation at the lesion site (Fig. 4A and Fig. 7A). Treated animals in which collagen biosynthesis has been inhibited by injection of BPY or anti-collagen type IV antibodies showed a massive axonal regeneration across the lesion site (Fig. 7B).

The regenerating fornix fibers grew up to their former target, the mammillary body, developed chemical synapses and became remyelinated (Fig. 8).

We further investigated this therapeutic approach in another, clinically more relevant animal model, the injured spinal cord of the adult rat.

Transection of Rat Dorsal Corticospinal Tract

The dorsal corticospinal tract (CST) of the adult rat was transected at the level of Th8 using a Scouten wire knife (Fig. 9). We chose this kind of lesioning mainly because the partial wire knife transection created a cavity that allowed controlled application of solutions.

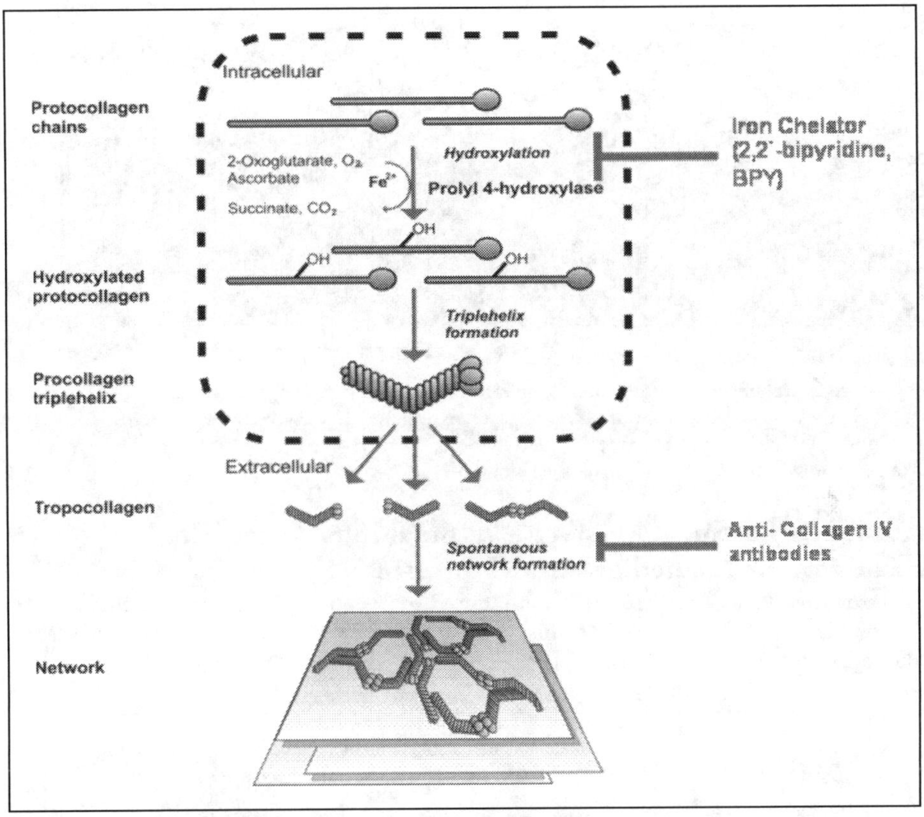

Figure 5. Biosynthesis of collagen and formation of the collagen IV network, a major component of the basement membrane acting as a scaffold to which numerous other glycoproteins and proteoglycans attach.

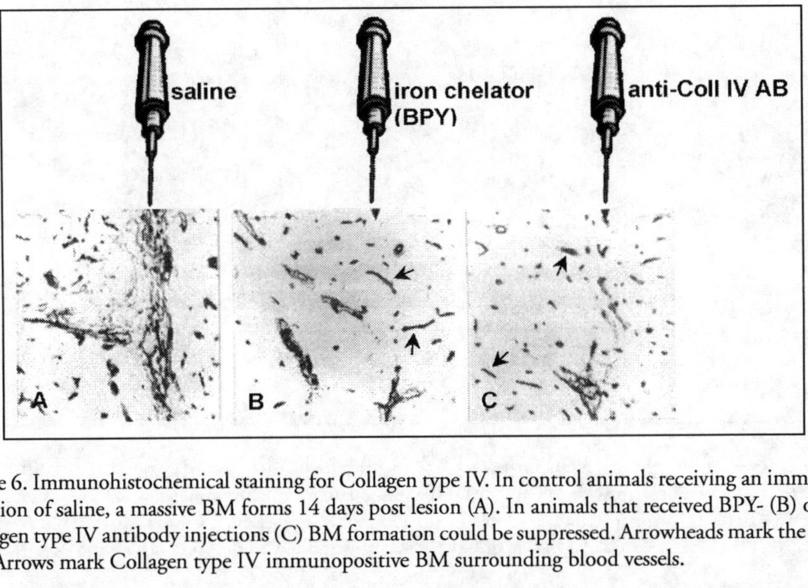

Figure 6. Immunohistochemical staining for Collagen type IV. In control animals receiving an immediate injection of saline, a massive BM forms 14 days post lesion (A). In animals that received BPY- (B) or anti Collagen type IV antibody injections (C) BM formation could be suppressed. Arrowheads mark the lesion site. Arrows mark Collagen type IV immunopositive BM surrounding blood vessels.

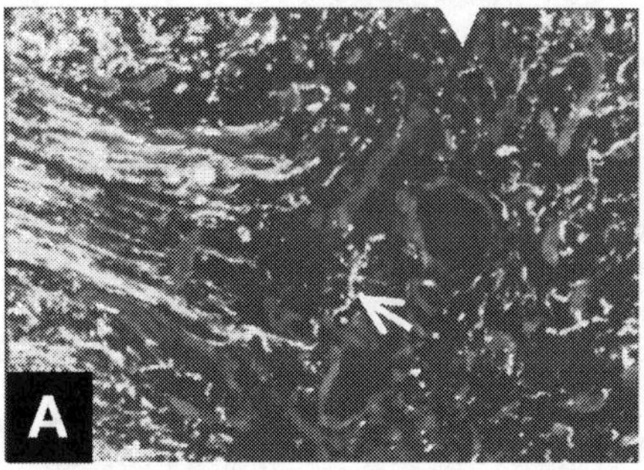

Figure 7. In control animals A) neurofilament stained fornix fibers stop growing when reaching the collagenous BM in the lesion core (arrowhead). Regrowing fibers even turn at the growth barrier (arrow). After pharmacological inhibition of BM formation severed fornix fibers traverse the lesion site and elongate in their former pathway B). Arrowheads mark the lesion site.

Detection of Basement Membrane in the Injured Spinal Cord Depends on the Method of Tissue Processing

Apparently, the detection of the lesion-induced collagenous BM was more difficult in spinal cord tissue than in the lesioned fornix. Using fresh frozen spinal cord tissue specimen, as we did in the fornix lesion, the basement membrane was completely washed out at the lesion

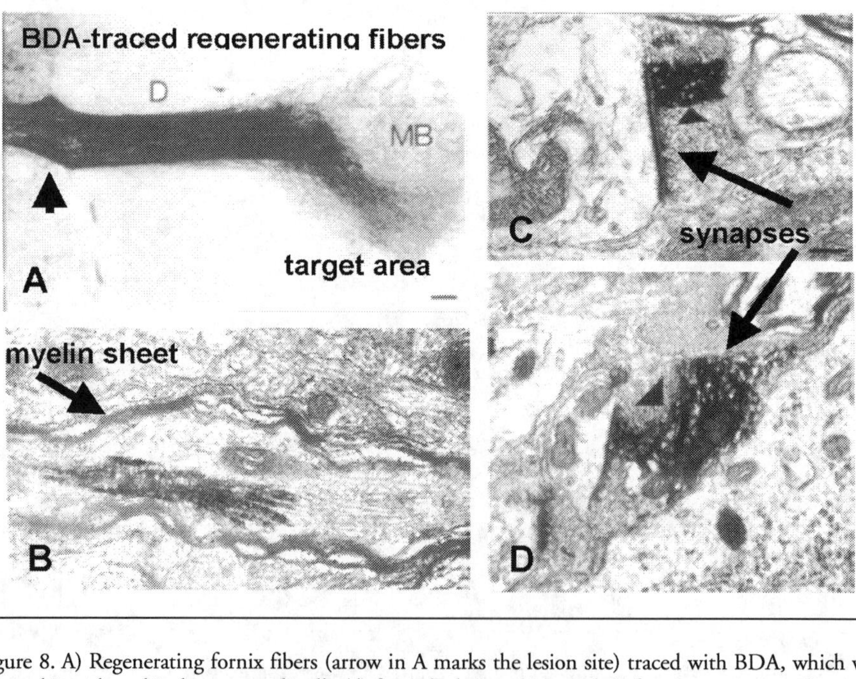

Figure 8. A) Regenerating fornix fibers (arrow in A marks the lesion site) traced with BDA, which was injected into the subiculum, grow distally (d) from the lesion site into their former target area, the mammillary body (MB). B) Ultrastructural evidence for myelination of regenerated traced axon distal to the lesion site. (C, D) Ultrastructural images of traced presynaptic endings in the mammillary body.

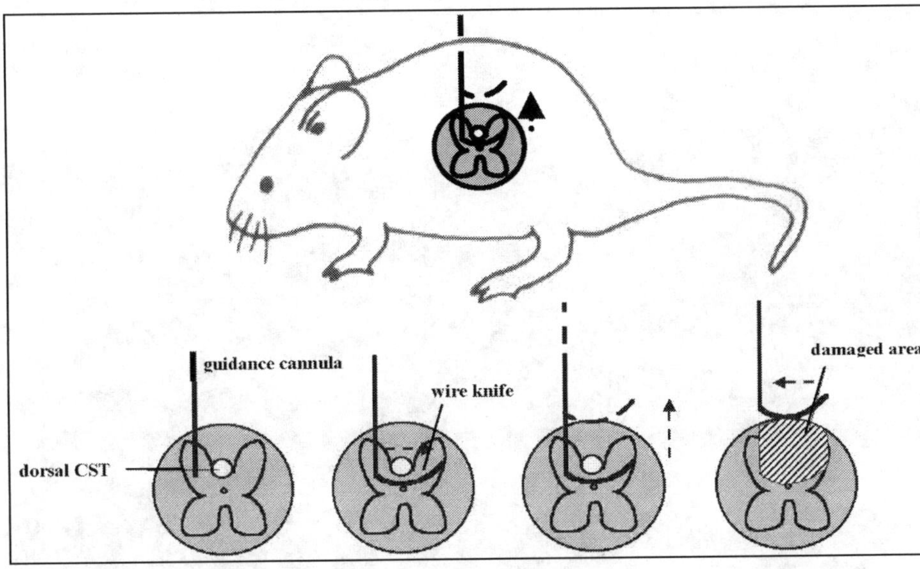

Figure 9. Transection of the dorsal corticospinal tract and dorsal columns in rat with a Scouten wire knife.

site during tissue processing. While BM of blood vessels was nicely stained, the fibrous scar in the lesion core could not be detected (Fig. 10A). In perfused tissue, stained free floating like most laboratories do, the lesion site was immunopositive for Collagen IV, but no blood-vessel staining could be detected in the surrounding spinal cord parenchyma. In order to determine suppression of BM formation at the lesion site, it was indispensable to use an internal reference for staining intensity. On the other hand, the fixative paraformaldehyde masked the collagen IV-antigen in regions where protein expression was not upregulated. To reveal a correct staining pattern we had to retrieve the antigen either by protease digestion or microwave treatment, which was problematic in free-floating sections. This is did not work well in free-floating sections. Additionally, we always observed some tissue loss in free-floating sections (Fig. 10B).

Therefore, we changed the tissue processing procedure to paraffin-embedding. This method required antigen retrieval by protease but led to reliable results with respect to tissue and basement membrane preservation as well as the detection of all Collagen type IV positive structures (Fig. 10 C, Ref. 22).

The Lesion-Induced BM Is an Extracellular Structure Not Associated with Blood Vessels

The collagenous scar or lesion-induced BM represents an unique structure of wound healing scars and should not be confounded with the glial scar, glia limitans or BM of newly formed blood vessels. In traumatic CNS lesions we find extracellular sheet-like structures that are not associated with blood vessels and which are immunopositive for the typical BM proteins collagen IV, laminin and nidogen. The endothelial marker von Willebrandt factor, however, does not stain extracellular sheet-like BM (Fig. 11).[23]

Extent of the Collagenous Scar Is Much Larger in Spinal Cord as Compared to Brain (Fornix) Lesions

Due to the close proximity of meningeal cells that invade spinal lesions[24,25] and secrete large amounts of collagen type IV,[26] the lesion induced BM is much more extensive in the

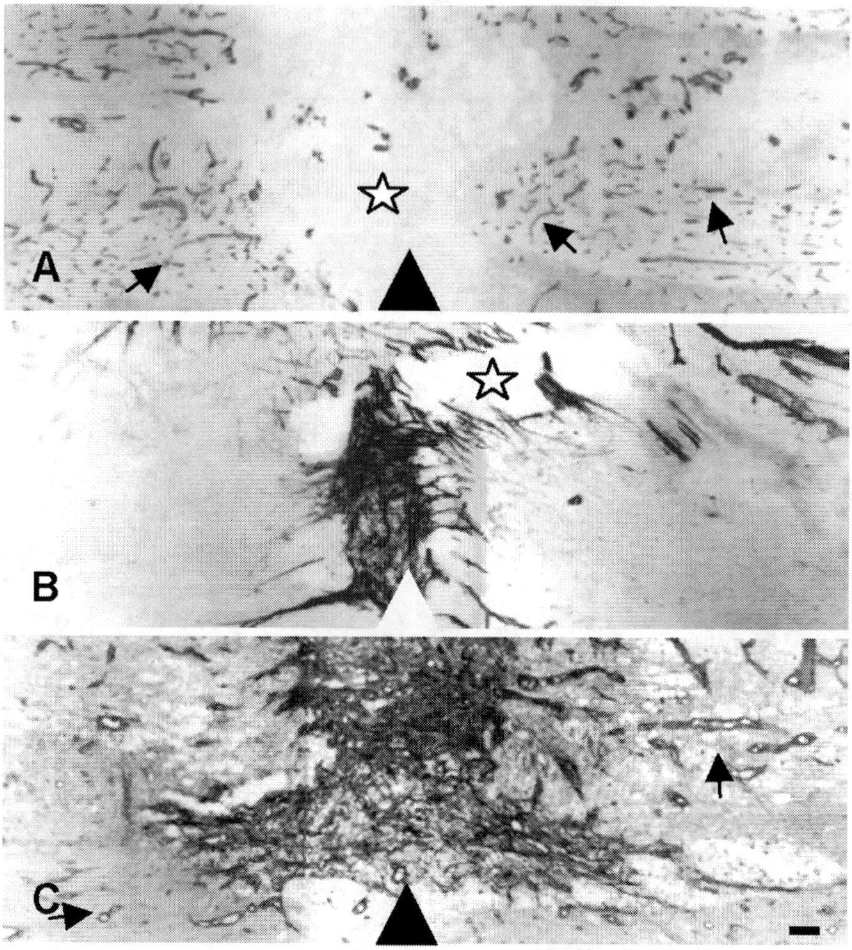

Figure 10. Collagen type IV immunostaining in transected spinal cord: A) fresh frozen tissue, B) paraformaldehyde fixed followed by free-floating immunostaining, and C) paraformaldehyde fixed and paraffin embedding. Arrowheads mark the lesion site, arrows point to stained BM of blood vessels and asterisks indicate tissue loss. Magnification bar in C for A-C: 100 μm.

spinal cord than the collagenous scar in fornix lesions (Fig. 12).[23,27] The fornix lesion was performed deeply in the brain (hypothalamus) which could not be invaded by meningeal cells from the brain surface. The collagenous BM in the postcommissural fornix lesion was probably produced by proliferating endothelial cells and/or astrocytes that have recently been described to express collagen type IV following CNS injury.[28] In lesioned spinal cord, however, leptomeningeal cells contribute significantly to lesion scarring.[29]

Because of this massive collagen expression, a single immediate injection of the formerly used iron chelator BPY turned out not to be successful in preventing scar formation after spinal cord lesions. Even continuous application of BPY by substance-soaked gelfoam and osmotic minipumps failed to reduce scar formation in the injured spinal cord (Fig. 13).[27]

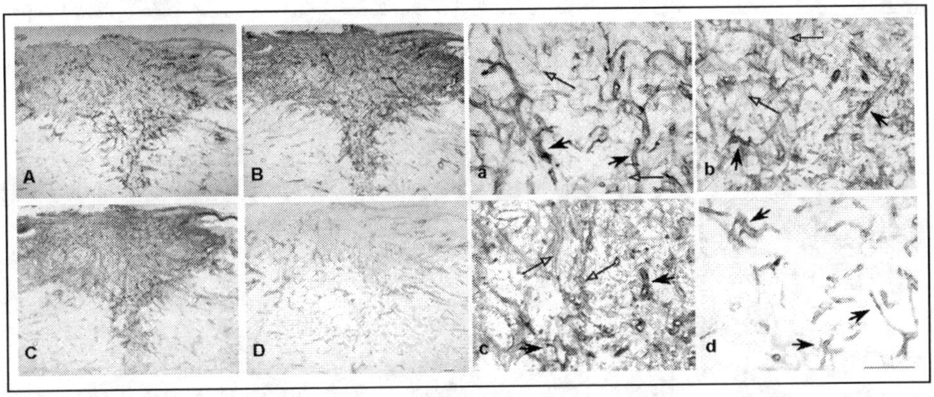

Figure 11. Immunohistochemical staining of the lesion scar for collagen type IV (A,a), Laminin (B,b), Nidogen (C,c) and von Willebrandt factor (D,d). The BM markers stain BM of blood vessels (black arrows in a-c) but also sheet-like immunopositive structures that are not associated with blood vessels (open arrows in a-c). The endothelial cell marker von Willebrandt factor solely stains blood vessels. Magnification bar in D for A-D and in d for a-d: 100 μm

Fibroblast proliferation and extracellular matrix production following traumatic injuries can be prevented by elevation of intracellular cAMP levels

Traumatic injuries lead to an inflammatory response at the lesion site which involves release of certain cytokines in the lesion area. Transforming growth factor β (TGFβ) is known as a primary regulatory factor in the formation of connective tissue in wound healing processes.[30] Its secretion results in an increased proliferation of fibroblasts, an increased ECM production, and a decreased ECM degradation mediated by direct inhibition of protease activity and stimulation of protease-inhibitor synthesis.[31,32] The release of the cytokine TGFβ by immunocompetent cells results in transcription of the CTGF gene in fibroblasts (Fig. 14A). CTGF or connective tissue growth factor is a mitogen and a chemotactic agent for fibroblasts[33] and stimulates their collagen- and fibronectin synthesis.[34] The protein is secreted by fibroblasts

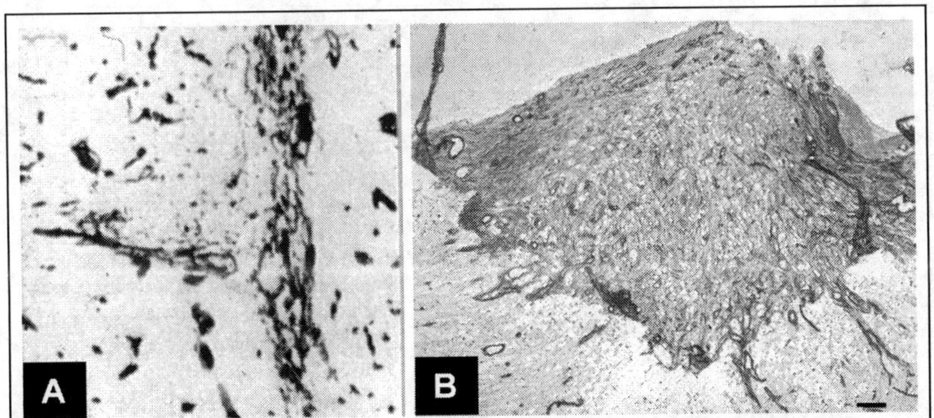

Figure 12. Immunohistochemical staining against Collagen type IV showing the much greater extent of the lesion-induced BM in spinal cord (B) as compared to fornix lesions (A). Magnification bar in B: for A 50 μm for B 100 μm.

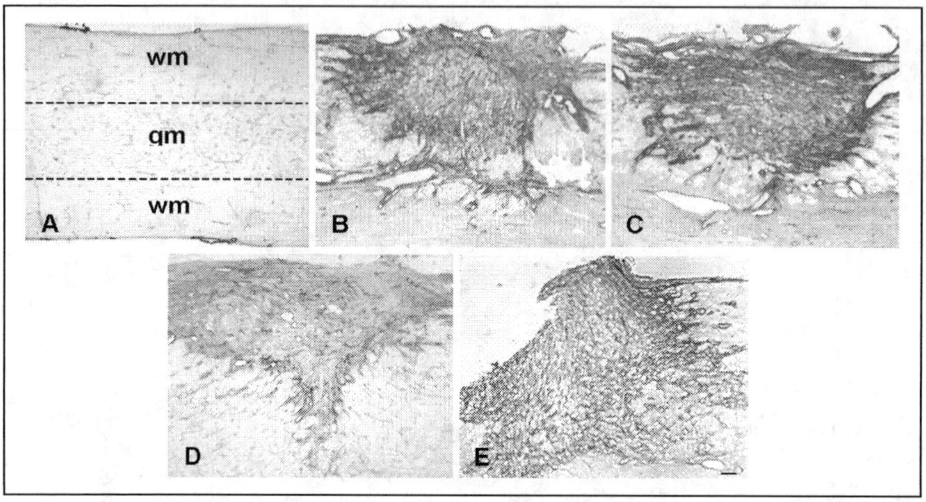

Figure 13. Immunohistochemical staining for collagen type IV. In the unlesioned cord A)Only BM of blood vessels are stained. Note that blood vessel B) distribution is more dense in gray matter (gm) than in white matter (wm). Application of BPY does not reduce BM formation: vehicle injection B), immediate injection of BPY 10 mM C), immediate injection of BPY 10 mM plus topical application of DPY-soaked gelfoam D), immediate injection of DPY 40mM plus continuous infusion of DPY (40 mM) by an osmotic minipump (E). Magnification bar in E for A-E: 100 µm.

in the presence of TGFβ and stimulates its own release via an autocrine mechanism. This TGFβ-mediated mechanism can be influenced by elevation of the intracellular cyclic AMP level (Fig. 14B).

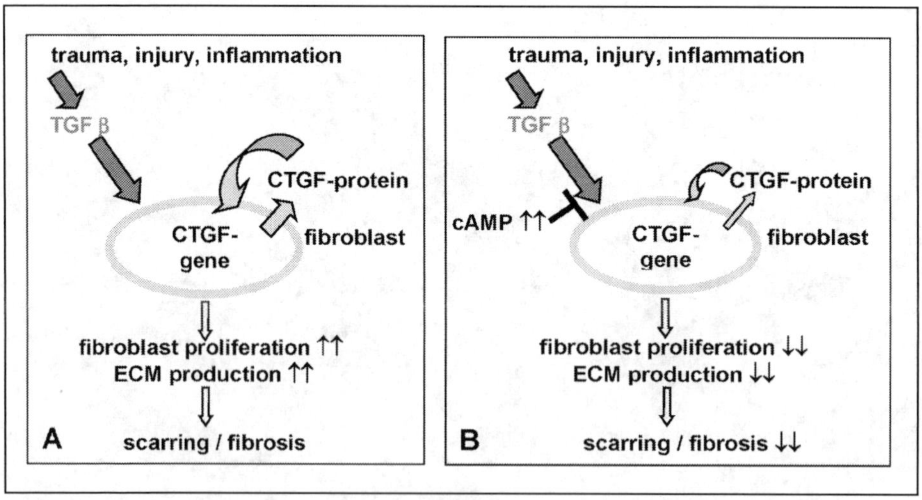

Figure 14. TGFβ release following traumatic injuries results in an increased fibroblast proliferation and ECM production that can be reduced by enhancing intracellular cAMP levels.

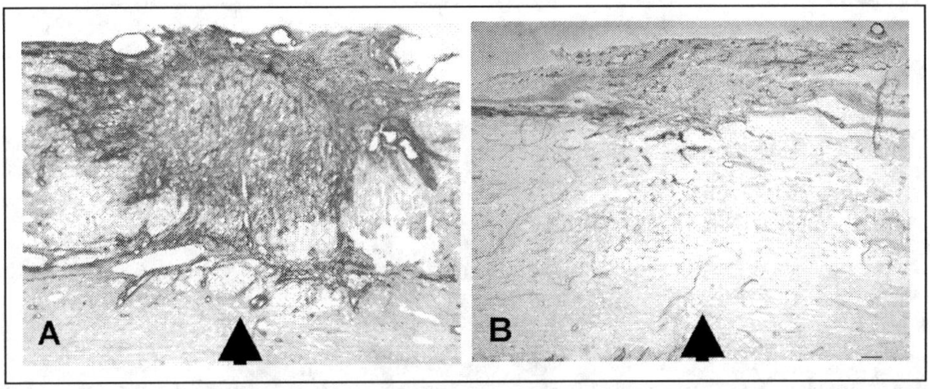

Figure 15. In control animals (A) a massive collagenous scar was formed 7 days post lesion. After combined application of an iron chelator and cAMP BM formation could be prevented (B). Arrowheads mark the lesion site and the asterix marks an extraspinal callus. Magnification bar in B for A + B: 100 μm.

An elevated cyclic AMP level in fibroblasts results in reduced transcription of the CTGF gene and, therefore, in reduced fibroblast proliferation[35] and extracellular matrix production.[30] The mechanism of this effect is not completely understood yet.

Combined Application of Iron Chelator and cAMP Prevents BM Formation in the Lesioned Spinal Cord

To reduce BM deposition in the spinal cord a combined treatment of more than one factor was required. The effective treatment consists of (1) multiple injections of dicarboxylated BPY (BPY-DCA), a very potent PH inhibitor, (2) topical application of solid 8-Br-cAMP to reduce fibroblast proliferation and ECM production and (3) continuous release of BPY-DCA from chelator-loaded Elvax copolymers. This treatment finally resulted in a complete reduction of lesion-induced BM in the spinal cord (Fig. 15).[23,27]

Suppression of BM Formation in the Lesioned Spinal Cord Promotes Corticospinal Fiber Outgrowth

We provide first evidence that suppression of BM formation in the injured spinal cord strongly supports corticospinal fiber regeneration. In treated animals anterogradely BDA traced corticospinal fibers could be detected within the lesion zone as well as in the distal cord (Fig. 16).[23]

The traced CST fibers differed in morphology compared to unlesioned fibers. While noninjured fibers appeared straight, regenerating axons showed a winding growth.

Numerous regenerating fibers grew along meningeal interfaces. To confirm that no false immunopositive staining at tissue borders occurred, we performed immunofluorescent double staining for BDA and neurofilament. Confocal images showing that BDA and neurofilament colocalize confirmed that regenerating corticospinal fibers are growing along the spinal cord / meningeal interface.[23] These experiments are still in progress and are currently analyzed in detail.

We showed that the collagenous lesion scar plays an important role as a hurdle for axonal regeneration in traumatic CNS lesions. Since it is well known that collagen IV is a good substrate for outgrowing fibers in vitro, it is unlikely that the collagen IV protein itself inhibits axonal regeneration. Further, the mere presence of fibroblasts / meningeal cells is probably not impeding axonal growth, because it has been shown that regenerating CST fibers grow along

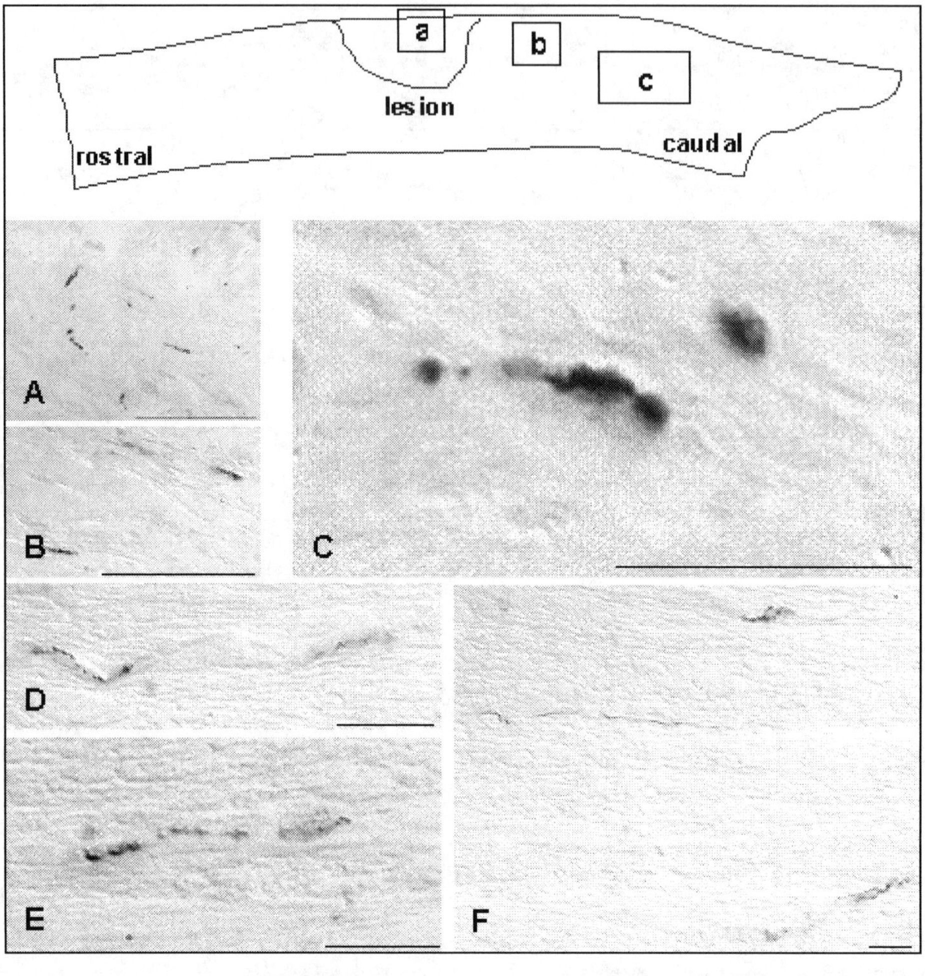

Figure 16. Regenerating BDA-traced corticospinal axons. A-C) Photomicrographs taken from the lesion site (region a in the schematic drawing on top). Higher magnification in (C) shows the winding morphology of a regenerating axon. D, E) Photomicrographs taken 2 mm distal to the lesion core (region b). F) Photomicrograph taken 5 mm distal to the lesion core (region c). Magnification bars in (A-C) and (F) 100 μm. Magnification bars in (D, E) 50 μm.

meningeal interfaces in vivo in different lesion and regeneration models.[35,23] However, it is much more likely that the lesion-induced collagenous basement membrane acts as a scaffold or anchor for a variety of growth inhibitory molecules including, e.g., certain proteoglycans and repulsive guidance molecules.

Acknowledgments

The authors are grateful for the excellent technical help of D. Steinhoff. Supported by the Deutsche Forschungsgemeinschaft, SFB 194/B5.

References

1. Stichel, CC and Müller, HW. Experimental strategies to promote axonal regeneration. Prog Neurobiol 1998; 56: 119-148.
2. Schnell L, Schwab ME. Axonal regeneration in the rat spinal cord produced by an antibody against myelin-associated neurite growth inhibitors. Nature 1990; 343: 269-272.
3. Cai D, Shen Y, De BM et al. Prior exposure to neurotrophins blocks inhibition of axonal regeneration by MAG and myelin via a cAMP-dependent mechanism. 1999; Neuron 22: 89-101.
4. Snow DM, Lemmon V, Carrino DA et al. Sulfated proteoglycans in astroglial barriers inhibit neurite outgrowth in vitro. Exp Neurol 1990; 109: 111-130.
5. Bovolenta P, Fernaud-Espinosa I, Mendez-Otero R et al. Neurite outgrowth inhibitor of gliotic brain tissue. Mode of action and cellular localization, studied with specific monoclonal antibodies. Eur J Neurosci 1997; 9: 977-989.
6. Moon LD, Brecknell JE, Franklin RJ et al. Robust regeneration of CNS axons through a track depleted of CNS glia. Exp Neurol 2000; 161: 49-66.
7. Pasterkamp RJ, De WF, Giger RJ et al. Role for semaphorin III and its receptor neuropilin-1 in neuronal regeneration and scar formation? Prog Brain Res 1998; 117: 151-170.
8. Bundesen LQ, Janis LS, Bregman BS et al. Alterations of EphB2 proteins following spinal cord injury in the adult rat. 2000; Abstract Book of the 30st annual meeting of the Society for Neuroscience.
9. Fidler PS, Schuette K, Asher RA et al. Comparing astrocytic cell lines that are inhibitory or permissive for axon growth: the major axon-inhibitory proteoglycan is NG2. J Neurosci 1999; 19: 8778-8788.
10. Bovolenta P, Fernaud-Espinosa I. Nervous system proteoglycans as modulators of neurite outgrowth. Prog Neurobiol 2000; 61: 113-132.
11. Probstmeier R, Stichel CC, Müller, HW Chondroitin sulfates expressed on oligodendrocyte-derived tenascin-R are involved in neural cell recognition—functional implications during CNS development and regeneration. J Neurosci Res 2000; 60: 21-36.
12. Becker T, Anliker B, Becker CG et al. Tenascin-R inhibits regrowth of optic fibers in vitro and persists in the optic nerve of mice after injury. GLIA 2000; 29: 330-346.
13. Asher RA, Morgenstern DA, Moon LD et al. Chondroitin sulphate proteoglycans: inhibitory components of the glial scar. Prog Brain Res 2001; 132: 611-619.
14. Bradbury EJ, Moon LD, Popat RJ et al. Chondroitinase ABC promotes functional recovery after spinal cord injury. Nature 2001; 416: 636-640.
15. Stichel CC, Niermann H, D'Urso D et al. Basal membrane-depleted scar in lesioned CNS: characteristics and relationships with regenerating axons. Neuroscience 1999; 93: 321-333.
16. Ramon y Cajal S, Degeneration and regeneration of the nervous system. 1928. New York, Hafner.
17. Matinian LA, Andreasian AS (1973) Enzyme therapy in organic lesions of the spinal cord. Akademia Nauk Armenian SSR, 94pp (English translation: Los Angeles, Brain Information Service, University of California, 1976, 156pp).
18. Feringa ER, Kowalski TF, Vahlsing HL et al. Enzyme treatment of spinal cord transected rats. Ann Neurol 1979; 5: 203-206.
19. Matthews MA, St Onge MF, Faciane CL et al. Spinal cord transection: a quantitative analysis of elements of the connective tissue matrix formed within the site of lesion following administration of piromen, cytoxan or trypsin. Neuropathol Appl Neurobiol 1979; 5: 161-180.
20. Guth L, Albuquerque EX, Deshpande SS et al. Ineffectiveness of enzyme therapy on regeneration in the transected spinal cord of the rat. J Neurosurg 1980; 52: 73-86.
21. Kagan HM. Intra- and extracellular enzymes of collagen biosynthesis as biological and chemical targets in the control of fibrosis. Acta Trop 2000; 77: 147-152.
22. Hermanns S, Müller HW. Preservation and detection of lesion-induced collagenous scar in the CNS depend on the method of tissue processing. Brain Res Protoc 2001; 7: 162-167.
23. Hermanns S, Klapka N, Müller HW. The collagenous lesion scar—an obstacle for axonal regeneration in brain and spinal cord injury. Restor Neurol Neurosci 2001a; 19: 139-148.
24. Carbonell AL, Boya J. Ultrastructural study on meningeal regeneration and meningo-glial relationships after cerebral stab wound in the adult rat. Brain Res 1988; 439: 337-344.

25. Halfter W, Liverani D, Vigny M et al. Deposition of extracellular matrix along the pathways of migrating fibroblasts. Cell Tissue Res 1990; 262: 467-481.
26. Eckes B, Zigrino P, Kessler D et al. Fibroblast-matrix interactions in wound healing and fibrosis. Matrix Biol 2000; 19: 325-332.
27. Hermanns S, Reiprich P, Müller HW. A reliable method to reduce collagen scar formation in the lesioned rat spinal cord. J Neurosci Methods 2001b; 110: 141-146.
28. Liesi P, Kauppila T. Induction of type IV collagen and other basement-membrane-associated proteins after spinal cord injury of the adult rat may participate in formation of the glial scar. Exp Neurol 2002; 173: 31-45.
29. Shearer MC, Fawcett JW. The astrocyte/meningeal cell interface—a barrier to successful nerve regeneration? Cell Tissue Res 2001; 305: 267-273.
30. Duncan MR, Frazier KS, Abramson S et al. Connective tissue growth factor mediates transforming growth factor beta-induced collagen synthesis: down-regulation by cAMP. FASEB J 1999; 13: 1774-1786.
31. Madri JA, Reidy MA, Kocher O et al. Endothelial cell behavior after denudation injury is modulated by transforming growth factor-beta1 and fibronectin. Lab Invest 1989; 60: 755-765.
32. Grotendorst GR. Connective tissue growth factor: a mediator of TGF-beta action on fibroblasts. Cytokine Growth Factor Rev 1997; 8: 171-179.
33. Bradham DM, Igarashi A, Potter RL et al. Connective tissue growth factor: a cysteine-rich mitogen secreted by human vascular endothelial cells is related to the SRC-induced immediate early gene product CEF-10. J Cell Biol 1992; 114: 1285-1294.
34. Frazier K, Williams S, Kothapalli D et al. Stimulation of fibroblast cell growth, matrix production, and granulation tissue formation by connective tissue growth factor. J Invest Dermatol 1996; 107: 404-411.
35. Kothapalli D, Hayashi N, Grotendorst GR. Inhibition of TGF-beta-stimulated CTGF gene expression and anchorage-independent growth by cAMP identifies a CTGF-dependent restriction point in the cell cycle. FASEB J 1998; 12: 1151-1161.
36. Ramon-Cueto A, Cordero MI, Santos-Benito FF et al. Functional recovery of paraplegic rats and motor axon regeneration in their spinal cords by olfactory ensheathing glia. Neuron 2000; 25: 425-435.

Chapter 12

Role of Endogenous Neural Stem Cells in Neurological Disease and Brain Repair

Jörg Dietrich and Gerd Kempermann

Introduction

There is abundant evidence that neural stem cells persist in the adult mammalian brain—including humans—throughout lifetime and support ongoing neurogenesis in restricted regions of the central nervous system (CNS). The potential role of neural stem cells not only in normal brain function, but also in neurological disease and repair now appears to be larger than anticipated only a few years ago. The question, however, remains whether the persistence of adult stem cells, their proliferation, and neurogenesis from these progenitors reflect the ability for self-repair in the mammalian brain. We here discuss recent advances in the understanding of the role of endogenous stem cells in normal brain function and under circumstances of neurological disease.

Neural Stem Cells in the Mammalian CNS

Neural stem cells (NSCs) are defined by their potential for theoretically unlimited self-renewal, and their ability to generate cells of both neuronal and glial lineages. During development, stem cells are found in the ventricular zone of the CNS.[1,2] In the adult brain, neural stem cells are primarily restricted to two brain regions, the subventricular zone of the lateral ventricles[3-8] and the subgranular zone of the dentate gyrus[9-11]—both regions in which neurogenesis persists throughout adulthood (Fig. 1). In low numbers, stem or progenitor cells have also been derived from many other brain regions, including septum, striatum,[12] cortex,[10] optic nerve,[12,13] spinal cord[14] and retina.[15,16] Apparently, these cells comprise a quiescent population of stem cells with as yet unknown functional relevance for the brain.

Stem cells of the adult brain have traditionally been classified as "multipotent". This term reflects their potential for differentiation into multiple neuroectodermal lineages, but not beyond this tissue-specificity. More recent evidence, however, suggests that cells with greater differentiation potential ("pluripotency") can be derived from the adult brain.[1,17-19] Moreover, stem cells from outside the brain can give rise to neurons in vivo (Fig. 2), at least under specific experimental conditions.[20-22]

The field of neural stem cell biology is currently undergoing dramatic changes in its concepts of "stemness", tissue-specificity, and developmental potential.[23,24] For the purpose of this review we adhere to the classical concepts of neural stem cell biology. However, much of what

Brain Repair, edited by Mathias Bähr. ©2006 Eurekah.com
and Kluwer Academic / Plenum Publishers.

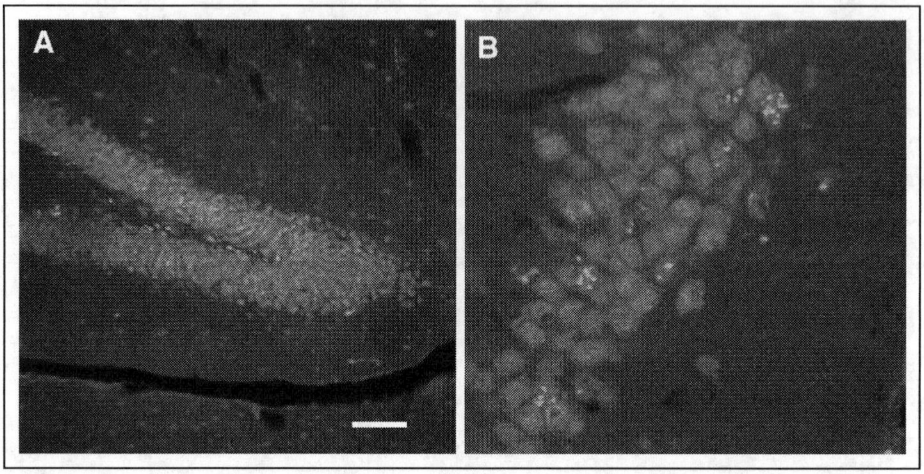

Figure 1. New neurons in the adult hippocampus. Confocal microscopic image of the hippocampus of an adult mouse, demonstrating neurogenesis in the granule cell layer of the dentate gyrus. A) Dividing cells in the subgranular zone (the border between granule cell layer and hilus) are labeled with proliferation marker bromodeoxyuridine (BrdU). B) Four to five weeks after cell division, newly generated neurons can be found throughout the granule cell layer. They are identified by their colocalization of immunoreactivity with antibodies against BrdU and neuronal marker NeuN. Astrocytes are identified by their expression of S100β. Scale bar (in A) equals 100μm for (A) and 20 μm for (B). Image from Kempermann, Bipolar Disorders 2002; 4:17-33, with kind permission of Munksgaard, Copenhagen, Denmark, ©2002.

will be discussed reflects a rather preliminary and probably simplifying view on the principles underlying stem cell biology in the adult brain.

Progeny from neural stem cells of the subventricular zone migrate along the rostral migratory stream to the olfactory bulb to differentiate into local interneurons.[25,26] In the hippocampus, neural stem cells give rise to new granule cells that extend their axons to area CA3 along the mossy fiber tract, as do all other granule cells of the dentate gyrus.[27-33] The new granule cells are electrophysiologically indistinguishable from older granule cells,[34] suggesting their functional integration.

Numerous factors that regulate adult hippocampal neurogenesis have been identified, but at present we are far from a unifying theory on which principles govern this regulation and which functional consequences it has (as reviewed in e.g., ref. 35).

Cells with stem-cell-like properties, dissected from diverse regions of the adult mammalian brain, can be induced to proliferate and differentiate in vitro in the presence of various growth factors, such as epidermal growth factor (EGF) or fibroblast growth factor (FGF-2).[2,11,36-38] Clonal analysis of these cells derived from the embryonic and adult brain has demonstrated their multipotency by giving rise to neurons, astrocytes and oligodendrocytes. This multipotency can also be detected in so-called neurospheres, a three-dimensional cell aggregates that are widely used to study neural stem cells in vitro.[37,39] With multiple neural stem cell populations loosely identified, questions arise where exactly these cells are located in the adult CNS and whether these stem cell populations are actually distinct cell types sharing similar potentials[40] or reflect different developmental stages that can be traced back to one unifying stem cell population. Interestingly, the isolation and characterization of neurospheres from different regions of the human embryonic CNS reveals a regionally specific pattern of growth and differentiation characteristics, suggesting the possible existence of distinct neural stem cell populations.[41] Consistent with these observations, there is evidence that stem cells

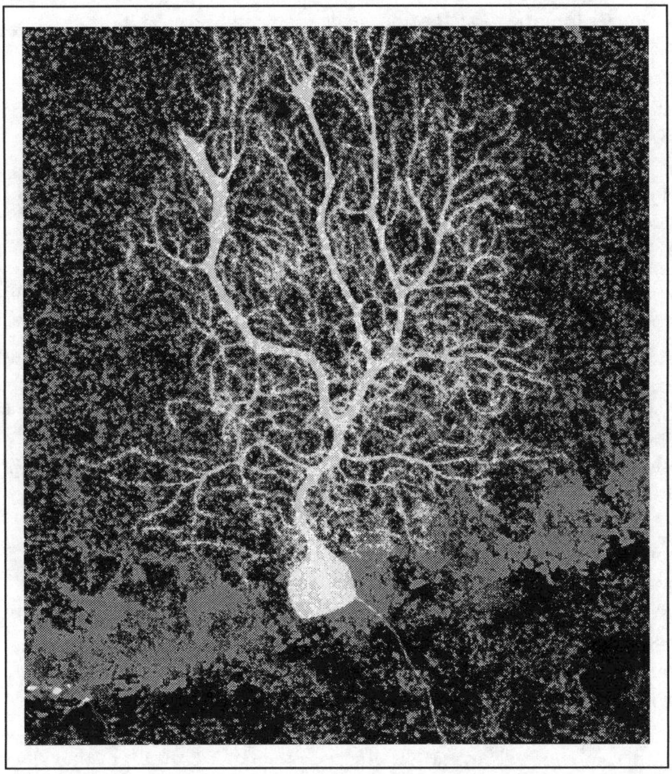

Figure 2. New neurons from bone marrow. Confocal microscopic image of bone marrow-derived Purkinje cells. One year after transplantation of bone marrow transduced with a retrovirus carrying green flourescent protein (GFP), donor-derived Purkinje cells were visualized by confocal microscopy. The surrounding staining is GAD (glutamate decarboxylase), identifying the neighboring GAD-expressing Purkinje cells (see Priller et al. J Cell Biol 2001; 155:733-738 for details. Image courtesy of Josef Priller, Berlin).

isolated from different brain regions maintain their regionally specific expression pattern of homeobox genes in vitro.[42] These results suggest that the identity of a particular stem or progenitor cell might be regionally and temporally specified depending on local environmental cues. However, in many respects, the behavior of the adult-derived stem and progenitor cells is indistinguishable from that of similar cells of the embryonic brain, suggesting some lineage continuity between the embryogenic and the adult CNS.[43]

Neural Stem Cells in Neurological Disease and Repair

The very existence of stem cells and neurogenesis in the adult brain throughout lifetime has shed new light on the potential of the brain for regeneration in the context of a variety of neurological diseases. In fact, several pathological conditions of the CNS have been associated with alterations in progenitor and stem cells – either as a consequence or as a cause of disease. In general, the following concepts for a stem cell-based therapy in the brain exist:

1. NSCs for direct replacement of lost cells from an identified neuronal population,
2. NSCs for replacement of glial or other cells with indirect effects on neurons (e.g., in spinal cord injury or multiple sclerosis),
3. NSCs for replacement of diffuse and complex cell losses (e.g., in stroke or trauma),

4. NSCs as vehicles for growth factor or gene delivery,
5. NSCs as basis of regeneration in situ, and
6. NSCs as target cells for other therapies based on the assumption that stem cells are involved in the pathology of a particular disease (e.g., depression, brain tumors).

In the following paragraphs we will discuss recent developments in which the concept of neural stem cells has been of special importance in changing the current understanding of neurological and psychiatric disease.

The Injured CNS

In contrast to other self-repairing tissues, such as the liver, skin or blood, the mammalian brain apparently lacks the regenerative potential to compensate adequately for neuronal and glial cell death, making this tissue particularly vulnerable to injury and disease.[44]

Repair strategies following CNS injury consist of different aspects of regeneration, including cellular replacement (by means of cell transplantation or endogenous stem cell activation), neurotrophic factor delivery, axon guidance and removal of growth inhibition, manipulation of intracellular signaling, bridging and artificial substrates, and modulation of the immune response (as reviewed in e.g., ref. 45).

For the purpose of this review we will specifically focus on recent findings in the contribution of endogenous neural stem cells to repair mechanisms.

Brain injury induced by traumatic lesions can cause a transient increase in proliferation of neural stem cells in the ventricle wall.[46-48] However, these studies could not clearly demonstrate any neuronal contribution of stem cells to the lesioned CNS.

As multipotent neural stem cells have been isolated from various regions of the adult mammalian brain, the failure of the normal brain to sufficiently regenerate under pathological conditions (e.g., traumatic brain injury) does not appear to be an intrinsic deficit of neural stem cells, but rather a characteristic feature of the damaged environment that does not sufficiently promote functional repair.

Nevertheless, the adult brain appears to be able to reorganize itself after *peripheral* injury and initial deficits in behavior, perception or cognition can be followed by a spontaneous recovery.[49-51] At least on a cortical level, this has been explained by the ability of the mammalian brain for cortical reorganization and plasticity.[52] From these studies, however, it cannot be concluded whether or to what extent adult neurogenesis contributes to such re-organizational processes.

In a more recent study Macklis and co-workers have demonstrated that neurogenesis can be induced in the lesioned neocortex of adult mice.[53] Endogenous precursor cells were stimulated by selective pyramidal cell apoptosis to generate cortical neurons that established appropriate corticothalamic connections. It has been speculated that either neural stem cells from the subventricular zone or resident cortical progenitors might have represented the source of these newborn neurons.[44] Thus, it seems possible that cells with stem-cell-like properties exist throughout the adult CNS. However, physiologically, these cells appear to give rise to neurons only in restricted neurogenic areas. Accordingly, it has also been suggested that adult neurogenesis represents the dormant capacity of the brain for a limited self-regeneration.[44] However, direct evidence is still missing that would clearly demonstrate the replacement of degenerated or dying neurons by newborn neurons. If this were the case, it would be important to know about the sequence of signals (e.g., released by apoptotic cell death) that are involved in the neurogenic response and that might direct newborn cells towards the lesioned area.

While the persistent neurogenesis in the brain with its apparent responsiveness to injury might reveal a possible endogenous repair program, the situation in the spinal cord seems to be somewhat different. A few studies report the existence of multipotent neural stem cells derived from adult spinal cord.[54-56] At present there is no convincing report on neurogenesis in the adult mammalian spinal cord. Progenitor cells in the adult rodent spinal cord produce glial cells.[57] Although there is increased proliferation of parenchymal progenitor cells[56] and nestin-expressing ependymal cells in the spinal cord after traumatic injury,[58] neurogenesis as a response to injury has not been found. This suggests that in case of spinal cord injury mobilizing endogenous neural stem cells to initiate neuronal repair remains a relatively distant possibility. However, at present no study on spinal cord injury has used such a highly specific and local induction of cell death as in the described study by Macklis and co-workers.[53]

Spinal cord injury is usually followed by a combination of neuronal and axonal damage, inflammation and demyelination. Thus, mobilizing more restricted progenitor cell pools (e.g., oligodendrocyte precursors) after spinal cord injury might—under appropriate conditions—contribute to myelin repair, regeneration of conduction velocity, and functional improvement (see ref. 59). Functional recovery frequently observed after spinal cord trauma in rodents and humans appears to be a consequence rather of axonal plasticity than of neural stem cell activation. [60] Nevertheless, the stimulation of local progenitor pools[59] and the enhancement of axonal plasticity, e.g., by local application of growth factors[61] and neuroprotective factor,[62-65] or other compounds[66,67] might become a useful approach to promote recovery after adult spinal cord injury.

Thus far, cell transplantation strategies for the injured brain and spinal cord have been performed using a variety of cell types and tissues, such as neuronal cell lines,[68] embryonic neuroblasts,[69,70] neural precursors,[71] oligodendrocyte precursors[72] and spinal cord tissue.[73]

In many of these studies, the expression of appropriate neurotransmitters by the grafted cells, the receiving of synaptic inputs from host neurons, or the establishment of long-distance projections could be demonstrated. In addition, functional improvements have been observed. Thus, cellular replacement of the injured CNS via transplantation might be possible, however, it seems to be critically dependent on the molecular host environment and the functional integration of the grafted tissue into the neuronal synaptic circuitry.

Taken together, in the field of spinal cord injury, stem cell research and the potential recruitment of endogenous cellular repair mechanisms hold great promises, but at present only few data exist that substantiate this optimism. However, the existence of neural stem cells in the adult mammalian brain and the positive effects of physical activity or the exposure to a complex environment on adult hippocampal neurogenesis suggest a potential practical impact of this research for neurorehabilitation.[52]

Neurodegenerative Diseases

Neurodegenerative diseases are characterized by a continuous loss of neurons with specific patterns of neuronal cell death associated with distinct disturbances in the neuronal network. Examples are the loss of dopaminergic input to the striatum from the substantia nigra in Parkinson's disease or the degeneration of cortical neurons with a cholinergic deafferentation in Alzheimer's disease.

In the light of our current understanding of the limited regenerative capacity of the adult mammalian CNS, the hypothesis has emerged that neurodegenerative diseases might actually reflect a failure of endogenous neural stem cells to replace lost neurons.[74] This "malfunction" of neuro-regeneration could be due to a primary failure of stem cell proliferation, migration, appropriate differentiation or a combination of all three, resulting in a lack of neurons at critically important topographical locations. At present there is no experimental evidence that a generalized theory of this kind could hold true, because with extremely few well-documented exceptions adult neurogenesis is restricted to the olfactory system and the hippocampal dentate gyrus.

Neurogenesis within neurogenic regions can be stimulated in vivo after exogenous administration of various growth factors and cytokines, including erythropoietin,[75] brain derived neurotrophic factor (BDNF),[76] Insulin-like growth factor I (IGF),[77] epidermal growth factor (EGF) and basic fibroblast growth factor (bFGF).[78,79] Moreover, cytokine infusion has been shown to stimulate neurogenesis in animal models of neurological disease. For example, infusion of transforming growth factor alpha (TGFα) to the forebrain of 6-hydroxydopamine lesioned rats (a model for Parkinson's disease) resulted in increased cell proliferation, directed migration of newly generated cells toward the infusion site and increased numbers of neurons in the striatum. This increase of neurons was associated with improvements in apomorphine-induced rotations of the animals (indicative of motor improvement).[80] This result fits well with data from the first study using transplantation of embryonic stem cells into a model of Parkinson's disease.[81]

Although very little is known about what signals direct newborn cells towards a particular CNS lesion and what factors orchestrate the appropriate site-specific differentiation of neural progenitors, the activation of endogenous stem cells to induce neurogenesis might be a possible means to overcome neuronal cell loss that occurs during the course of neurodegenerative disease. Despite encouraging first findings, this strategy remains speculative at the present time.

Most experiences in using neural stem cells for treatment of neurological disease have been made in Parkinson's disease.[82] Fetal cells have already been transplanted to severely impaired patients in clinical trials[83-86] and demonstrate that fetal human mesencephalic cells containing dopaminergic neurons can survive after transplantation, restore striatal dopamine release, and ameliorate motor behavior impairments. However, clinical and experimental studies have shown that functional integration of the grafted neurons within the host brain is necessary to produce substantial recovery.[87-89]

In Huntington's disease, a genetic disease characterized by a progressive neurodegeneration in the striatum and cerebral cortex, transplantation of fetal neural tissue has also offered a therapeutic opportunity.[82,90] In animal models of Huntington's disease, transplantation of fetal striatal neuroblasts to the striatum have been shown to be functionally integrated into the host environment and to restore striatal connections.[91] Reconstruction of neuronal circuitry by grafted tissue into the striatum could also be demonstrated in primates.[92,93] Moreover, transplantation of fetal striatal tissue has also been applied in patients with Huntington's disease.[90,94] In a clinical study, five patients with Huntington's disease who received fetal grafts were assessed for therapeutic outcome one year after transplantation. Patients with presumed surviving grafts (demonstrated by positron-emission tomography) showed improvements in motor and cognitive functions, and functional benefits were seen in daily-life activities.[94]

Due to a number of limitations using fetal grafts (e.g., ethical problems, survival of grafted cells, and problems in standardization and quality control), future efforts will also focus on the in vitro expansion and differentiation of neural stem cells as alternative sources to primary fetal CNS tissue for replacement therapy in neurodegenerative diseases.[82,95,96]

A cell-mediated gene therapy of the diseased CNS offers an alternative approach for the treatment of neurodegenerative diseaes.[97-100] For example, specific neurotransmitter release (e.g., γ-aminobutyric acid or dopamine) by transplanted neurotransmitter-synthesizing cells into the affected regions of the CNS has been shown to improve disease-related symptoms.[101,102]

Taken together, neurodegenerative diseases in which a defined cell type is being lost or damaged, such as Parkinson's disease, might be good candidates for a targeted stem cell therapy. In contrast, owing to the diffuse nature of neuronal and glial cell death that is associated with other neurodegenerative diseases such as Alzheimer's disease, repair of such disorders represents a potentially different category of problem than the repair of focal degeneration. Assuming that there is no primary deficit in the neural stem cells themselves as a cause of disease, a direct stimulation of endogenous neural stem cells through pharmaceutical or behavioral manipulation might increase brain plasticity and repair. Regardless, the continued cell loss during the course of neurodegenerative diseases will be challenging to overcome. It seems substantial, however, to define more clearly the cellular compartments affected in those disorders as a prelude to the analysis of cell protection or cell replacement strategies. It seems more likely that future therapies of neurodegenerative diseases will be advanced by general lessons on cellular development, survival and plasticity learned from stem cell biology than by the direct application of stem cells in these conditions.

Brain Tumors

Compelling theories are linking neural stem cell biology to neurological disease in the field of neuro-oncology. An increasing knowledge about neural stem and progenitor cells has started to shed light on the potential role of these cells in respect to tumorigenesis, brain tumor classification, and the treatment of brain tumors. Moreover, understanding the vulnerability of CNS precursor cells towards drug toxicity or irradiation might help to reveal the biological basis for brain damage frequently associated with cancer treatments.

Consistent with experiences in the treatment of tumors of the hematopoietic system, the diagnosis of brain tumors based on cell lineage appears to be of great potential value. In hemato-oncology, the normal cellular lineage to which a tumor is related seems to be closely correlated with treatment response and prognosis.[103-109]

At present, various precursor cell populations have been identified in the developing and/or mature CNS: EGF-dependent and FGF-dependent neuroepithelial stem cells (NSCs),[3,37,54,110-113] lineage-restricted precursor cells,[114-116] including neuron-restricted precursor (NRP) cells[114,117] and glial-restricted precursor (GRP) cells,[115] and oligodendrocyte-type-2 astrocyte (O2A) progenitor cells.[118,119] In addition, a number of astrocyte precursor cells [120-123] and a pre-O2A progenitor cell have been reported in the literature.[124] Both multipotent neural stem cells and lineage restricted precursor cells have also been identified in the adult mammalian brain, including humans.[11,125-131]

To make matters even more complex, there have been recent reports on stem cells with an overlapping developmental potential between the CNS and the hematopoetic system, which could give rise to both neural cells and blood cells.[23,132-136] Bartlett and colleages have isolated a more-than-multipotent stem cell from the adult mouse brain, whose developmental potential approached that of pluripotent embryonic stem cells.[1] Similar data exist for stem cells from bone marrow[137] and skin.[138] These data question the validity of the term "tissue specificity" of stem or progenitor cells and elucidate that we are far from a final systematic of stem cell biology in the adult.

All of the described cells could be potential targets of transformation events to initiate the development of a tumor in the CNS.[139] Thus, a lineage-based classification system for brain tumors might lead to the establishment of better prognostic criteria and might also help to define patient populations that would benefit from a particular treatment.[139-142]

Establishment of brain tumor models on the basis of CNS precursor cells will not only increase our understanding of potential genomic alterations during tumorigenesis, but will also provide helpful information on the relation between transformed precursor cells and the (heterogeneous) tumors they create in vivo.[140,143-145] In this regard, the cell lineage appears to be important in determining whether or not a particular genetic lesion will have functional consequence. For example, a specific genetic alteration could result in a different tumor-forming ability or tumor phenotype, depending on the precursor cell that was targeted.[146]

Several studies have described the expression of glial and neuronal markers in brain tumors, including astrocytomas, oligodendrogliomas, medulloblastomas and primitive neuroectodermal tumors (PNETS).[147-151] To date no markers are available that specifically and unambiguously label neural stem or precursor cells, only the development of antibodies to new cell type specific antigens, e.g., by gene expression analysis and microarray technology, might help to assign particular brain tumors to their lineages of origin. [152-154] For some tumor populations such analysis has already been successfully applied in order to cluster tumors with a similar prognosis.[155-157]

Neural stem cells might become useful as possible means of brain tumor treatment. To achieve this it will be important to understand to what extend endogenous stem and precursor cells influence or modulate tumor cells in the CNS. The most straightforward approach is to use stem cells as vehicle cells in gene therapy, building on the readiness of these cells to develop tissue specificity and become functionally integrated in the host organism. For example, neural stem cells genetically modified to produce interleukin-4 promoted tumor regression and prolonged survival in mice injected intracranially with a mouse glioma cell line.[158] In another study, implantation of a neural stem cell line, engineered to produce cytosine deaminase (which converts 5-fluorocytosine to the oncolytic drug 5-fluorouracil), into the CNS of mice harboring a tumor resulted in a reduction of the tumor mass in vivo, when animals were treated with 5-fluorocytosine.[159] Interestingly, implanted neural stem cells were detected in the primary tumor mass when injected at a distance from the tumor and were also seen to co-cluster with tumor cells at distances remote from the tumor injection site. This result is exciting since one of the most important impediments to the treatment of malignant brain tumors has been the invasion of tumor cells into the surrounding normal brain tissue, which makes their treatment particularly challenging.[160,161] Many more experiments will be required and many questions will have to be answered before it is clear whether neural stem cells can be used for the treatment of brain tumors (as reviewed in refs. 162,163). For example, do neural stem cells really migrate towards dispersed tumor cells or are they simply using the same migratory substrates leading to an occasional juxtaposition? Do endogenous or transplanted neural stem cells change their biological properties (e.g., differentiation and proliferation) in case of a present tumor mass? What are potential adverse effects of transplanting neural stem cells into the human brain? And would it be possible to visualize and monitor transplanted stem cells in vivo to design controlled clinical studies? In this regard, there have already been promising results with labeling neural stem or progenitor cells in order to make them detectable by magnetic resonance imaging after transplantation.[164-166]

A different aspect relating neural stem cells to oncology has been raised in regard to the vulnerability of the CNS to conventional cancer treatment. In fact, traditional approaches to cancer therapy are often associated with severe neurotoxicity. For example, radiation-induced neurological complications include leukoencephalopathy, radionecrosis, myelopathy, cranial nerve damage and cognitive impairment.[167-169]

Moreover, it has been well known that many chemotherapeutic regimens may be associated with severe neurotoxicity. For example, multiple reports have confirmed cognitive impairment in children and adults after chemotherapy. Neurotoxicity of chemotherapy may be particularly hazardous when combined with radiotherapy.[169,170] For example, computed tomography (CT) studies of patients receiving both brain radiation and chemotherapy showed that all patients surviving a malignant glioma for more than 4 years developed leukencephalopathy and brain atrophy.[171]

Thus, improvements in survival for children with leukemias or brain tumors treated with radiotherapy and chemotherapy have led to increasing concerns on quality-of-life issues for long-term survivors, in which neuropsychological testing has revealed a high frequency of cognitive deficits.[168,172-175]

Potential clues to the biological basis of neurotoxicity, such as cognitive impairment, have come from studies on the effects of radiation on the brain. On a cellular basis, radiation appears to cause damage to both dividing and non-dividing CNS cells. Irradiation has been shown to cause apoptosis in the subgranular zone of the hippocampal dentate gyrus [176,177] and in the subependymal zone,[178] both of which are sites of continuing stem or progenitor cell proliferation in the adult CNS. Such damage also is associated with long-term impairment of subependymal repopulation,[177] indicating that surviving stem cells are unable to regenerate the subependymal zone. This could in fact lead to a profound impairment of the overall cellular plasticity in the CNS.

Furthermore, non-dividing cells, such as oligodendrocytes, are also killed by irradiation,[179] which seems to be consistent with clinical evidence, where irradiation induces diffuse myelin and axonal loss, tissue necrosis and diffuse spongiosis of the white matter.[180] There is considerable discussion whether the damage caused by irradiation represents a direct or indirect effect on the brain. Although some damage in vivo might be secondary to vascular damage, radiation is also directly damaging important CNS populations, such as oligodendrocyte precursor cells.[181]

Recent studies have indicated that carmustine (BCNU), a lipophilic alkylating agent, is toxic for oligodendrocytes at doses that would be routinely achieved in patients. In contrast, astrocytes appear to be relatively resistant to BCNU, and O-2A cells showed intermediate levels of sensitivity.[182] The sensitivity of oligodendrocytes to BCNU raises the disturbing issue of whether the normal cells of the brain are damaged by exposure to chemotherapeutic agents. Preliminary results raise the concern that multiple types of neural precursor cells are at least as sensitive to death induced by chemotherapeutic agents as are cancer cells themselves (J. Dietrich and M. Noble, unpublished observations).

Taken together, in the field of neuro-oncology various intersections between neural stem cells and tumors of the CNS seem to emerge. An increasing knowledge about the lineage relationships and biological properties of different neural stem and precursor cells might help to better understand the process of tumorigenesis in the CNS and might also help to develop novel treatment options for future cancer therapies.

Demyelinating Diseases

Demyelination is a common feature of various neurological diseases with different underlying causes such as inflammation, automimmune reactions, degeneration, and trauma.[183-186] Multiple sclerosis (MS), as the most prominent example of a demyelinating disease, is characterized by chronic inflammatory focal demyelination associated with a variable degree of axonal loss.[187-189]

The appearance of demyelination and axonal loss very early in the course of the disease[190,191] suggests that strategies of myelin repair might be a possible means of protecting axons from further immunological insults.

In general, therapeutic strategies to promote myelin repair have focused on two major avenues: (1) cell transplantation to provide an exogenous source of cells which are competent to form myelin producing cells, and (2) recruitment of endogenous cell populations that are capable to produce myelin.[192]

Experimentally, the transplantation of certain types of cells, including oligodendrocyte precursor cells or multipotent stem cells, which are able to generate myelin-producing oligodendrocytes, can lead to remyelination of chronic demyelinated tissue.[193-197] Promising results in myelin repair and re-establishment of nerve conduction have come from the use of embryonic neural stem cells that were expanded in vitro and induced to the oligodendroglial lineage prior to transplantation.[192,193,198] However, regardless of the potential of transplanted cells to produce myelin, poor survival of grafted cells, lack of migration of these cells beyond the lesion site and therefore an unpredictable therapeutic outcome are current limitations to this approach.[199]

In demyelinated CNS regions a certain amount of remyelination occurs,[200-204] but remyelination in the adult damaged brain remains incomplete.[202,205-207]

The identity and origin of cells, that participate in endogenous remyelination has been unraveled to some degree. Multipotential stem cells, oligodendrocytes, or oligodendrocyte precursor cells are possible candidates involved in the remyelination process. Multipotent neural stem cells have been implicated in myelin repair.[208] For example, in a lysolecithin-induced demyelination of the corpus callosum progenitors of the rostral subventricular zone (SVZ) expressing the polysialylated form of the neural cell adhesion molecule (PSA-NCAM) proliferate and seem to migrate towards the lesion site and differentiate into glia but not neurons.[209]

Whereas oligodendrocytes of the adult forebrain are primarily postmitotic,[210] persistent cycling oligodendrocyte precursor cells (OPCs) might represent the most likely population and source for myelin repair.[72,210-212]

OPCs are not only present in the developing but also in the adult mammalian brain.[118,213-215] Their specific function in the normal brain is largely unknown. It has been hypothesized that they play a role in influencing neuronal activity[216,217] and synaptic growth.[218,219]

In the injured brain or in case of demyelination, oligodendrocyte precursor cells might form a glial population that can be activated to proliferate and to become involved in repair mechanisms by giving rise to myelinating oligodendrocytes.[209,220] This exciting potential has initiated research on the exact meachnisms that induce oligodendrocyte precursor cells migration, proliferation and differentiation to promote myelin repair.

Several studies have also indicated the persistence of mitotic oligodendrocyte precursor cells in the *adult human* subcortical white matter.[221-224] However, only small numbers of oligodendrocytes are generated in the intact adult mammalian brain.[225] Thus, subcortical white matter progenitors appear to be a quiescent population, and oligodendrocyte differentiation from these cells to a myelinating stage is considered to be a rare event.[215]

In contrast, glial progenitor proliferation can be found after injury and in several animal studies of induced demyelination.[192,220,226] Moreover, oligodendrocyte precursor cells have also been identified in multiple sclerosis lesions.[227,228]

However, they apparently fail to proliferate and to differentiate during chronic stages of the disease.[203] Reasons for the incomplete repair might lie in a profound axonal loss, the lack of sufficient precursor migration towards a lesion, insufficient precursor pools that could be mobilized or the lack of permissive environmental cues (e.g., growth factors and cytokines) to activate precursor cells.

For example, immature cycling progenitor cells of the adult subcortical white matter can be recruited to give rise to myelin-producing oligodendrocytes in response to experimental focal demyelination by lysolecithin,[229] but do not migrate towards the lesioned area so that only cells present at the site of demyelination can participate in remyelination.[229] Consequently, severe demyelination might damage all resident progenitors at one particular lesion site—as it is to be assumed in chronic MS lesions—and thus profoundly reduce the capacity for myelin repair in that region.

Therefore it would be a useful approach to direct the migration of cycling precursor cells towards a demyelinated area that has suffered depletion of its own precursor population.

Since oligodendrocyte precursors fail to survive and migrate when transplanted into the *intact* mammalian brain,[230,231] multiple environmental factors might be important to trigger progenitor cells to proliferate and differentiate. In fact, the balance between cell proliferation and differentiation appears to be mediated by local environmental cues, such as growth factors locally synthesized by surrounding neurons and glia. For example, glial growth factor 2 (a neuroregulin isoform) or Insulin-Growth-Factor 1, have been shown to promote remyelination in animal models of inflammatory demyelination.[232]

The immune system itself is likely to influence myelin repair.[233-235] For example, there is evidence that circulating immunoglobulins bind to oligodendrocyte surface antigens to promote remyelination,[236] and that antigen-antibody binding may facilitate the opsonization of myelin debris allowing repair to proceed.[236]

Taken together, it remains to be determined to what extent cell types such as neural stem or progenitor cells can contribute to myelin repair. It is a substantial scientific challenge to determine the signals involved in the activation of oligodendrocyte precursor cells and in the induction of multipotent CNS precursor cells to proliferate and differentiate into migratory oligodendrocyte precursor cells to enhance remyelination and to improve neurological deficits.

Seizures

Epilepsy is a common neurological condition that is characterized by recurrent seizures due to hyperactivity and synchronization of activity within neuronal populations (as reviewed e.g., in refs. 237-240). Animal models of induced seizure activity attempt to mimic these characteristic features of epilepsy. For example, *kindling* is a widely used model, in which repeated electrical stimulation of limbic areas leads to a stimulus-induced stable seizure activity, that resembles the temporal lobe epilepsy in humans.[241,242] Other commonly used seizure models are based on the stimulation of glutamatergic or cholinergic receptors by drugs such as kainate and pilocarpine.[243-245] For review on other widely used seizure models, see e.g., ref. 245.

Epileptic activity has been reported to result in a number of long-term alterations, such as altered gene and growth factor expressions,[246-249] neuronal cell loss in the hippocampus,[250-253] Ammon's horn sclerosis,[254-258] dendritic abnormalities of pyramidal cells,[259] and synaptic reorganizations within the hippocampus,[260-262] all of which have a potential impact on the neuronal circuits. These effects might induce a cascade of consequences, including alterations of glutamate receptor expression, glial hypertrophy, axonal growth and formation of new synapses that might contribute to an increased susceptibility to further seizures.[263]

In patients with temporal lobe epilepsy, nests of ectopic granule cells have been described.[264,265] Similarly, Houser and co-workers found aberrant sprouting of mossy fibers, the axons of granule cells, in the brains of these subjects.[266] These findings were first interpreted as reflecting a deranged development and thus a cause of the seizures. Research on adult hippocampal neurogenesis has allowed the alternate hypothesis that these changes are a direct consequence of seizures.[267-271] Parent et al were the first to report that pilocarpine induced seizures in rats lead to a transient increase in cell proliferation in the dentate gyrus.[272] Others have extended on these findings with similar models and similar findings.[273-278] Such increased proliferation corresponds to the up-regulation of several cytokines and mitogens, as described elsewhere.[279-283] Interestingly, it appears that stimulation of neurogenesis following kainate-induced seizures requires endogenously synthesized FGF-2, since this result cannot be seen in FGF-2 knockout mice. [284] While the induction of cell proliferation has been convincingly documented in several studies, less attention has been given to the question, whether a greater number of mature, functionally relevant neurons develop from these dividing cells and what their ultimate fate is. Parent et al have initially speculated that it might be the new cells that produce aberrant connections considered to sustain seizure activity,[272] but later provided arguments that this might not be the case (see below).[285]

Activity-induced cell proliferation in the dentate gyrus—and in some cases neurogenesis—has been demonstrated in both electrical[286,287] and chemically induced seizures,[272,273,277,285,288] suggesting a fundamental response mechanism as a result of synchronized neuronal activity. However, it remains to be shown, whether altered neurogenesis is a cause or consequence of increased seizure activity.

Blümcke et al reported an increased proliferation index as assessed by Ki-67 immunoreactivity in the dentate gyrus of children with early-onset temporal lobe epilepsy who had undergone surgery to remove the epileptogenic focus in the hippocampus.[289] They found an increase in nestin-labeled cells as a putative marker of progenitor cells in the dentate gyrus. Although the further development of the proliferating cells remains unclear, these finding are suggestive that the response of the human hippocampus is similar to the rodent hippocampus. Scharfman et al demonstrated that pilocarpine- and kainate-induced seizures in rats cause proliferation of hippocampal neural stem or progenitor cells whose progeny can migrate to the CA3 region of the hippocampus to give rise to ectopic $NeuN^+$ and $Calbindin^+$ granule cells.[277] This would be consistent with the observation by Houser et al in humans.[290] However, despite exhibiting granule cell specific intrinsic properties (e.g., membrane properties, firing behavior and morphology), these cells seem to be abnormally integrated into the CA3 network.[277]

Interestingly, conditions known to be able to induce neurogenesis, such as living in an enriched environment,[291] have not only been shown to be associated with reduced spontaneous apoptotic cell death in the rat hippocampus, but also to protect from kainate-induced seizure activity itself.[292] The relation between these two separate observations that is suggested by Young et al remains to be proven, but raises the question whether seizure-induced structural changes in the brain might be linked to altered stem cell activity.

It appears, however, that there are fundamental differences between the immature and the aged brain in regard to susceptibility to seizures and the functional consequences of seizures.[293-297] Despite the evidence that seizures result in a more profound cell death in aged animals compared to young animals,[295,298] seizures might have deleterious effects in the neonatal brain. For example, seizures in the developing brain can result in irreversible alterations in neuronal connectivity, as reviewed in 295. It has been reported that newborn animals receiving 10 daily electroconvulsive seizures have significantly smaller brains than controls.[298] Furthermore, seizures in the neonatal brain result in a reduced neurogenesis in the dentate gyrus, measured by BrdU incorporation and phenotypical characterization of newborn cells by the neuronal marker NeuN. In contrast, aged animals exposed to the same number of seizures show a significant increase in hippocampal neurogenesis.[299] While the underlying cause of the age-related differences is not exactly known, it has been speculated that increased glutamate release following seizures or a pronounced level of sensitivity to hypoxia in the neonatal brain might be partly responsible for altering the balance between cell death and birth.[299]

These observations are intriguing, because recurrent neonatal seizures could therefore—even in the absence of cell loss—have profound effects on brain development and might explain some of the late neurological impairments following recurrent seizure activity.[294,300-302]

In addition, recurrent seizures in the neonatal brain cause sprouting of mossy fibers into the inner molecular layer of the dentate gyrus and pyramidal layer of CA3 in rats. [303] While seizure-induced progenitor proliferation in the dentate gyrus can be inhibited by irradiation, synaptic remodeling of the mossy fiber pathway appears not to be altered.[285] Thus, it seems likely that mossy fiber synaptic reorganization is independent of neurogenesis, suggesting that sprouting arises from mature granule cells.[299]

A completely different question is whether the use of neural stem cells (either as transplanted cells or as recruitment of endogenous cells) might provide a possible means for the treatment of epilepsy.[82] For example, there have been several studies trying to circumvent the imbalance between excitatory and inhibitory neurotransmission in seizures by transplanting embryonic cells that release inhibitory neurotransmitters such as GABA. Although successful transplants resulted in seizure suppression, the underlying mechanisms of the graft action are mostly unclear and seizure suppression has so far only been transient.[304-307]

It remains to be established whether there is any potential therapeutic benefit to be derived from endogenous stem cell activity in response to seizures, and whether the seen effects are part of the epileptogenic pathology or attempts of endogenous regeneration.

In summary, repetitive seizures have been shown to lead to well-described neuropathological changes such as neuronal cell death, reactive gliosis, enhanced neurogenesis and axonal sprouting.[308] Most of these damages seen in animal models are similar to those seen in humans, e.g., in cases of intractable temporal lobe epilepsy. Many questions about the molecular mechanisms involved in these changes remain to be elucidated, as reviewed in e.g., ref. 309).

Better understanding the exact mechanisms that modulate proliferation and differentiation of neural stem and progenitor cells following seizure activity might offer potential targets for future therapies.

Ischemia

Cerebrovascular insults are a major cause for permanent neurological impairment. After cerebral infarction, necrotic tissue is usually not replaced and functional recovery of the patient is very limited. As in other areas, neural stem cell activity and neurogenesis in the adult brain might play a role in the clinical outcome of CNS disease such as cerebral infarction. Unfortunately, much of this general optimism is as yet backed by only limited experimental evidence. Nevertheless, the use of neural stem cells might offer future treatment options either as vehicle systems to deliver neurotrophic factors or as cell replacement therapy via transplantation. Recent studies have described a significant increase of cell proliferation in the subventricular zone and the dentate gyrus of the mammalian brain in response to vascular injury, such as global and focal ischemia in experimental models.[310-315]

Thus, transient ischemia could be a potent signal for inducing neural stem cell proliferation (as measured by BrdU-incorporation) and differentiation into neuronal and astroglial phenotypes (by co-labeling BrdU-positive cells with lineage specific markers). However, it is not known, what is actually reflected by the increase in the number of BrdU-labeled cells, which is interpreted as increased cell proliferation, because most conclusions are indirect. No specific positive markers for neural stem cells in vivo are known and most studies lack the examination of long survival periods in order to assess the net and long-term effects. Also, at least in the hippocampus, prolonged periods of global ischemia (> 2 minutes) seem to be necessary to significantly increase BrdU incorporation.[312] Despite several reports describing the proliferation and differentiation of hippocampal progenitors following global ischemia, further quantitative studies will be required to determine whether this also results in an sustained increase of granule neurons. Interestingly, BrdU incorporation in the subgranular zone of the dentate gyrus has been described to a more pronounced degree on the ipsilateral side, in case focal cerebral ischemia was applied,[310] suggesting that signals associated with cell death might locally stimulate cell proliferation.

Increased cell proliferation has also been described in the rat neocortex following transient global[316] and focal ischemia.[317] Newborn cells were distributed randomly in cortical layers II-VI with highest densities in the ischemic boundary zone.[316] Reactive neurogenesis in a photothrombotic stroke model has been reported,[317] which seems to fit well with cortical neurogenesis after phototoxic lesions [53] and the controversial report of spontaneous neurogenesis in cortical areas of the primate brain.[318] Photothrombotic stroke is an interesting model system that however lacks several features characterizing normal embolic or thrombotic ischemia. It remains generally conceivable that some ischemia-induced neurogenesis might also be present in the human brain in various brain areas and might even be a potential means for brain repair after stroke. However, the unequivocal demonstration of functional and lasting neurogenesis following ischemia has still to be made.

Mechanisms that have been shown to reduce vascular damage and ischemia-induced cell death, such as glutamate receptor blockade, have also been demonstrated to positively influence stem or progenitor cell proliferation.[319] Several mechanisms have been controversially discussed that might influence hippocampal neurogenesis after ischemia. Potential signals include changes of NMDA receptor signaling,[319-323] death of glutamatergic neurons that project into the granule cell layer,[322,324] dying hippocampal neurons,[272,273,325-329] growth factors or mitogenic factors such as FGF[330,331] and erythropoietin.[75]

In addition, age-related differences in stem or progenitor cell activity following ischemic insults appear to be important. While neurogenesis in the dentate gyrus following global ischemia seems to be less accelerated in aged and more pronounced in young animals,[332] young animals seem also to be more vulnerable to ischemic insults. Neonatal rats exposed to cerebral ischemia show a severe and sustained damage of the subventricular zone (SVZ) with necrotic and apoptotic cell death.[333] More specifically, oligodendrocyte precursor cells and neural stem cells appear to be particularly vulnerable resulting in subcortical white matter demyelination and profound cell loss in the SVZ three weeks after the insult.[333] Moreover, there is evidence for a maturation-dependent vulnerability of oligodendrocyte precursors to hypoxia and ischemia, which might explain the selective vulnerability of the periventricular white matter to hypoxia and ischemia seen in the premature infant.[334]

Thus, neurological impairment (e.g., cognitive and motor dysfunction) caused by asphyxia of the newborn might be due to damage to progenitors and stem cells of the CNS.

It has been hypothesized that grafted neural tissue may be a possible means for therapy of brain ischemia by either direct cell replacement or by releasing neurotrophic factors to the damaged brain, as reviewed e.g., in refs. 82,335. A variety of grafted cell types have been studied in ischemic brain models, including fetal cells and tissues,[336] immortalized cells[337] and genetically modified cells, as reviewed in e.g., 338. Numerous reports have demonstrated that transplanted cells were able to survive, to migrate preferentially toward the lesioned area, to differentiate into neuronal cells, to re-establish functional connections within the host animal, and to restore functional deficits.[335-337,339] A first clinical study by Kondziolka et al is an initial indication that transplanting cultured neuronal cells into the brains of humans after stroke is safe and could have functional benefits.[340] In general, however, clinical application of this strategy appears premature, because risks and potential benefits cannot yet be reasonably judged.

The establishment of a functional neuronal circuitry between the host and the grafted tissue is dependent on many variables, including the availability of trophic factors. Neurotrophic factors are essential to maintain the physiological function of glia and neuronal cells.[341-344] Furthermore, proliferation and differentiation of endogenous stem and progenitor cells is also dependent on appropriate neurotrophic signals.[345,346]

In pathological situations, as in the ischemic brain, neurotrophic factors protect brain tissue from experimentally induced damage.[347-351] For example, gene delivery of the *glial cell line-derived neurotrophic factor* (GDNF) into the rat brain one day before a transient middle cerebral artery occlusion resulted in a significantly smaller infarct volume and was associated with a reduction of apoptosis. [352] Consistent with these reports, neurotrophin receptors are up-regulated in cholinergic striatal interneurons after global cerebral ischemia, suggesting that neurotrophin signaling might be important for the survival and function of these cells. [325]

In summary, recent findings have raised hopes for novel treatment approaches of ischemic brain damages, including activation of endogenous neural stem cells and transplantation of neural grafts. However, the use of neural transplants for the treatment of CNS ischemia has to be considered with caution and further pre-clinical studies are needed to validate the safety and efficacy of such an approach before neural stem cells could be applied to stroke patients. Alternatively, there is evidence that endogenous progenitors and stem cells are activated and might be involved in repair mechanisms following ischemic brain injury. As of yet, the plasticity of the adult human brain in acute and chronic ischemic conditions is poorly understood. For example, compensatory reactions and functional recovery (as it can be seen in stroke patients) that have been thus far explained by synaptic or functional plasticity might in fact include a limited neuronal replacement, potentially far from the injury site.

Mood Disorders

Adult hippocampal neurogenesis affects hippocampal function and is thus potentially involved in higher cognitive functions. Some of the known factors that are able to induce neuronal cell death and to potently suppress hippocampal neurogenesis are psychosocial stress[353,354] and glucocorticoid hormones.[355-357]

Major depression, although not a hippocampal disease in a strict sense, shows hippocampal impairment, for example with regard to symptoms of dementia and memory loss.[358,359] Recent studies have indicated that the pathogenesis and treatment of depression is likely to involve the impaired plasticity of neuronal circuits within the hippocampal formation. Thus, stress-induced impairment of dentate gyrus neurogenesis has been linked to the onset of clinical phases of depression [360]

In accordance with this hypothesis, studies using magnetic resonance imaging demonstrated selective atrophies in the *limbic-cortical-striatal-pallidal-thalamic tract*, which consistently includes a volume loss of the hippocampus in various psychiatric disorders, like in long-standing depression.[361] These findings were complemented by postmortem observations of hippocampal atrophy and cell loss in patients with mood disorders.[362-364] These structural changes correlate with deficits in declarative, spatial and contextual memory performance, supporting a link between hippocampal dysfunctions and the development and clinical appearance of certain psychiatric conditions.[365]

This hypothesis has led to the assumption that remodeling the hippocampal network, e.g., by increased neurogenesis, might be a possible means to influence the outcome of stress-related mood disorders.[355,366-368] Hence, circumstantial evidence to support this hypothesis has come from several studies showing that drugs used for treatment of depression, including tricyclic antidepressants and serotonin re-uptake inhibitors, as well as electroconvulsive therapy and physical activity stimulate adult hippocampal neurogenesis.[287,369-372]

It has been suggested that antidepressants might therefore exert their therapeutic effects by stimulating changes in neuronal systems, such as by an increase in neurogenesis – possibly by enhancing the expression of growth and survival promoting factors like neurotrophins.[373] Interestingly, in the case of serotonin re-uptake inhibitors, stimulation of neurogenesis requires long-term treatment,[374] which is consistent with the clinical experience of a long latency period before onset of an anti-depressive effect.

Furthermore, lithium (in clinical use for the treatment of bipolar disorders) has an effect on adult hippocampal neurogenesis, too.[375] This effect is possibly mediated through the upregulation of the anti-apoptotic protein bcl-2.[375]

Stimulation of neurogenesis (e.g., by antidepressants) might thus inhibit or reverse the effects of stress-induced downregulation of hippocampal neurogenesis and hippocampus atrophy.[370,374,376]

At present, however, it remains unknown, whether a disturbance in adult hippocampal neurogenesis is a consequence, cause or correlate of major depression and bipolar disorders.[35] Several recent reviews have speculated about this potential pathogenic link.[355,360,377] As of yet, this relation remains an interesting hypothesis that still has to be substantiated by empiric and experimental evidence. Intriguingly, major depression and schizophrenia share some characteristic features such as hippocampal involvement. Whether an impairment of cellular plasticity within the hippocampus is involved in schizophrenia has been suggested,[378-380] but at present, this hypothesis is even more speculative than in the case of major depression.

Summary

These examples show that stem-cell-based therapy of neuro-psychiatric disorders will not follow a single scheme, but rather include widely different approaches. This is in accordance with the notion that the impact of stem cell biology on neurology will be fundamental, providing a shift in perspective, rather than introducing just one novel therapeutic tool. Stem cell biology, much like genomics and proteomics, offers a "view from within" with an emphasis on a theoretical or real potential and thereby the inherent openness, which is central to the concept of stem cells. Thus, stem cell biology influences many other, more traditional therapeutic approaches, rather than introducing one distinct novel form of therapy.

Substantial advances have been made in neural stem cell research during the past few years. With the identification of stem and progenitor cells in the adult brain and the complex interaction of different stem cell compartments in the CNS—both, under physiological and pathological conditions—new questions arise: What is the lineage relationship between the different progenitor cells in the CNS and how much lineage plasticity exists? What are the signals controlling proliferation and differentiation of neural stem cells and can these be utilized to allow repair of the CNS? Insights in these questions will help to better understand the role of stem cells during development and aging and the possible relation of impaired or disrupted stem cell function and their impact on both the development and treatment of neurological disease. A number of studies have indicated a limited neuronal and glial regeneration in certain pathological conditions. These fundamental observations have already changed our view on understanding neurological disease and the brain's capacity for endogenous repair. The following years will have to show how we can influence and modulate endogenous repair mechanisms by increasing the cellular plasticity in the young and aged CNS.

References

1. Rietze RL, Valcanis H, Brooker GF et al. Purification of a pluripotent neural stem cell from the adult mouse brain. Nature 2001; 412(6848):736-9.
2. Temple S. The development of neural stem cells. Nature 2001; 414(6859):112-7.
3. Reynolds BA, Weiss S. Generation of neurons and astrocytes from isolated cells of the adult mammalian central nervous system. Science 1992; 255(5052):1707-10.
4. Alvarez-Buylla A, Garcia-Verdugo JM. Neurogenesis in adult subventricular zone. J Neurosci 2002; 22(3):629-34.
5. Alvarez-Buylla A, Herrera DG, Wichterle H. The subventricular zone: source of neuronal precursors for brain repair. Prog Brain Res 2000; 127:1-11.
6. Goldman SA, Nedergaard M, Crystal RG et al. Neural precursors and neuronal production in the adult mammalian forebrain. Ann N Y Acad Sci 1997; 835:30-55.
7. Martens DJ, Tropepe V, van Der Kooy D. Separate proliferation kinetics of fibroblast growth factor-responsive and epidermal growth factor-responsive neural stem cells within the embryonic forebrain germinal zone. J Neurosci 2000; 20(3):1085-95.
8. Chiasson BJ, Tropepe V, Morshead CM et al. Adult mammalian forebrain ependymal and subependymal cells demonstrate proliferative potential, but only subependymal cells have neural stem cell characteristics. J Neurosci 1999; 19(11):4462-71.
9. Ray J, Peterson DA, Schinstine M et al. Proliferation, differentiation, and long-term culture of primary hippocampal neurons. Proc Natl Acad Sci USA 1993; 90(8):3602-6.
10. Palmer TD, Markakis EA, Willhoite AR et al. Fibroblast growth factor-2 activates a latent neurogenic program in neural stem cells from diverse regions of the adult CNS. J Neurosci 1999; 19(19):8487-97.
11. Gage FH. Mammalian neural stem cells. Science 2000; 287(5457):1433-8.
12. Palmer TD, Ray J, Gage FH. FGF-2-responsive neuronal progenitors reside in proliferative and quiescent regions of the adult rodent brain. Mol Cell Neurosci 1995; 6(5):474-86.
13. Kondo T, Raff M. Oligodendrocyte precursor cells reprogrammed to become multipotential CNS stem cells. Science 2000; 289(5485):1754-7.
14. Shihabuddin LS, Horner PJ, Ray J et al. Adult spinal cord stem cells generate neurons after transplantation in the adult dentate gyrus. J Neurosci 2000; 20(23):8727-35.
15. Tropepe V, Coles BL, Chiasson BJ et al. Retinal stem cells in the adult mammalian eye. Science 2000; 287(5460):2032-6.
16. Reh TA, Levine EM. Multipotential stem cells and progenitors in the vertebrate retina. J Neurobiol 1998; 36(2):206-20.
17. Bjornson CR, Rietze RL, Reynolds BA et al. Turning brain into blood: a hematopoietic fate adopted by adult neural stem cells in vivo. Science 1999; 283(5401):534-7.
18. Johansson CB, Momma S, Clarke DL et al. Identification of a neural stem cell in the adult mammalian central nervous system. Cell 1999; 96(1):25-34.

19. Morshead CM, Benveniste P, Iscove NN et al. Hematopoietic competence is a rare property of neural stem cells that may depend on genetic and epigenetic alterations. Nat Med 2002; 8(3):268-73.
20. Mezey E, Chandross KJ, Harta G et al. Turning blood into brain: cells bearing neuronal antigens generated in vivo from bone marrow. Science 2000; 290(5497):1779-82.
21. Brazelton TR, Rossi FM, Keshet GI et al. From marrow to brain: expression of neuronal phenotypes in adult mice. Science 2000; 290(5497):1775-9.
22. Priller J, Persons DA, Klett FF et al. Neogenesis of cerebellar Purkinje neurons from gene-marked bone marrow cells in vivo. J Cell Biol 2001; 155(5):733-8.
23. Anderson DJ, Gage FH, Weissman IL. Can stem cells cross lineage boundaries? Nat Med 2001; 7(4):393-5.
24. Anderson DJ. Stem cells and pattern formation in the nervous system: the possible versus the actual. Neuron 2001; 30(1):19-35.
25. Lois C, Garcia-Verdugo JM, Alvarez-Buylla A. Chain migration of neuronal precursors. Science 1996; 271(5251):978-81.
26. Luskin MB. Restricted proliferation and migration of postnatally generated neurons derived from the forebrain subventricular zone. Neuron 1993; 11(1):173-89.
27. Altman J, Das GD. Autoradiographic and histological evidence of postnatal hippocampal neurogenesis in rats. J Comp Neurol 1965; 124(3):319-35.
28. Kaplan MS, Hinds JW. Neurogenesis in the adult rat: electron microscopic analysis of light radioautographs. Science 1977; 197(4308):1092-4.
29. Cameron HA, McKay RD. Adult neurogenesis produces a large pool of new granule cells in the dentate gyrus. J Comp Neurol 2001; 435(4):406-17.
30. Kuhn HG, Dickinson-Anson H, Gage FH. Neurogenesis in the dentate gyrus of the adult rat: age-related decrease of neuronal progenitor proliferation. J Neurosci 1996; 16(6):2027-33.
31. Stanfield BB, Trice JE. Evidence that granule cells generated in the dentate gyrus of adult rats extend axonal projections. Exp Brain Res 1988; 72(2):399-406.
32. Markakis EA, Gage FH. Adult-generated neurons in the dentate gyrus send axonal projections to field CA3 and are surrounded by synaptic vesicles. J Comp Neurol 1999; 406(4):449-60.
33. Hastings NB, Gould E. Rapid extension of axons into the CA3 region by adult-generated granule cells. J Comp Neurol 1999; 413(1):146-54.
34. van Praag H, Schinder AF, Christie BR et al. Functional neurogenesis in the adult hippocampus. Nature 2002; 415(6875):1030-4.
35. Kempermann G. Regulation of adult hippocampal neurogenesis – implications for novel theories of major depression. Bipolar disorders 2002; 4:17-33.
36. Palmer TD, Takahashi J, Gage FH. The adult rat hippocampus contains primordial neural stem cells. Mol Cell Neurosci 1997; 8(6):389-404.
37. Reynolds BA, Weiss S. Clonal and population analyses demonstrate that an EGF-responsive mammalian embryonic CNS precursor is a stem cell. Dev Biol 1996; 175(1):1-13.
38. Ciccolini F, Svendsen CN. Fibroblast growth factor 2 (FGF-2) promotes acquisition of epidermal growth factor (EGF) responsiveness in mouse striatal precursor cells: identification of neural precursors responding to both EGF and FGF-2. J Neurosci 1998; 18(19):7869-80.
39. Caldwell MA, He X, Wilkie N et al. Growth factors regulate the survival and fate of cells derived from human neurospheres. Nat Biotechnol 2001; 19(5):475-9.
40. Temple S. CNS development: The obscure origins of adult stem cells. Curr Biol 1999;9(11):R397-9.
41. Ostenfeld T, Joly E, Tai YT et al. Regional specification of rodent and human neurospheres. Brain Res Dev Brain Res 2002; 134(1-2):43-55.
42. Hitoshi S, Tropepe V, Ekker M et al. Neural stem cell lineages are regionally specified, but not committed, within distinct compartments of the developing brain. Development 2002; 129(1):233-44.
43. Johe KK, Hazel TG, Muller T et al. Single factors direct the differentiation of stem cells from the fetal and adult central nervous system. Genes Dev 1996; 10(24):3129-40.
44. Bjorklund A, Lindvall O. Self-repair in the brain. Nature 2000; 405(6789):892-3, 895.
45. Horner PJ, Gage FH. Regenerating the damaged central nervous system. Nature 2000; 407(6807):963-70.

46. Reznikov K. [Incorporation of 3H-thymidine into glial cells of the parietal region and cells of the subependymal zone of two-week and adult mice under normal conditions and following brain injury]. Ontogenez 1975; 6(2):169-76.
47. Szele FG, Chesselet MF. Cortical lesions induce an increase in cell number and PSA-NCAM expression in the subventricular zone of adult rats. J Comp Neurol 1996; 368(3):439-54.
48. Tzeng SF, Wu JP. Responses of microglia and neural progenitors to mechanical brain injury. Neuroreport 1999; 10(11):2287-92.
49. Merzenich MM, Nelson RJ, Stryker MP et al. Somatosensory cortical map changes following digit amputation in adult monkeys. J Comp Neurol 1984; 224(4):591-605.
50. Kaas JH, Merzenich MM, Killackey HP. The reorganization of somatosensory cortex following peripheral nerve damage in adult and developing mammals. Annu Rev Neurosci 1983; 6:325-56.
51. Wall PD, Egger MD. Formation of new connexions in adult rat brains after partial deafferentation. Nature 1971; 232(5312):542-5.
52. Taub E, Uswatte G, Elbert T. New treatments in neurorehabilitation founded on basic research. Nat Rev Neurosci 2002; 3:228-236.
53. Magavi SS, Leavitt BR, Macklis JD. Induction of neurogenesis in the neocortex of adult mice. Nature 2000; 405(6789):951-5.
54. Weiss S, Dunne C, Hewson J et al. Multipotent CNS stem cells are present in the adult mammalian spinal cord and ventricular neuroaxis. J Neurosci 1996; 16(23):7599-609.
55. Shihabuddin LS, Ray J, Gage FH. FGF-2 is sufficient to isolate progenitors found in the adult mammalian spinal cord. Exp Neurol 1997; 148(2):577-86.
56. Yamamoto S, Yamamoto N, Kitamura T et al. Proliferation of parenchymal neural progenitors in response to injury in the adult rat spinal cord. Exp Neurol 2001; 172(1):115-27.
57. Horner PJ, Power AE, Kempermann G et al. Proliferation and differentiation of progenitor cells throughout the intact adult rat spinal cord. J Neurosci 2000; 20(6):2218-28.
58. Namiki J, Tator CH. Cell proliferation and nestin expression in the ependyma of the adult rat spinal cord after injury. J Neuropathol Exp Neurol 1999; 58(5):489-98.
59. McTigue DM, Horner PJ, Stokes BT et al. Neurotrophin-3 and brain-derived neurotrophic factor induce oligodendrocyte proliferation and myelination of regenerating axons in the contused adult rat spinal cord. J Neurosci 1998; 18(14):5354-65.
60. Weidner N, Ner A, Salimi N et al. Spontaneous corticospinal axonal plasticity and functional recovery after adult central nervous system injury. Proc Natl Acad Sci USA 2001; 98(6):3513-8.
61. Rubio F, Kokaia Z, Arco A et al. BDNF gene transfer to the mammalian brain using CNS-derived neural precursors. Gene Ther 1999; 6(11):1851-66.
62. Grill R, Murai K, Blesch A et al. Cellular delivery of neurotrophin-3 promotes corticospinal axonal growth and partial functional recovery after spinal cord injury. J Neurosci 1997; 17(14):5560-72.
63. Liu Y, Kim D, Himes BT et al. Transplants of fibroblasts genetically modified to express BDNF promote regeneration of adult rat rubrospinal axons and recovery of forelimb function. J Neurosci 1999; 19(11):4370-87.
64. Himes BT, Liu Y, Solowska JM et al. Transplants of cells genetically modified to express neurotrophin-3 rescue axotomized Clarke's nucleus neurons after spinal cord hemisection in adult rats. J Neurosci Res 2001; 65(6):549-64.
65. Kim DH, Gutin PH, Noble LJ et al. Treatment with genetically engineered fibroblasts producing NGF or BDNF can accelerate recovery from traumatic spinal cord injury in the adult rat. Neuroreport 1996; 7(13):2221-5.
66. Teng YD, Lavik EB, Qu X et al. Functional recovery following traumatic spinal cord injury mediated by a unique polymer scaffold seeded with neural stem cells. Proc Natl Acad Sci USA 2002; 99(5):3024-3029.
67. Benowitz LI, Goldberg DE, Madsen JR et al. Inosine stimulates extensive axon collateral growth in the rat corticospinal tract after injury. Proc Natl Acad Sci USA 1999; 96(23):13486-90.
68. Whittemore SR. Neuronal replacement strategies for spinal cord injury. J Neurotrauma 1999; 16(8):667-73.
69. Shin JJ, Fricker-Gates RA, Perez FA et al. Transplanted neuroblasts differentiate appropriately into projection neurons with correct neurotransmitter and receptor phenotype in neocortex undergoing targeted projection neuron degeneration. J Neurosci 2000; 20(19):7404-16.

70. McDonald JW, Liu XZ, Qu Y et al. Transplanted embryonic stem cells survive, differentiate and promote recovery in injured rat spinal cord. Nat Med 1999; 5(12):1410-2.
71. Akiyama Y, Honmou O, Kato T et al. Transplantation of clonal neural precursor cells derived from adult human brain establishes functional peripheral myelin in the rat spinal cord. Exp Neurol 2001; 167(1):27-39.
72. Groves AK, Barnett SC, Franklin RJ et al. Repair of demyelinated lesions by transplantation of purified O-2A progenitor cells. Nature 1993; 362(6419):453-5.
73. Reier PJ. The Neurobiology of Central Nervous System Trauma. New York: Oxford University Press, 1994.
74. Armstrong RJ, Barker RA. Neurodegeneration: a failure of neuroregeneration? Lancet 2001; 358(9288):1174-6.
75. Shingo T, Sorokan ST, Shimazaki T et al. Erythropoietin regulates the in vitro and in vivo production of neuronal progenitors by mammalian forebrain neural stem cells. J Neurosci 2001; 21(24):9733-43.
76. Pencea V, Bingaman KD, Wiegand SJ et al. Infusion of brain-derived neurotrophic factor into the lateral ventricle of the adult rat leads to new neurons in the parenchyma of the striatum, septum, thalamus, and hypothalamus. J Neurosci 2001; 21(17):6706-17.
77. Aberg MA, Aberg ND, Hedbacker H et al. Peripheral infusion of IGF-I selectively induces neurogenesis in the adult rat hippocampus. J Neurosci 2000; 20(8):2896-903.
78. Wagner JP, Black IB, DiCicco-Bloom E. Stimulation of neonatal and adult brain neurogenesis by subcutaneous injection of basic fibroblast growth factor. J Neurosci 1999; 19(14):6006-16.
79. Kuhn HG, Winkler J, Kempermann G et al. Epidermal growth factor and fibroblast growth factor-2 have different effects on neural progenitors in the adult rat brain. J Neurosci 1997; 17(15):5820-9.
80. Fallon J, Reid S, Kinyamu R et al. In vivo induction of massive proliferation, directed migration, and differentiation of neural cells in the adult mammalian brain. Proc Natl Acad Sci USA 2000; 97(26):14686-91.
81. Bjorklund LM, Sanchez-Pernaute R, Chung S et al. Embryonic stem cells develop into functional dopaminergic neurons after transplantation in a Parkinson rat model. Proc Natl Acad Sci USA 2002; 99(4):2344-9.
82. Bjorklund A, Lindvall O. Cell replacement therapies for central nervous system disorders. Nat Neurosci 2000; 3(6):537-44.
83. Freed CR, Greene PE, Breeze RE et al. Transplantation of embryonic dopamine neurons for severe Parkinson's disease. N Engl J Med 2001; 344(10):710-9.
84. Wenning GK, Odin P, Morrish P et al. Short- and long-term survival and function of unilateral intrastriatal dopaminergic grafts in Parkinson's disease. Ann Neurol 1997; 42(1):95-107.
85. Lindvall O, Sawle G, Widner H et al. Evidence for long-term survival and function of dopaminergic grafts in progressive Parkinson's disease. Ann Neurol 1994; 35(2):172-80.
86. Lindvall O, Brundin P, Widner H et al. Grafts of fetal dopamine neurons survive and improve motor function in Parkinson's disease. Science 1990; 247(4942):574-7.
87. Dobrossy MD, Dunnett SB. The influence of environment and experience on neural grafts. Nat Rev Neurosci 2001; 2(12):871-9.
88. Piccini P, Lindvall O, Bjorklund A et al. Delayed recovery of movement-related cortical function in Parkinson's disease after striatal dopaminergic grafts. Ann Neurol 2000; 48(5):689-95.
89. Hagell P, Schrag A, Piccini P et al. Sequential bilateral transplantation in Parkinson's disease: effects of the second graft. Brain 1999; 122(Pt 6):1121-32.
90. Mazurova Y. New therapeutic approaches for the treatment of Huntington's disease. Acta Medica (Hradec Kralove) 2001; 44(4):119-23.
91. Dunnett SB, Nathwani F, Bjorklund A. The integration and function of striatal grafts. Prog Brain Res 2000; 127:345-80.
92. Kendall AL, Rayment FD, Torres EM et al. Functional integration of striatal allografts in a primate model of Huntington's disease. Nat Med 1998; 4(6):727-9.
93. Palfi S, Conde F, Riche D et al. Fetal striatal allografts reverse cognitive deficits in a primate model of Huntington disease. Nat Med 1998; 4(8):963-6.

94. Bachoud-Levi AC, Remy P, Nguyen JP et al. Motor and cognitive improvements in patients with Huntington's disease after neural transplantation. Lancet 2000; 356(9246):1975-9.
95. Svendsen CN, Smith AG. New prospects for human stem-cell therapy in the nervous system. Trends Neurosci 1999; 22(8):357-64.
96. Dietrich J, Easterday MC. Developing concepts in neural stem cells. Trends Neurosci 2002; 25(3):129-31.
97. Dumas TC, Sapolsky RM. Gene therapy against neurological insults: sparing neurons versus sparing function. Trends Neurosci 2001; 24(12):695-700.
98. Zurn AD, Widmer HR, Aebischer P. Sustained delivery of GDNF: towards a treatment for Parkinson's disease. Brain Res Brain Res Rev 2001; 36(2-3):222-9.
99. Hoffer B, Olson L. Treatment strategies for neurodegenerative diseases based on trophic factors and cell trasnplantation techniques. In: Mizuno Y YM, Calne DB, Horowski R et al, eds. Advances in Research on Neurodegeneration. Wien, New York: Springer Verlag, 1997:1-10.
100. Silani V, Braga M, Cardin V et al. The pathogenesis of ALS: implications for treatment strategies. Neurol Neurochir Pol 2001; 35(1 Suppl):25-39.
101. Martin JB. Gene therapy and pharmacological treatment of inherited neurological disorders. Trends Biotechnol 1995; 13(1):28-35.
102. Fisher LJ. Neural precursor cells: applications for the study and repair of the central nervous system. Neurobiol Dis 1997; 4(1):1-22.
103. Reya T, Morrison SJ, Clarke MF et al. Stem cells, cancer, and cancer stem cells. Nature 2001; 414(6859):105-11.
104. Moe PJ, Holen A. High-dose methotrexate in childhood all. Pediatr Hematol Oncol 2000; 17(8):615-22.
105. Rubnitz JE. Molecular diagnostics in the treatment of childhood acute lymphoblastic leukemia. J Biol Regul Homeost Agents 2000; 14(3):182-6.
106. Alizadeh AA, Staudt LM. Genomic-scale gene expression profiling of normal and malignant immune cells. Curr Opin Immunol 2000; 12(2):219-25.
107. Knuutila S, Teerenhovi L, Larramendy ML et al. Cell lineage involvement of recurrent chromosomal abnormalities in hematologic neoplasms. Genes Chromosomes Cancer 1994; 10(2):95-102.
108. Nilsson L, Astrand-Grundstrom I, Arvidsson I et al. Isolation and characterization of hematopoietic progenitor/stem cells in 5q-deleted myelodysplastic syndromes: evidence for involvement at the hematopoietic stem cell level. Blood 2000; 96(6):2012-21.
109. Russell NH. Biology of acute leukaemia. Lancet 1997; 349(9045):118-22.
110. McKay R. Stem cells in the central nervous system. Science 1997; 276:66-71.
111. Temple S, Qian X. Vertebrate neural progenitor cells: subtypes and regulation. Current Opinion in Neurobiology 1996; 6:11-17.
112. Reynolds BA, Tetzlaff W, Weiss S. A Mulitpotent EGF-Responsive Striatal Embryonic Progenitor Cell Produces Neurons and Astrocytes. J Neurosci 1992; 12(11):4565-4574.
113. Vescovi A, Gritti A, Galli R et al. Isolation and intracerebral grafting of nontransformed multipotential embryonic human CNS stem cells. J Neurotrauma 1999; 16:689-693.
114. Mayer-Proschel M, Kalyani AJ, Mujtaba T et al. Isolation of lineage-restricted neuronal precursors from multipotent neuroepithelial stem cells. Neuron 1997; 19(4):773-85.
115. Rao MS, Noble M, Mayer-Proschel M. A tripotential glial precursor cell is present in the developing spinal cord. Proc Natl Acad Sci USA 1998; 95(7):3996-4001.
116. Rao MS, Mayer-Proschel M. Glial-restricted precursors are derived from multipotent neuroepithelial stem cells. Dev Biol 1997; 188(1):48-63.
117. Kalyani A, Piper D, Mujitaba T et al. E-NCAM immunoreactive neuronal precursor cells differentiate into multiple neurotransmitter phenotypes. J Neurosci 1998; 18:7856-7868.
118. Raff MC, Miller RH, Noble M. A glial progenitor cell that develops in vitro into an astrocyte or an oligodendrocyte depending on culture medium. Nature 1983; 303(5916):390-6.
119. Noble M, Murray K. Purified astrocytes promote the in vitro division of a bipotential glial progenitor cell. Embo J 1984; 3(10):2243-7.
120. Fok-Seang J, Miller HR. Astrocyte precursors in neonatal rat spinal cord cultures. J Neurosci 1992; 12:2751-2764.

121. Fok-Seang J, Miller RH. Distribution and differentiation of A2B5+ glial precursors in the developing rat spinal cord. J Neurosci Res 1994; 37:219-235.
122. Seidman K, Teng A, Rosenkopf R et al. Isolation, cloning and characterization of a putative type-1 astrocyte cell line. Brain Res 1997; 753:18-26.
123. Mi H, Barres BA. Purification and characterization of astrocyte precursor cells in the developing rat optic nerve. J Neurosci 1999; 19:1049-1061.
124. Grinspan JB, Stern JL, Pustilnik SM et al. Cerebral white matter contains PDGF-responsive precursors to O-2A cells. J Neuroscience 1990; 10:1866-1873.
125. Palmer TD, Schwartz PH, Taupin P et al. Cell culture. Progenitor cells from human brain after death. Nature 2001; 411(6833):42-3.
126. Laywell ED, Rakic P, Kukekov VG et al. Identification of a multipotent astrocytic stem cell in the immature and adult mouse brain. Proc Natl Acad Sci USA 2000; 97(25):13883-8.
127. Roy NS, Wang S, Jiang L et al. In vitro neurogenesis by progenitor cells isolated from the adult human hippocampus. Nat Med 2000; 6(3):271-7.
128. Arsenijevic Y, Villemure JG, Brunet JF et al. Isolation of multipotent neural precursors residing in the cortex of the adult human brain. Exp Neurol 2001; 170(1):48-62.
129. Piper DR, Mujtaba T, Keyoung H et al. Identification and characterization of neuronal precursors and their progeny from human fetal tissue. J Neurosci Res 2001; 66(3):356-68.
130. Reubinoff BE, Itsykson P, Turetsky T et al. Neural progenitors from human embryonic stem cells. Nat Biotechnol 2001; 19(12):1134-40.
131. Zhang SC, Wernig M, Duncan ID et al. In vitro differentiation of transplantable neural precursors from human embryonic stem cells. Nat Biotechnol 2001; 19(12):1129-33.
132. Shih CC, Weng Y, Mamelak A et al. Identification of a candidate human neurohematopoietic stem-cell population. Blood 2001; 98(8):2412-22.
133. Scheffler B, Horn M, Blumcke I et al. Marrow-mindedness: a perspective on neuropoiesis. Trends Neurosci 1999; 22(8):348-57.
134. Parati EA, Bez A, Ponti D et al. Human neural stem cells express extra-neural markers. Brain Res 2002; 925(2):213-21.
135. Price J, Williams BP. Neural stem cells. Curr Opin Neurobiol 2001; 11(5):564-7.
136. Weissman IL, Anderson DJ, Gage F. Stem and progenitor cells: origins, phenotypes, lineage commitments, and transdifferentiations. Annu Rev Cell Dev Biol 2001; 17:387-403.
137. Krause DS, Theise ND, Collector MI et al. Multi-organ, multi-lineage engraftment by a single bone marrow-derived stem cell. Cell 2001; 105(3):369-77.
138. Toma JG, Akhavan M, Fernandes KJ et al. Isolation of multipotent adult stem cells from the dermis of mammalian skin. Nat Cell Biol 2001; 3(9):778-84.
139. Noble M, Dietrich J. Intersections between neurobiology and oncology: tumor origin, treatment and repair of treatment-associated damage. Trends Neurosci 2002; 25(2):103-7.
140. Noble M. The oligodendrocyte-type-2 astrocyte lineage: in vitro and in vivo studies on development, tissue repair and neoplasia. In: Gage FH, Christen Y., eds. Isolation, characterization and utilization of CNS stem cells. Berlin, Heidelberg: Springer, 1997: 101-128.
141. Linskey ME. Developmental glial biology: the key to understanding glial tumors. Clin Neurosurg 2000; 47:46-71.
142. Linskey ME. Glial ontogeny and glial neoplasia: the search for closure. J Neurooncol 1997; 34(1):5-22.
143. Barnett SC, Robertson L, Graham D et al. Oligodendrocyte-type-2 astrocyte (O-2A) progenitor cells transformed with c-myc and H-ras form high-grade glioma after stereotactic injection into the rat brain. Carcinogenesis 1998; 19(9):1529-37.
144. Holland EC, Celestino J, Dai C et al. Combined activation of Ras and Akt in neural progenitors induces glioblastoma formation in mice. Nat Genet 2000; 25(1):55-7.
145. Seyfried TN. Perspectives on brain tumor formation involving macrophages, glia, and neural stem cells. Perspect Biol Med 2001; 44(2):263-82.
146. Noble M, Gutowski N, Bevan K et al. From rodent glial precursor cell to human glial neoplasia in the oligodendrocyte-type-2 astrocyte lineage. Glia 1995; 15(3):222-30.
147. Kashima T, Tiu SN, Merrill JE et al. Expression of oligodendrocyte-associated genes in cell lines derived from human gliomas and neuroblastomas. Cancer Res 1993; 53(1):170-5.

148. Seyfried TN, Yu RK. Ganglioside GD3: structure, cellular distribution, and possible function. Mol Cell Biochem 1985; 68(1):3-10.
149. Tohyama T, Lee VM, Rorke LB et al. Nestin expression in embryonic human neuroepithelium and in human neuroepithelial tumor cells. Lab Invest 1992; 66(3):303-13.
150. Li J, Pearl DK, Pfeiffer SE et al. Patterns of reactivity with anti-glycolipid antibodies in human primary brain tumors. J Neurosci Res 1994; 39(2):148-58.
151. Yang HY, Lieska N, Shao D et al. Proteins of the intermediate filament cytoskeleton as markers for astrocytes and human astrocytomas. Mol Chem Neuropathol 1994; 21(2-3):155-76.
152. Shoshan Y, Nishiyama A, Chang A et al. Expression of oligodendrocyte progenitor cell antigens by gliomas: implications for the histogenesis of brain tumors. Proc Natl Acad Sci USA 1999; 96(18):10361-6.
153. Lu QR, Park JK, Noll E et al. Oligodendrocyte lineage genes (OLIG) as molecular markers for human glial brain tumors. Proc Natl Acad Sci USA 2001; 98(19):10851-6.
154. Marie Y, Sanson M, Mokhtari K et al. OLIG2 as a specific marker of oligodendroglial tumour cells. Lancet 2001; 358(9278):298-300.
155. Pomeroy SL, Tamayo P, Gaasenbeek M et al. Prediction of central nervous system embryonal tumour outcome based on gene expression. Nature 2002; 415(6870):436-42.
156. Van't Veer LJ, De Jong D. The microarray way to tailored cancer treatment. Nat Med 2002; 8(1):13-4.
157. Armstrong SA, Staunton JE, Silverman LB et al. MLL translocations specify a distinct gene expression profile that distinguishes a unique leukemia. Nat Genet 2002; 30(1):41-7.
158. Benedetti S, Pirola B, Pollo B et al. Gene therapy of experimental brain tumors using neural progenitor cells. Nat Med 2000; 6(4):447-50.
159. Aboody KS, Brown A, Rainov NG et al. From the cover: neural stem cells display extensive tropism for pathology in adult brain: evidence from intracranial gliomas. Proc Natl Acad Sci USA 2000; 97(23):12846-51.
160. Berens ME, Giese A. "...those left behind." Biology and oncology of invasive glioma cells. Neoplasia 1999; 1(3):208-19.
161. Prados MD, Levin V. Biology and treatment of malignant glioma. Semin Oncol 2000; 27(3 Suppl 6):1-10.
162. Noble M. Can neural stem cells be used to track down and destroy migratory brain tumor cells while also providing a means of repairing tumor-associated damage? Proc Natl Acad Sci USA 2000; 97(23):12393-5.
163. Noble M. Can neural stem cells be used as therapeutic vehicles in the treatment of brain tumors? Nat Med 2000; 6(4):369-70.
164. Lewin M, Carlesso N, Tung CH et al. Tat peptide-derivatized magnetic nanoparticles allow in vivo tracking and recovery of progenitor cells. Nat Biotechnol 2000; 18(4):410-4.
165. Franklin RJ, Blaschuk KL, Bearchell MC et al. Magnetic resonance imaging of transplanted oligodendrocyte precursors in the rat brain. Neuroreport 1999; 10(18):3961-5.
166. Bulte JW, Zhang S, van Gelderen P et al. Neurotransplantation of magnetically labeled oligodendrocyte progenitors: magnetic resonance tracking of cell migration and myelination. Proc Natl Acad Sci USA 1999; 96(26):15256-61.
167. Keime-Guibert F, Napolitano M, Delattre JY. Neurological complications of radiotherapy and chemotherapy. J Neurol. 1998; 245:695-708.
168. Grill J, Renaux VK, Bulteau C et al. Long-term intellectual outcome in children with posterior fossa tumors according to radiation doses and volumes. Int. J Radiat Oncol Biol Phys 1999; 45:137-145.
169. Schlegel U, Pels H, Oehring R et al. Neurologic sequelae of treatment of primary CNS lymphomas. J Neurooncol 1999; 43:277-286.
170. Cetingul N, Aydinok Y, Kantar M et al. Neuropsychologic sequelae in the long-term survivors of childhood acute lymphoblastic leukemia. Pediatr. Hematol. Oncol 1999; 16:213-220.
171. Stylopoulos LA, George AE, de Leon MJ et al. Longitudinal CT study of parenchymal brain changes in glioma survivors. AJNR Am J Neuroradiol 1988; 9(3):517-22.
172. Riva D, Giorgi C. The neurodevelopmental price of survival in children with malignant brain tumours. Childs Nerv Syst 2000; 16(10-11):751-4.

173. Waber DP, Tarbell NJ. Toxicity of CNS prophylaxis for childhood leukemia. Oncology (Huntingt) 1997; 11(2):259-64; discussion 264-5.
174. Glauser TA, Packer RJ. Cognitive deficits in long-term survivors of childhood brain tumors. Childs Nerv Syst 1991; 7(1):2-12.
175. Appleton RE, Farrell K, Zaide J et al. Decline in head growth and cognitive impairment in survivors of acute lymphoblastic leukaemia. Arch Dis Child 1990; 65(5):530-4.
176. Peissner W, Kocher M, Treuer H et al. Ionizing radiation-induced apoptosis of proliferating stem cells in the dentate gyrus of the adult rat hippocampus. Brain Res Mol. Brain Res 1999; 71:61-68.
177. Tada E, Parent JM, Lowenstein DH et al. X-irradiation causes a prolonged reduction in cell proliferation in the dentate gyrus of adult rats. Neuroscience 2000; 99(1):33-41.
178. Bellinzona M, Gobbel GT, Shinohara C et al. Apoptosis is induced in the subependyma of young adult rats by ionizing irradiation. Neurosci Lett 1996; 208:163-166.
179. Li YQ, Wong CS. Apoptosis and its relationship with cell proliferation in the irradiated rat spinal cord. Int. J Radiat Biol 1998; 74:405-417.
180. Vigliani MC, Duyckaerts C, Hauw JJ et al. Dementia following treatment of brain tumors with radiotherapy administered alone or in combination with nitrosourea-based chemotherapy: a clinical and pathological study. J Neurooncol 1999; 41:137-149.
181. Hopewell JW, van der Kogel AJ. Pathophysiological mechanisms leading to the development of late radiation-induced damage to the central nervous system. Front Radiat Ther Oncol 1999; 33:265-75.
182. Nutt CL, Noble M, Chambers AF et al. Differential expression of drug resistance genes and chemosensitivity in glial cell lineages correlate with differential response of oligodendrogliomas and astrocytomas to chemotherapy. Cancer Res 2000; 60(17):4812-8.
183. Lassmann H. Classification of demyelinating diseases at the interface between etiology and pathogenesis. Curr Opin Neurol 2001; 14(3):253-8.
184. McTigue DM, Popovich PG, Jakeman LB et al. Strategies for spinal cord injury repair. Prog Brain Res 2000; 128:3-8.
185. Beattie MS, Farooqui AA, Bresnahan JC. Review of current evidence for apoptosis after spinal cord injury. J Neurotrauma 2000; 17(10):915-25.
186. Kakulas BA. A review of the neuropathology of human spinal cord injury with emphasis on special features. J Spinal Cord Med 1999; 22(2):119-24.
187. Prineas JW, MacDonald WI. Demyelinating diseases. In: Graham DI, Lantos PL, eds. Greenfield's neuropathology. 6th ed. London: Arnold, 1997: 813-889.
188. Lassmann H, Bruck W, Lucchinetti C. Heterogeneity of multiple sclerosis pathogenesis: implications for diagnosis and therapy. Trends Mol Med 2001; 7(3):115-21.
189. Kornek B, Storch MK, Weissert R et al. Multiple sclerosis and chronic autoimmune encephalomyelitis: a comparative quantitative study of axonal injury in active, inactive, and remyelinated lesions. Am J Pathol 2000; 157(1):267-76.
190. Bitsch A, Schuchardt J, Bunkowski S et al. Acute axonal injury in multiple sclerosis. Correlation with demyelination and inflammation. Brain 2000; 123(Pt 6):1174-83.
191. Ferguson B, Matyszak MK, Esiri MM et al. Axonal damage in acute multiple sclerosis lesions. Brain 1997; 120(Pt 3):393-9.
192. Duncan ID, Grever WE, Zhang SC. Repair of myelin disease: strategies and progress in animal models. Mol Med Today 1997; 3(12):554-61.
193. Blakemore WF, Franklin RJ. Transplantation options for therapeutic central nervous system remyelination. Cell Transplant 2000; 9(2):289-94.
194. Blakemore WF, Olby NJ, Franklin RJ. The use of transplanted glial cells to reconstruct glial environments in the CNS. Brain Pathol 1995; 5(4):443-50.
195. Brustle O, McKay RD. Neuronal progenitors as tools for cell replacement in the nervous system. Curr Opin Neurobiol 1996; 6(5):688-95.
196. Cattaneo E, McKay R. Identifying and manipulating neuronal stem cells. Trends Neurosci 1991; 14(8):338-40.
197. Duncan ID. Glial cell transplantation and remyelination of the central nervous system. Neuropathol Appl Neurobiol 1996; 22(2):87-100.

198. Brustle O, Jones KN, Learish RD et al. Embryonic stem cell-derived glial precursors: a source of myelinating transplants. Science 1999; 285(5428):754-6.
199. Rao MS, Mayer-Proschel M. Precursor cells for transplantation. Prog Brain Res 2000; 128:273-92.
200. Prineas JW, Barnard RO, Kwon EE et al. Multiple sclerosis: remyelination of nascent lesions. Ann Neurol 1993; 33(2):137-51.
201. Prineas JW, Kwon EE, Goldenberg PZ et al. Multiple sclerosis. Oligodendrocyte proliferation and differentiation in fresh lesions. Lab Invest 1989; 61(5):489-503.
202. Ludwin SK. Remyelination in demyelinating diseases of the central nervous system. Crit Rev Neurobiol 1987; 3(1):1-28.
203. Wolswijk G. Chronic stage multiple sclerosis lesions contain a relatively quiescent population of oligodendrocyte precursor cells. J Neurosci 1998; 18(2):601-9.
204. Blakemore WF, Franklin, R.J.M., Noble, M. Glial cell transplantation and the repair of demyelinating lesions. In: Jessen KR, Richardson WD, eds. Glial cell development: Basic principles and relevance. Oxford: BIOS Scientific Publishers Ltd., 1996: 209-220.
205. Allen IV, Kirk, J. Demyelinating diseases. In: Adams JH, Duchen LW, eds. Greenfield's neuropathology. New York: Oxford University Press, 1992: 447-520.
206. Raine CS. Demyelinating Diseases. In: Davis RL, Robertson DM, eds. Textbook of Neuropathology. Baltimore: Williams and Wilkins, 1991: 535-620.
207. Nait-Oumesmar B, Lachapelle F, Decker L et al. Do central nervous system axons remyelinate? Pathol Biol (Paris) 2000; 48(1):70-9.
208. Decker L, Picard N, Lachapelle F et al. Neural precursors and demyelinating diseases. Prog Brain Res 2001; 132:175-84.
209. Nait-Oumesmar B, Decker L, Lachapelle F et al. Progenitor cells of the adult mouse subventricular zone proliferate, migrate and differentiate into oligodendrocytes after demyelination. Eur J Neurosci 1999; 11(12):4357-66.
210. Norton WT. Do oligodendrocytes divide? Neurochem Res 1996; 21(4):495-503.
211. Carroll WM, Jennings AR, Ironside LJ. Identification of the adult resting progenitor cell by autoradiographic tracking of oligodendrocyte precursors in experimental CNS demyelination. Brain 1998; 121(Pt 2):293-302.
212. Keirstead HS, Blakemore WF. Identification of post-mitotic oligodendrocytes incapable of remyelination within the demyelinated adult spinal cord. J Neuropathol Exp Neurol 1997; 56(11):1191-201.
213. Ffrench-Constant C, Raff MC. Proliferating bipotential glial progenitor cells in adult rat optic nerve. Nature 1986; 319(6053):499-502.
214. Wolswijk G, Noble M. Identification of an adult-specific glial progenitor cell. Development 1989; 105(2):387-400.
215. Gensert JM, Goldman JE. In vivo characterization of endogenous proliferating cells in adult rat subcortical white matter. Glia 1996; 17(1):39-51.
216. Bergles DE, Roberts JD, Somogyi P et al. Glutamatergic synapses on oligodendrocyte precursor cells in the hippocampus. Nature 2000; 405(6783):187-91.
217. Butt AM, Duncan A, Hornby MF et al. Cells expressing the NG2 antigen contact nodes of Ranvier in adult CNS white matter. Glia 1999; 26(1):84-91.
218. Levine JM, Reynolds R, Fawcett JW. The oligodendrocyte precursor cell in health and disease. Trends Neurosci 2001; 24(1):39-47.
219. Dou CL, Levine JM. Inhibition of neurite growth by the NG2 chondroitin sulfate proteoglycan. J Neurosci 1994; 14(12):7616-28.
220. Levine JM, Reynolds R. Activation and proliferation of endogenous oligodendrocyte precursor cells during ethidium bromide-induced demyelination. Exp Neurol 1999; 160(2):333-47.
221. Gogate N, Verma L, Zhou JM et al. Plasticity in the adult human oligodendrocyte lineage. J Neurosci 1994; 14(8):4571-87.
222. Armstrong RC, Dorn HH, Kufta CV et al. Pre-oligodendrocytes from adult human CNS. J Neurosci 1992; 12(4):1538-47.
223. Scolding NJ, Rayner PJ, Sussman J et al. A proliferative adult human oligodendrocyte progenitor. Neuroreport 1995; 6(3):441-5.

224. Roy NS, Wang S, Harrison-Restelli C et al. Identification, isolation, and promoter-defined separation of mitotic oligodendrocyte progenitor cells from the adult human subcortical white matter. J Neurosci 1999; 19(22):9986-95.
225. McCarthy GF, Leblond CP. Radioautographic evidence for slow astrocyte turnover and modest oligodendrocyte production in the corpus callosum of adult mice infused with 3H-thymidine. J Comp Neurol 1988; 271(4):589-603.
226. Redwine JM, Armstrong RC. In vivo proliferation of oligodendrocyte progenitors expressing PDGFalphaR during early remyelination. J Neurobiol 1998; 37(3):413-28.
227. Maeda Y, Solanky M, Menonna J et al. Platelet-derived growth factor-alpha receptor-positive oligodendroglia are frequent in multiple sclerosis lesions. Ann Neurol 2001; 49(6):776-85.
228. Scolding N, Franklin R, Stevens S et al. Oligodendrocyte progenitors are present in the normal adult human CNS and in the lesions of multiple sclerosis. Brain 1998; 121(Pt 12):2221-8.
229. Gensert JM, Goldman JE. Endogenous progenitors remyelinate demyelinated axons in the adult CNS. Neuron 1997; 19(1):197-203.
230. O'Leary MT, Blakemore WF. Oligodendrocyte precursors survive poorly and do not migrate following transplantation into the normal adult central nervous system. J Neurosci Res 1997; 48(2):159-67.
231. Franklin RJ, Bayley SA, Blakemore WF. Transplanted CG4 cells (an oligodendrocyte progenitor cell line) survive, migrate, and contribute to repair of areas of demyelination in X-irradiated and damaged spinal cord but not in normal spinal cord. Exp Neurol 1996; 137(2):263-76.
232. Dubois-Dalcq M, Murray K. Why are growth factors important in oligodendrocyte physiology? Pathol Biol (Paris) 2000; 48(1):80-6.
233. Hohlfeld R, Kerschensteiner M, Stadelmann C et al. The neuroprotective effect of inflammation: implications for the therapy of multiple sclerosis. J Neuroimmunol 2000; 107(2):161-6.
234. Molina-Holgado E, Vela JM, Arevalo-Martin A et al. LPS/IFN-gamma cytotoxicity in oligodendroglial cells: role of nitric oxide and protection by the anti-inflammatory cytokine IL-10. Eur J Neurosci 2001; 13(3):493-502.
235. Kil K, Zang YC, Yang D et al. T cell responses to myelin basic protein in patients with spinal cord injury and multiple sclerosis. J Neuroimmunol 1999; 98(2):201-7.
236. Warrington AE, Asakura K, Bieber AJ et al. Human monoclonal antibodies reactive to oligodendrocytes promote remyelination in a model of multiple sclerosis. Proc Natl Acad Sci USA 2000; 97(12):6820-5.
237. Browne TR, Holmes GL. Epilepsy. N Engl J Med 2001; 344(15):1145-51.
238. Stefan H. Pathophysiology of human epilepsy: imaging and physiologic studies. Curr Opin Neurol 2000; 13(2):177-81.
239. Hosford DA. Models of primary generalized epilepsy. Curr Opin Neurol 1995; 8(2):121-5.
240. Brodie MJ, French JA. Management of epilepsy in adolescents and adults. Lancet 2000; 356(9226):323-9.
241. Kalynchuk LE. Long-term amygdala kindling in rats as a model for the study of interictal emotionality in temporal lobe epilepsy. Neurosci Biobehav Rev 2000; 24(7):691-704.
242. Sutula TP. Experimental models of temporal lobe epilepsy: new insights from the study of kindling and synaptic reorganization. Epilepsia 1990; 31(Suppl 3):S45-54.
243. Meldrum BS, Akbar MT, Chapman AG. Glutamate receptors and transporters in genetic and acquired models of epilepsy. Epilepsy Res 1999; 36(2-3):189-204.
244. Kupferberg H. Animal models used in the screening of antiepileptic drugs. Epilepsia 2001; 42(Suppl 4):7-12.
245. Fisher RS. Animal models of the epilepsies. Brain Res Brain Res Rev 1989; 14(3):245-78.
246. Golarai G, Greenwood AC, Feeney DM et al. Physiological and structural evidence for hippocampal involvement in persistent seizure susceptibility after traumatic brain injury. J Neurosci 2001; 21(21):8523-37.
247. Whitaker WR, Faull RL, Dragunow M et al. Changes in the mRNAs encoding voltage-gated sodium channel types II and III in human epileptic hippocampus. Neuroscience 2001; 106(2):275-85.
248. Hendriksen H, Datson NA, Ghijsen WE et al. Altered hippocampal gene expression prior to the onset of spontaneous seizures in the rat post-status epilepticus model. Eur J Neurosci 2001; 14(9):1475-84.

249. Hughes PE, Alexi T, Walton M et al. Activity and injury-dependent expression of inducible transcription factors, growth factors and apoptosis-related genes within the central nervous system. Prog Neurobiol 1999; 57(4):421-50.
250. Osawa M, Uemura S, Kimura H et al. Amygdala kindling develops without mossy fiber sprouting and hippocampal neuronal degeneration in rats. Psychiatry Clin Neurosci 2001; 55(6):549-57.
251. Zhang LX, Smith MA, Li XL et al. Apoptosis of hippocampal neurons after amygdala kindled seizures. Brain Res Mol Brain Res 1998; 55(2):198-208.
252. Thompson K, Holm AM, Schousboe A et al. Hippocampal stimulation produces neuronal death in the immature brain. Neuroscience 1998; 82(2):337-48.
253. Sloviter RS, Dean E, Sollas AL et al. Apoptosis and necrosis induced in different hippocampal neuron populations by repetitive perforant path stimulation in the rat. J Comp Neurol 1996; 366(3):516-33.
254. Harding B, Thom M. Bilateral hippocampal granule cell dispersion: autopsy study of 3 infants. Neuropathol Appl Neurobiol 2001; 27(3):245-51.
255. Briellmann RS, Newton MR, Wellard RM et al. Hippocampal sclerosis following brief generalized seizures in adulthood. Neurology 2001; 57(2):315-7.
256. Fisher PD, Sperber EF, Moshe SL. Hippocampal sclerosis revisited. Brain Dev 1998; 20(8):563-73.
257. Liu Z, Mikati M, Holmes GL. Mesial temporal sclerosis: pathogenesis and significance. Pediatr Neurol 1995; 12(1):5-16.
258. Sloviter RS. The functional organization of the hippocampal dentate gyrus and its relevance to the pathogenesis of temporal lobe epilepsy. Ann Neurol 1994; 35(6):640-54.
259. Swann JW, Al-Noori S, Jiang M et al. Spine loss and other dendritic abnormalities in epilepsy. Hippocampus 2000; 10(5):617-25.
260. Schauwecker PE, Ramirez JJ, Steward O. Genetic dissection of the signals that induce synaptic reorganization. Exp Neurol 2000; 161(1):139-52.
261. Liu Z, Yang Y, Silveira DC et al. Consequences of recurrent seizures during early brain development. Neuroscience 1999; 92(4):1443-54.
262. Sloviter RS. Status epilepticus-induced neuronal injury and network reorganization. Epilepsia 1999; 40(Suppl 1):S34-9; discussion S40-1.
263. Ben-Ari Y. Cell death and synaptic reorganizations produced by seizures. Epilepsia 2001; 42(Suppl 3):5-7.
264. Houser CR. Granule cell dispersion in the dentate gyrus of humans with temporal lobe epilepsy. Brain Res 1990; 535(2):195-204.
265. Lurton D, El Bahh B, Sundstrom L et al. Granule cell dispersion is correlated with early epileptic events in human temporal lobe epilepsy. J Neurol Sci 1998; 154(2):133-6.
266. Houser CR. Morphological changes in the dentate gyrus in human temporal lobe epilepsy. Epilepsy Res Suppl 1992; 7:223-34.
267. Gorter JA, van Vliet EA, Aronica E et al. Progression of spontaneous seizures after status epilepticus is associated with mossy fibre sprouting and extensive bilateral loss of hilar parvalbumin and somatostatin-immunoreactive neurons. Eur J Neurosci 2001; 13(4):657-69.
268. Pierce JP, Milner TA. Parallel increases in the synaptic and surface areas of mossy fiber terminals following seizure induction. Synapse 2001; 39(3):249-56.
269. Pitkanen A, Nissinen J, Lukasiuk K et al. Association between the density of mossy fiber sprouting and seizure frequency in experimental and human temporal lobe epilepsy. Epilepsia 2000; 41(Suppl 6):S24-9.
270. Cantallops I, Routtenberg A. Kainic acid induction of mossy fiber sprouting: dependence on mouse strain. Hippocampus 2000; 10(3):269-73.
271. Wenzel HJ, Woolley CS, Robbins CA et al. Kainic acid-induced mossy fiber sprouting and synapse formation in the dentate gyrus of rats. Hippocampus 2000; 10(3):244-60.
272. Parent JM, Yu TW, Leibowitz RT et al. Dentate granule cell neurogenesis is increased by seizures and contributes to aberrant network reorganization in the adult rat hippocampus. J Neurosci 1997; 17(10):3727-38.
273. Bengzon J, Kokaia Z, Elmer E et al. Apoptosis and proliferation of dentate gyrus neurons after single and intermittent limbic seizures. Proc Natl Acad Sci USA 1997; 94(19):10432-7.

274. Gray WP, Sundstrom LE. Kainic acid increases the proliferation of granule cell progenitors in the dentate gyrus of the adult rat. Brain Res 1998; 790(1-2):52-9.
275. Nakagawa E, Aimi Y, Yasuhara O et al. Enhancement of progenitor cell division in the dentate gyrus triggered by initial limbic seizures in rat models of epilepsy. Epilepsia 2000; 41(1):10-8.
276. Parent JM, Janumpalli S, McNamara JO et al. Increased dentate granule cell neurogenesis following amygdala kindling in the adult rat. Neurosci Lett 1998; 247(1):9-12.
277. Scharfman HE, Goodman JH, Sollas AL. Granule-like neurons at the hilar/CA3 border after status epilepticus and their synchrony with area CA3 pyramidal cells: functional implications of seizure-induced neurogenesis. J Neurosci 2000; 20(16):6144-58.
278. Scott BW, Wang S, Burnham WM et al. Kindling-induced neurogenesis in the dentate gyrus of the rat. Neurosci Lett 1998; 248(2):73-6.
279. Van Der Wal EA, Gomez-Pinilla F, Cotman CW. Seizure-associated induction of basic fibroblast growth factor and its receptor in the rat brain. Neuroscience 1994; 60(2):311-23.
280. Humpel C, Wetmore C, Olson L. Regulation of brain-derived neurotrophic factor messenger RNA and protein at the cellular level in pentylenetetrazol-induced epileptic seizures. Neuroscience 1993; 53(4):909-18.
281. Humpel C, Lippoldt A, Chadi G et al. Fast and widespread increase of basic fibroblast growth factor messenger RNA and protein in the forebrain after kainate-induced seizures. Neuroscience 1993; 57(4):913-22.
282. Bugra K, Pollard H, Charton G et al. aFGF, bFGF and flg mRNAs show distinct patterns of induction in the hippocampus following kainate-induced seizures. Eur J Neurosci 1994; 6(1):58-66.
283. Ballabriga J, Pozas E, Planas AM et al. bFGF and FGFR-3 immunoreactivity in the rat brain following systemic kainic acid administration at convulsant doses: localization of bFGF and FGFR-3 in reactive astrocytes, and FGFR-3 in reactive microglia. Brain Res 1997; 752(1-2):315-8.
284. Yoshimura S, Takagi Y, Harada J et al. FGF-2 regulation of neurogenesis in adult hippocampus after brain injury. Proc Natl Acad Sci USA 2001; 98(10):5874-9.
285. Parent JM, Tada E, Fike JR et al. Inhibition of dentate granule cell neurogenesis with brain irradiation does not prevent seizure-induced mossy fiber synaptic reorganization in the rat. J Neurosci 1999; 19(11):4508-19.
286. Scott BW, Wojtowicz JM, Burnham WM. Neurogenesis in the dentate gyrus of the rat following electroconvulsive shock seizures. Exp Neurol 2000; 165(2):231-6.
287. Madsen TM, Treschow A, Bengzon J et al. Increased neurogenesis in a model of electroconvulsive therapy. Biol Psychiatry 2000; 47(12):1043-9.
288. Sankar R, Shin D, Liu H et al. Granule cell neurogenesis after status epilepticus in the immature rat brain. Epilepsia 2000; 41(Suppl 6):S53-6.
289. Blumcke I, Schewe JC, Normann S et al. Increase of nestin-immunoreactive neural precursor cells in the dentate gyrus of pediatric patients with early-onset temporal lobe epilepsy. Hippocampus 2001; 11(3):311-21.
290. Houser CR. Neuronal loss and synaptic reorganization in temporal lobe epilepsy. Adv Neurol 1999; 79:743-61.
291. Kempermann G, Kuhn HG, Gage FH. More hippocampal neurons in adult mice living in an enriched environment. Nature 1997; 386(6624):493-5.
292. Young D, Lawlor PA, Leone P et al. Environmental enrichment inhibits spontaneous apoptosis, prevents seizures and is neuroprotective. Nat Med 1999; 5(4):448-53.
293. Kubova H, Moshe SL. Experimental models of epilepsy in young animals. J Child Neurol 1994; 9(Suppl 1):S3-11.
294. Wasterlain CG, Shirasaka Y. Seizures, brain damage and brain development. Brain Dev 1994; 16(4):279-95.
295. Holmes GL, Ben-Ari Y. The neurobiology and consequences of epilepsy in the developing brain. Pediatr Res 2001; 49(3):320-5.
296. Haas KZ, Sperber EF, Opanashuk LA et al. Resistance of immature hippocampus to morphologic and physiologic alterations following status epilepticus or kindling. Hippocampus 2001; 11(6):615-25.
297. Lado FA, Sankar R, Lowenstein D et al. Age-dependent consequences of seizures: relationship to seizure frequency, brain damage, and circuitry reorganization. Ment Retard Dev Disabil Res Rev 2000; 6(4):242-52.

298. Wasterlain CG, Plum F. Vulnerability of developing rat brain to electroconvulsive seizures. Arch Neurol 1973; 29(1):38-45.
299. McCabe BK, Silveira DC, Cilio MR et al. Reduced neurogenesis after neonatal seizures. J Neurosci 2001; 21(6):2094-103.
300. Hoch DB, Hill RA, Oas KH. Epilepsy and mental decline. Neurol Clin 1994; 12(1):101-13.
301. Battaglia D, Rando T, Deodato F et al. Epileptic disorders with onset in the first year of life: neurological and cognitive outcome. Europ J Paediatr Neurol 1999; 3(3):95-103.
302. Koh S, Storey TW, Santos TC et al. Early-life seizures in rats increase susceptibility to seizure-induced brain injury in adulthood. Neurology 1999; 53(5):915-21.
303. Holmes GL, Gairsa JL, Chevassus-Au-Louis N et al. Consequences of neonatal seizures in the rat: morphological and behavioral effects. Ann Neurol 1998; 44(6):845-57.
304. Loscher W, Ebert U, Lehmann H et al. Seizure suppression in kindling epilepsy by grafts of fetal GABAergic neurons in rat substantia nigra. J Neurosci Res 1998; 51(2):196-209.
305. Kokaia M, Cenci MA, Elmer E et al. Seizure development and noradrenaline release in kindling epilepsy after noradrenergic reinnervation of the subcortically deafferented hippocampus by superior cervical ganglion or fetal locus coeruleus grafts. Exp Neurol 1994; 130(2):351-61.
306. Kokaia M, Aebischer P, Elmer E et al. Seizure suppression in kindling epilepsy by intracerebral implants of GABA- but not by noradrenaline-releasing polymer matrices. Exp Brain Res 1994; 100(3):385-94.
307. Thompson K, Anantharam V, Behrstock S et al. Conditionally immortalized cell lines, engineered to produce and release GABA, modulate the development of behavioral seizures. Exp Neurol 2000; 161(2):481-9.
308. Cole AJ. Is epilepsy a progressive disease? The neurobiological consequences of epilepsy. Epilepsia 2000; 41(Suppl 2):S13-22.
309. Jankowsky JL, Patterson PH. The role of cytokines and growth factors in seizures and their sequelae. Prog Neurobiol 2001; 63(2):125-49.
310. Jin K, Minami M, Lan JQ et al. Neurogenesis in dentate subgranular zone and rostral subventricular zone after focal cerebral ischemia in the rat. Proc Natl Acad Sci USA 2001; 98(8):4710-5.
311. Kee NJ, Preston E, Wojtowicz JM. Enhanced neurogenesis after transient global ischemia in the dentate gyrus of the rat. Exp Brain Res 2001; 136(3):313-20.
312. Liu J, Solway K, Messing RO et al. Increased neurogenesis in the dentate gyrus after transient global ischemia in gerbils. J Neurosci 1998; 18(19):7768-78.
313. Iwai M, Hayashi T, Zhang WR et al. Induction of highly polysialylated neural cell adhesion molecule (PSA-NCAM) in postischemic gerbil hippocampus mainly dissociated with neural stem cell proliferation. Brain Res 2001; 902(2):288-93.
314. Takagi Y, Nozaki K, Takahashi J et al. Proliferation of neuronal precursor cells in the dentate gyrus is accelerated after transient forebrain ischemia in mice. Brain Res 1999; 831(1-2):283-7.
315. Kitagawa K, Matsumoto M, Hori M. Protective and regenerative response endogenously induced in the ischemic brain. Can J Physiol Pharmacol 2001; 79(3):262-5.
316. Jiang W, Gu W, Brannstrom T et al. Cortical neurogenesis in adult rats after transient middle cerebral artery occlusion. Stroke 2001; 32(5):1201-7.
317. Gu W, Brannstrom T, Wester P. Cortical neurogenesis in adult rats after reversible photothrombotic stroke. J Cereb Blood Flow Metab 2000; 20(8):1166-73.
318. Gould E, Reeves AJ, Graziano MS et al. Neurogenesis in the neocortex of adult primates. Science 1999; 286(5439):548-52.
319. Bernabeu R, Sharp FR. NMDA and AMPA/kainate glutamate receptors modulate dentate neurogenesis and CA3 synapsin-I in normal and ischemic hippocampus. J Cereb Blood Flow Metab 2000; 20(12):1669-80.
320. Arvidsson A, Kokaia Z, Lindvall O. N-methyl-D-aspartate receptor-mediated increase of neurogenesis in adult rat dentate gyrus following stroke. Eur J Neurosci 2001; 14(1):10-8.
321. Westerberg E, Monaghan DT, Kalimo H et al. Dynamic changes of excitatory amino acid receptors in the rat hippocampus following transient cerebral ischemia. J Neurosci 1989; 9(3):798-805.
322. Cameron HA, McEwen BS, Gould E. Regulation of adult neurogenesis by excitatory input and NMDA receptor activation in the dentate gyrus. J Neurosci 1995; 15(6):4687-92.

323. Gould E, Cameron HA, McEwen BS. Blockade of NMDA receptors increases cell death and birth in the developing rat dentate gyrus. J Comp Neurol 1994; 340(4):551-65.
324. Gould E, McEwen BS, Tanapat P et al. Neurogenesis in the dentate gyrus of the adult tree shrew is regulated by psychosocial stress and NMDA receptor activation. J Neurosci 1997; 17(7):2492-8.
325. Larsson E, Lindvall O, Kokaia Z. Stereological assessment of vulnerability of immunocytochemically identified striatal and hippocampal neurons after global cerebral ischemia in rats. Brain Res 2001; 913(2):117-32.
326. Mody I, Otis TS, Bragin A et al. GABAergic inhibition of granule cells and hilar neuronal synchrony following ischemia-induced hilar neuronal loss. Neuroscience 1995; 69(1):139-50.
327. Sugimoto A, Shozuhara H, Kogure K et al. Exposure to sub-lethal ischemia failed to prevent subsequent ischemic death of dentate hilar neurons, as estimated by laminin immunohistochemistry. Brain Res 1993; 629(1):159-62.
328. Li Y, Chopp M, Powers C. Granule cell apoptosis and protein expression in hippocampal dentate gyrus after forebrain ischemia in the rat. J Neurol Sci 1997; 150(2):93-102.
329. Bering R, Draguhn A, Diemer NH et al. Ischemia changes the coexpression of somatostatin and neuropeptide Y in hippocampal interneurons. Exp Brain Res 1997; 115(3):423-9.
330. Kawamata T, Dietrich WD, Schallert T et al. Intracisternal basic fibroblast growth factor enhances functional recovery and up-regulates the expression of a molecular marker of neuronal sprouting following focal cerebral infarction. Proc Natl Acad Sci USA 1997; 94(15):8179-84.
331. Endoh M, Pulsinelli WA, Wagner JA. Transient global ischemia induces dynamic changes in the expression of bFGF and the FGF receptor. Brain Res Mol Brain Res 1994; 22(1-4):76-88.
332. Yagita Y, Kitagawa K, Ohtsuki T et al. Neurogenesis by progenitor cells in the ischemic adult rat hippocampus. Stroke 2001; 32(8):1890-6.
333. Levison SW, Rothstein RP, Romanko MJ et al. Hypoxia/ischemia depletes the rat perinatal subventricular zone of oligodendrocyte progenitors and neural stem cells. Dev Neurosci 2001; 23(3):234-47.
334. Back SA, Han BH, Luo NL et al. Selective vulnerability of late oligodendrocyte progenitors to hypoxia-ischemia. J Neurosci 2002; 22(2):455-63.
335. Park KI. Transplantation of neural stem cells: cellular & gene therapy for hypoxic-ischemic brain injury. Yonsei Med J 2000; 41(6):825-35.
336. Fukunaga A, Uchida K, Hara K et al. Differentiation and angiogenesis of central nervous system stem cells implanted with mesenchyme into ischemic rat brain. Cell Transplant 1999; 8(4):435-41.
337. Sinden JD, Stroemer P, Grigoryan G et al. Functional repair with neural stem cells. Novartis Found Symp 2000; 231:270-83; discussion 283-8, 302-6.
338. Nishino H, Borlongan CV. Restoration of function by neural transplantation in the ischemic brain. Prog Brain Res 2000; 127:461-76.
339. Hodges H, Nelson A, Virley D et al. Cognitive deficits induced by global cerebral ischaemia: prospects for transplant therapy. Pharmacol Biochem Behav 1997; 56(4):763-80.
340. Kondziolka D, Wechsler L, Goldstein S et al. Transplantation of cultured human neuronal cells for patients with stroke. Neurology 2000; 55(4):565-9.
341. Semkova I, Krieglstein J. Neuroprotection mediated via neurotrophic factors and induction of neurotrophic factors. Brain Res Brain Res Rev 1999; 30(2):176-88.
367. Manji HK, Moore GJ, Chen G. Clinical and preclinical evidence for the neurotrophic effects of mood stabilizers: implications for the pathophysiology and treatment of manic-depressive illness. Biol Psychiatry 2000; 48(8):740-54.
368. Duman RS, Malberg J, Nakagawa S. Regulation of adult neurogenesis by psychotropic drugs and stress. J Pharmacol Exp Ther 2001; 299(2):401-7.
369. Blows WT. The neurobiology of antidepressants. J Neurosci Nurs 2000; 32(3):177-80.
370. Malberg JE, Eisch AJ, Nestler EJ et al. Chronic antidepressant treatment increases neurogenesis in adult rat hippocampus. J Neurosci 2000; 20(24):9104-10.
371. Kempermann G, van Praag H, Gage FH. Activity-dependent regulation of neuronal plasticity and self repair. Prog Brain Res 2000; 127:35-48.
372. van Praag H, Kempermann G, Gage FH. Running increases cell proliferation and neurogenesis in the adult mouse dentate gyrus. Nat Neurosci 1999; 2(3):266-70.

373. Vaidya VA, Duman RS. Depresssion–emerging insights from neurobiology. Br Med Bull 2001; 57:61-79.
374. Duman RS, Nakagawa S, Malberg J. Regulation of adult neurogenesis by antidepressant treatment. Neuropsychopharmacology 2001; 25(6):836-44.
375. Chen G, Rajkowska G, Du F et al. Enhancement of hippocampal neurogenesis by lithium. J Neurochem 2000; 75(4):1729-34.
376. Czeh B, Michaelis T, Watanabe T et al. Stress-induced changes in cerebral metabolites, hippocampal volume, and cell proliferation are prevented by antidepressant treatment with tianeptine. Proc Natl Acad Sci USA 2001; 98(22):12796-801.
377. Duman RS, Malberg J, Thome J. Neural plasticity to stress and antidepressant treatment. Biol Psychiatry 1999; 46(9):1181-91.
378. Finamore TL, Seybold KS, Noble M et al. Contributions of hippocampal cellular damage and NMDA receptor dysfunction to behavioral markers of schizophrenia. Int J Neurosci 2001; 109(1-2):61-70.
379. Weinberger DR. Cell biology of the hippocampal formation in schizophrenia. Biol Psychiatry 1999; 45(4):395-402.
380. Port RL, Seybold KS. Hippocampal synaptic plasticity as a biological substrate underlying episodic psychosis. Biol Psychiatry 1995; 37(5):318-24.

Chapter 13

Transplantation in Parkinson's Disease:

The Future Looks Bright

Gesine Paul, Young Hwan Ahn, Jia-Yi Li and Patrik Brundin

Introduction

Parkinson's disease (PD) is the second most common neurodegenerative disorder and affects almost 1% of the population above the age of 50. Early in the course of the disease, patients primarily display motor symptoms, including bradykinesia, rigidity and resting tremor.[1,2] These symptoms worsen over time and, in addition, most patients eventually exhibit vegetative disturbances, and almost 40% of PD patients are affected by depression.[3] Dementia is also increasingly recognized as an important feature of PD, especially in the elderly cases[4,5] and approximately 25-35% of the patients exhibit cognitive decline.[6]

The most conspicious neuropathologic finding in PD is the loss of dopaminergic (DA) neurons in the substantia nigra pars compacta, which leads to a reduction of striatal DA levels.[7,8] As expected, there is also a loss of striatal DA transporter (DAT), a marker for nigrostriatal nerve terminals, which correlates inversely with the severity of motor symptoms.[9,10]

It is unlikely that the pathological event that triggers neurodegeneration is identical in all patients classified as idiopathic PD cases. Thus, there might be several different causes underlying PD, but despite differing etiologies, crucial components of the pathogenetic process may be similar in all patients.[11] Several interesting hypotheses regarding pathogenetic mechanisms in PD such as mitochondrial dysfunction, oxidative stress, exogenous toxins, intracellular accumulation of toxic metabolites, viral infections, excitotoxicity, deficient trophic support and immune mechanisms have been proposed (for review see refs. 12-15). The neuropathological hallmark of PD is an intracytoplasmic hyaline inclusion called "Lewy body", which is found in some of the remaining degenerating DA neurons. The inclusions are rich in the synaptic protein alpha-synuclein.[16,17] The observation that a small number of cases of autosomal dominant PD exhibit mutations in the alpha-synuclein gene[16,17] has implicated this protein as a potential key player in the pathogenesis of PD, but its role is clearly not yet understood.[18,19]

Two further genes (encoding for ubiquitin carboxy terminal hydroxylase 1 and a ubiquitin protein ligase named parkin) have been found to be mutant in some PD families. These proteins are both involved in the ubiquitin pathway of protein degradation[20,21] and it has been suggested that neurons in PD patients may be impaired in their ability to handle misfolded and damaged proteins.[22] Additionally, at least two other chromosomal loci (2p and 4p) that may harbour mutations in rare forms of inherited PD have been identified (for review see ref. 23).

In the absence of an understanding of PD etiology and pathogenesis, there is no treatment that can prevent or retard progression of the disease. Since the late 1960s, the main approach to

Brain Repair, edited by Mathias Bähr. ©2006 Eurekah.com
and Kluwer Academic / Plenum Publishers.

treating PD has been the pharmacological alleviation of the striatal DA deficit. This can be achieved by administration of the DA precursor L-Dopa in its standard or controlled release formulations in association with drugs that inhibit L-Dopa breakdown (dopa-decarboxylase inhibitors, carbidopa and benzeraside) and drugs enhancing dopaminergic transmission by blocking the breakdown of DA, monoamine-oxidase-B (MAO-B) and catechol-O-methyltransferase (COMT) inhibitors. There are also several DA agonists that act by direct binding on postsynaptic receptor sites. Nondopaminergic medications are less frequently used today and typically of limited benefit. They include anticholinergic agents, and amantadine which is a blocker of a class of glutamatergic receptors (for review see Ref. 24).

On the whole, the pharmacological treatment of PD has been remarkably successful in reducing clinical symptoms in PD. This motivated the award of the Nobel prize in medicine and physiology to Arvid Carlsson in 2000 for his work on signal transduction in the nervous system, in particular his suggestion that L-dopa could be used as a treatment for PD. L-Dopa treatment is usually highly effective early in the course of the disease, but at later stages patients may develop one or more of several treatment complications: The loss of drug efficacy is apparent in "wearing-off" symptoms, which are related to a loss of capacity to store L-Dopa presynaptically. As a result, increasing and more frequent doses of medication are required. Later, psychiatric disturbances such as hallucinations may develop, probably as a consequence of stimulation of DA receptors outside the striatum. Another major problem are the partly unpredictable fluctuations between immobility and an increased ability to move (the so called "on-off phenomenon"). During the "on" phases, when the drug allows the patients to move, the patients typically exhibit disturbing involuntary movements, termed dyskinesias.

Due to the problems associated with pharmacological DA replacement therapy, it is clear that new treatments are needed. Recent years have seen the development of novel neurosurgical methods for PD. Three different neurosurgical approaches have been successfully tested in PD patients. First, ablative procedures, thalamotomy and pallidotomy, and recently also subthalamotomy have been employed. Second, lesions using electrical currents (deep brain stimulation, DBS), aiming at the ventral intermediate nucleus (VIM) of the thalamus for the treatment of tremor,[25] the subthalamic nucleus (STN) to reduce tremor, bradykinesia and rigidity; or the internal segment of the globus pallidus (Gpi) for the reduction of mainly L-Dopa-induced dyskinesias besides other motor symptoms.[26,27] Third, restorative strategies, i.e., transplantation of different cell types to the brain have been tested during the past 15 years.[28]

Ablative and functional interventions attempt to restore functional imbalance in the basal ganglia and are based on observations that imply a hyperactivity of structures such as the STN and the GPi as a result of the dopaminergic deficit.[29,30] Neural transplantation, in contrast, aims at the replacement of the lost dopaminergic cells by grafting dopaminergic neurons or precursors into the target site, the striatum.[31] The ultimate role of these surgical methods in clinical practice is not firmly established and essentially they all still remain experimental.

In this chapter, we focus on cell replacement therapies in PD. We describe the methodology and efficacy of grafts of human embryonic nigral tissue in patients with PD, and highlight some of the clinical problems the procedure presents. We also discuss the technical aspects of neural transplantation in patients in more detail, including the importance and difficulties in achieving good graft survival. Finally, we will consider future alternative sources of donor tissue, including xenografts and various forms of stem cells.

Neurotransplantation: What History Has Taught Us

The adult central nervous system has a limited capacity for endogenous cell replacement. Cell therapies in PD are currently aimed at the replacement of lost dopaminergic neurons by grafting immature neurons into the target striatum. This treatment option for PD is especially promising since the cell loss in PD is primarily focused, albeit not exclusively, on the substantia nigra. Cell-based therapies for experimental and clinical PD have undergone a considerable development over the past three decades.[31]

Olson and coworkers were the first to graft *embryonic* dopaminergic neurons into the anterior eye chamber of the rat.[32,33] These studies revealed that developing neurons are needed to achieve survival and neurite outgrowth from grafted neurons; cells therefore have to be dissected during a critical stage of development.

A few years later, Björklund and colleagues demonstrated that embryonic DA neurons could also be grafted to the brain, although these pioneering studies only focused on morphological aspects of the grafts and did not address their possible functional implications.[34,35]

At the end of the 1970s the first reports on functional recovery due to neural grafts in hemiparkinsonian rats appeared.[36,37] The rats were injected unilaterally into the nigrostriatal pathway with 6-hydroxydopamine (6-OHDA) causing an ipsilateral depletion of striatal DA due to cell loss in the substantia nigra. As a consequence, such rats exhibit motor asymmetry that can be amplified into an overt rotational behavior following administration of drugs affecting the DA system. Thus, administration of the DA-releasing agent amphetamine causes DA release in the intact striatum and rotation ipsilateral to the lesion. In contrast, systemic injection of a low dose of apomorphine, a directly-acting DA receptor agonist, results in stimulation of supersensitive DA receptors in the denervated striatum and rotation contralateral to the lesion. In these pioneering studies, graft tissue was dissected from the ventral mesencephalon (VM) of rat embryos, a region containing large numbers of immature dopaminergic neurons in the developing substantia nigra and ventral tegmental area. The rats received grafts of solid nigral tissue pieces into the lateral ventricle adjacent to the denervated striatum[36] or premade cortical cavities, that had been allowed to develop well vascularized pial scar tissue, overlying the striatum.[37] These early landmark papers described that the transplants of nigral tissue reversed the motor asymmetry in the rats and that they extended axons into the denervated host striatum. Shortly thereafter a series of papers used a novel cell suspension grafting technology, emulating the dissociated neuronal culture technique. With this method, the donor tissue is enzymatically digested and mechanically dissociated resulting in a cell suspension that can be drawn into a microsyringe and stereotactically injected into any desired brain region. Using this grafting technique it was easier to reveal the functional importance of placing the cells into different subregions of the striatum.[38-40] Reinnervation of different parts of the striatum is essential to reverse specific parts of the behavioral syndrome that is the consequence of unilateral striatal DA depletion.[41,42] Thus, reversal of drug-induced rotation was associated with grafts innervating the central and dorsal parts of the striatum, while reversal of "sensory neglect" occurred when the implants innervated the ventrolateral striatum.[38,41,43] The grafted DA neurons were shown to make synaptic contacts with the host striatum[44,45] and to release DA in a regulated fashion.[46,47] Moreover, transplanted neurons can receive inputs from host neurons[48,49] and partially reverse several different motor deficits in animal models of PD. While the grafts are able to reverse many of the simple sensorimotor deficits that follow unilateral striatal DA depletion in the rat, more complex motor behaviors have not been as readily ameliorated. The details of these behavioral studies have been the subject of several review articles that can be recommended for further reading.[28,50-52]

Before the first systematic transplantation trials using embryonic dopaminergic neurons in patients with PD were performed in the mid-1980s,[53-57] adrenal medulla chromaffin cells had been used as autografts in attempts to provide a source of catecholamine in the striatum. Initial reports on stereotactic implants of chromaffin cells, performed in Sweden, indicated that the procedure was relatively safe, but did not report of any significant beneficial effects.[58,59] In contrast, the Mexican group headed by Madrazo used "open microsurgery" to access the caudate nucleus via the lateral ventricle and implanted adrenal medulla.[57,60] They reported significant amelioration of symptoms in one patient and the paper was followed, within a couple of years, by surgery using a similar technique in several hundred patients worldwide. However, these surgeries not only failed to reproduce the reported beneficial effects, but were also accompanied by an unacceptable level of mortality and morbidity.[61] In conclusion, adrenal medullary graft tissue proved ineffective and the procedure is no longer considered an acceptable option.[62]

Other sources of donor cells, such as, carotid body cells,[63,64] superior cervical ganglion[65,66] and retinal pigment epithelium[67] have been investigated in human experiments. However, the data available so far do not suggest that there are major clinical benefits.

At present, results obtained using grafts of embryonic ventral mesencephalic neurons have been most impressive. The first surgeries reported in the scientific literature are from 1988.[54,55] Two years later, the first evidence for graft survival and function was reported.[68] It is estimated that over 300 PD patients have been operated with grafts of nigral tissue.[69] Only a minority of these cases has been reported in the scientific literature. Nonetheless, there are multiple reports of long lasting symptomatic improvement.[70-76] Functional improvement seems to correlate with neuroradiological evidence of surviving graft tissue.[77-79] A more detailed description of the clinical outcome follows later in the "Graft Effects" section.

Methodology for Nigral Tissue Transplantation in Parkinson's Disease Patients

The neural transplantation surgery protocol varies significantly between different centers regarding several parameters, e.g., age of donor tissue; tissue storage and preparation; numbers and location of injections; use of immunosuppression etc. The protocol described in this review is based on the one employed at Lund University. Cells are taken from the ventral midbrain during embryonic development at a time when the neurons are undergoing terminal differentiation between the 5th and 8th week post-fertilization.[80,81] The embryos are collected from routine elective abortions using ultrasound-guided suction technique with informed consent of the woman undergoing abortion, in accordance with strict ethical guidelines. A set of ethical guidelines recommended by a transnational European organization has been published.[82] Soon after the abortion the embryo is rinsed in sterile balanced salt solution and the VM dissected. Some centers have employed a tissue storage step, to separate donor tissue procurement and transplantation surgery in time. This storage has either taken place in explant culture for 1-4 weeks[83] or at 4°C in a hibernation medium for 1-8 days.[84] Typically the VM from multiple embryonic donors are collected, each cut into 6-16 pieces, and enzymatically digested (0.1% trypsin and 0.05% deoxyribonuclase) in the balanced salt solution at 37°C for 20 min.[80] They are then mechanically dissociated, using a firepolished Pasteur pipette, into a suspension consisting of single cells and small aggregates. Using MRI-guided stereotactic neurosurgery, grafts are placed along multiple (3-7) trajectories, in the coronal plane, in the putamen and in some cases along two trajectories into the head of the caudate nucleus.

Depending on the design of the surgical program and the expected availability of the donor tissue, different centers have opted for different policies on the use of immunosuppression (see "Is Immunosuppression Necessary?" section) (Fig. 1).

Technical Aspects of Graft Preparation: Donor Age and Cell Number

Several lines of evidence suggest that survival of dopaminergic neurons is crucial for nigral grafts to be effective. The age of the embryonic donor tissue constitutes one vital parameter that affects survival of grafted DA neurons (for review see refs. 79,80). The optimal donor age is governed by the tissue preparation method. When the embryonic mesencephalon is mechanically dissociated, tissue from younger embryos has to be used in contrast to when solid tissue pieces are transplanted.[80,85] Based on data from human-to-rat xenograft trials, donor

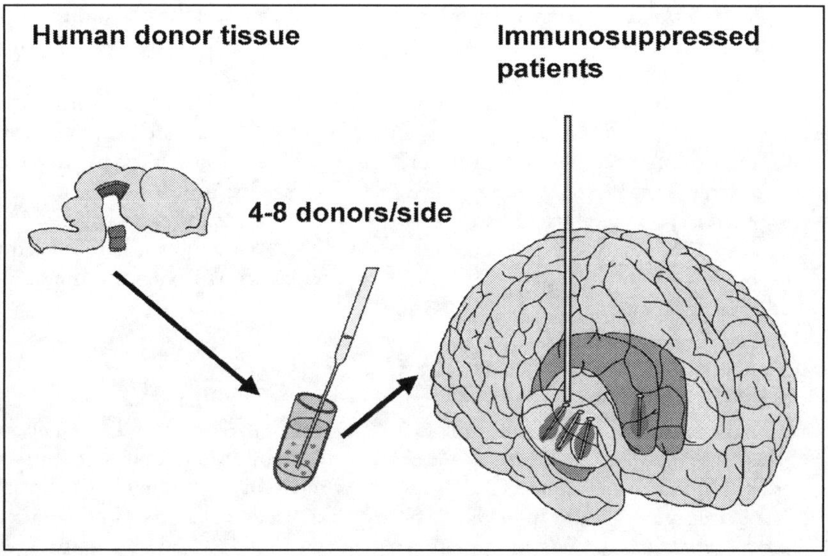

Figure 1. Methodology of nigral transplants: The ventral mesencephalon (VM) from human embryos between the 5th and 8th week of gestation is dissected, enzymatically and mechanically dissociated into a mixture of single cells and small aggregates and grafted along several trajectories into the caudate and putamen of an immunosuppressed patient. Multiple donors, according to some sources as many as 4-8 donors per side of the brain, are necessary to obtain sufficient tissue.

tissue from early embryonic stages of around 5.5 to 8 weeks post conception is most suitable when the tissue is subjected to dissociation.[80] Slightly older tissue can probably be used successfully if it is not dissociated into a cell supension, but cut into small pieces before surgery instead.[85]

Careful donor tissue dissection is another vital aspect of the graft surgery. If the mesenchymal membranes covering the embryonic central nervous tissue are not removed, they may be the source of undesirable growth of nonneuronal tissue in the host brain. Indeed, in two post-mortem cases nonneuronal tissue was found in the grafted area[86,87] and in at least one of the cases it was probably directly related to the cause of death of the patient (for review see Ref. 79).

Each human mesencephalon has been estimated to contain about 500.000-1.000.000 dopaminergic neurons[88] and it is likely that these cells have already begun differentiation in the embryonic mesencephalon at the time of donor tissue procurement.[89] Only approximately 20.000-40.000 of DA neurons from a single human embryo survive when xenografted to immunosuppressed rats.[89] A similar survival rate seems to apply also when the cells are grafted to PD patients with a similar protocol.[90] Therefore, to achieve complete replacement of DA-containing neurons it may be necessary to transplant mesencephalic tissue from multiple donors into each putamen.[73,79] Results from rat experiments clearly illustrate that the recovery in drug-induced rotation is greater when more grafted DA neurons survive. Eventually a ceiling effect is attained, and higher cell numbers have no additional impact on the assessed function[91] (for review see Ref. 50). So far, different surgical centers have implanted different amounts of donor tissue, ranging from one to seven donors per side of the brain (for review see Ref. 79). Due to the fact that treatment of grafted tissue (donor age, tissue preparation and storage, and surgical technique) differs considerably between centers, definite conclusions are currently difficult to make regarding the importance of the amount of implanted tissue. When 35 patients from six different centers were compared regarding clinical outcome and the amount of tissue transplanted, there was a trend favouring better clinical outcome when more tissue was grafted. However, this trend did not reach statistical significance.[92]

Tissue storage is another issue of debate. At Lund University we have recently introduced storage at 4°C prior to transplantation up to 8 days in hibernation medium supplemented with a lipid peroxidation inhibitor and a growth factor.[84] Other centers also subject the tissue to cool storage[93] or alternatively grow the tissue in explant culture for up to 4 weeks.[76,94] Although there are reports on the impact of such tissue storage on graft survival in rats,[84,95,96] it is not clear whether a tissue storage step could affect the clinical outcome in grafted patients and whether there is a greater risk of development of dyskinesias in patients when the donor tissue is stored prior to surgery.[97]

Technical Aspects of Graft Implantation: Graft Placement

The human striatum is approximately 200 times bigger in volume than that of rats.[98] It is believed that the diffusion of DA released from graft-derived terminals is limited.[99,100] Therefore it has been suggested that grafts need to be implanted along multiple injections tracts placed as close as 5 mm apart (6-8 needle tracks in the coronal plane) throughout the putamen.[73,75] Bilateral transplants ought to be more effective than unilateral grafts, although there are some *bilateral* graft effects in *unilaterally* grafted patients.[68,83,94,101,102] There is no prospective study comparing unilateral versus bilateral implants, but in patients with a functional graft on one side of the brain, implantation into the contralateral striatum provides some additional benefit.[75]

In PD, the reduction of DA in the putamen, rather than the caudate nucleus, seems to underlie the motor symptoms. Studies in normal individuals indicate that the dorsal and intermediate subdivisions of the putamen are primarily involved in motor function, whereas the caudate nucleus plays an important role for cognition and for visual cues important for motor function and correction of balance.[103,104] However, the caudate may still be important as an additional implantation target since motor symptoms are improved following transplantation into the caudate in a marmoset model of PD.[105] It has been suggested that better results should be obtained following implantation into both the putamen and the caudate, than putamen alone,[73,74] although there is still insufficient clinical proof.

Is Immunosuppression Necessary?

Little is known about immunological aspects of intracerebral grafting in patients. Some confusion has arisen from the concept of the brain being immunologically privileged. The "immune privilege" infers that histoincompatible grafts survive longer when implanted into the brain in comparison with another site, e.g., the skin. However, it does not necessarily imply that there is indefinite survival.[106] Many factors contribute to the immune privilege, including a complex set of barriers (e.g., blood-brain barrier) that restrict passage of immunocompetent cells into the brain. Only activated lymphocytes pass endothelial cells in brain capillaries. In addition, the antigen-presenting capacity of the brain is reduced in comparison to other sites. The regulation of the immune response is also dependent on the local response of cytokines and chemokines, which are produced in response to various stimuli and traumata, such as the implantation trauma.[106-109]

Allografted (same species, but different genetic background) nigral tissue seems to survive indefinitely in the striatum of experimental rodents.[106] However, several studies in rats have shown that allografted DA neurons can undergo rejection from the brain if the host immune system is challenged by an orthotopic skin allograft[110] or if the neural cells are cografted with allogeneic spleen cells.[111] So there is no doubt that, under certain conditions, the rat immune system can effectively reject neurons allografted to the brain. Interestingly, the same may also be true for humans. Brains obtained from two post-mortem cases of patients grafted with human allogeneic tissue were found to contain macrophages, B-cells and T-cells around the implant sites,[90,112] suggesting that human allografts can evoke an immune reaction.

Depending on the design of the surgical program and the expected availability of the donor tissue, different centers have opted for different policies on the use of immunosuppression. As an example, the Lund program has performed two surgeries separated by 2 weeks up to 3-5 years.[75] As the first graft may immunize the patient to human antigens, the current protocol in Lund uses a triple combination of cyclosporin, azathioprin and prednisolone for immunosuppression during the time between the surgeries. The immunosuppression has also been given for 12 months after the second and final implantation, and then tapered off over a 3-month period. There have been very few side effects of this treatment.[113] After withdrawal of the immunosuppression there has not been any sign of rejection, neither clinically nor as assessed by PK 11195 PET scanning (Widner H, personal communication).[113,114]

Despite the possibility for immune rejection of human allografts, clinical data obtained so far suggest that long-term, continuous immunosuppression may not be necessary for survival of intracerebral allografts.[71,77,115-117] Therefore, other surgical centers have opted for different policies, e.g., the Denver/New York group used no immunosuppression at all.[83] However, it is unclear if the survival is optimal under these conditions. For example, a recent report described that some grafted human neurons survived for at least 7 months-8 years in patients who had not received immunosuppression.[76,118] The number of surviving DA neurons reported in two of these patients[76] was considerably lower compared to other autopsy cases.[119,120] Because the amounts of implanted tissue and graft preparation also differed between these two sets of patients, it is difficult even to speculate whether the reduced survival is due to an immune rejection. The Tampa program uses an immunosuppressive treatment with cyclosporin A alone for 6 months after surgery.[93] As mentioned briefly in an earlier section, lymphocytes have been observed in an autopsy case after 18 months after grafting (12 months without immunosuppression) from this series.[121]

In summary, there is currently no consensus between centers on how and for how long allografted PD patients should be immunosuppressed to achieve lasting graft survival.

Graft Effects

Patient Selection

Patients have to be carefully selected and undergo detailed clinical evaluation before and after neurotransplantation. Preferably, this should be done according to the "Core Assessment Program for Surgical interventions and Transplantation" (CAPSIT),[122] which is a revised version of the "Core Assessment Program for Intracerebral Transplanation" CAPIT.[123] These protocols allow comparisons between clinical studies at different centers.[117] They includes diagnostic criteria, exclusion criteria and suggestions on preoperative and post-operative follow-ups. In order to reduce the level of placebo effects, CAPSIT suggests three preoperative investigations and neuro-imaging. It also allows for comparison between DBS and transplantation studies.

The clinical assessement part of CAPIT comprises

- Patient-derived "on/off diaries" to assess daily motor fluctuations.
- L-dopa-test which tests the effect of a single dose of L-dopa on motor symptoms.
- Activities of daily living, although the value of this measure has been questioned.[124,125]

Bradykinesia, rigidity and motor fluctuations are the most disabling motor PD symptoms of patients treated pharmacologically, and they are all effectively ameliorated by intrastriatal VM grafts.[78,117] Therefore, primarily patients in whom these symptoms predominate should be considered for grafting. A recent study suggested that the L-Dopa-response before transplantation is predictive for the outcome of the transplantation.[118] This finding, however, is not surprising, since L-Dopa-responsiveness is the main predictor for PD, irrespective of the stage of the disease. Unless a clear L-dopa response can be determined, transplantation surgery is not warranted.

Except for one single case,[120] no significant improvement of Parkinsonian tremor has been demonstrated following transplantation (for review see Ref. 117). Although most patients grafted so far have not had tremor as a dominant symptom, there is currently little support for selecting such patients for transplantation surgery.

Also patients with cognitive impairment are probably not suitable for grafting.

Only mild improvement of motor symptoms was observed in PD patients with impairment of cognitive function.[70,75] These patients also stand a greater risk for developing post-surgical complications with transient worsening of cognitive functions.[117]

Motor Assessment

In successful cases, transplants of embryonic nigral tissue markedly improve bradykinesia and rigidity contralateral, and to some extent also ipsilateral, to the implanted side. Motor fluctuations are also reduced markedly by the procedure (for review see refs. 113,117).

In most cases, there is a gradual increase of clinical benefit during the first 1-2 years (in some patients apparent already after 3-6 months), and the effects seem to plateau after 3-4 years at the latest.[77]

Most patients evaluated according to the CAPIT criteria have experienced long-lasting improvements of 30-50% on the motor score of the UPDRS scale (for review see refs. 113,117). The 18 patients transplanted in Lund in collaboration with centers in San José, London and Munich/Marburg[117] have been followed up for 10 years. The majority of these patients spent significantly more time in the "off"-phase and less time in the "on" phase before surgery, and after surgery the relationship was reversed. When patients were tested in defined "off" (12 hours without any anti-parkinson medication), the speed and accuracy of movement was evaluated in timed series of pronation/supination movements of the hand and found to improve significantly. The L-dopa medication could be reduced progressively or even terminated in some cases.[68,74,75,116,126,127]

As mentioned earlier, not all symptoms are clearly improved by transplantation. For example, impaired postural function, swallowing and speech do not seem to improve as dramatically as limb hypokinesia and rigidity (Fig. 2).[117,128]

In Vivo Imaging of Brains Containing Grafted Dopamine Neurons

Many clinical transplantation programs include assessment of graft survival using positron emission tomography (PET) with ^{18}Fluorodopa (FD) as a marker for dopaminergic neurons. The FD-PET scans provide an index of the number of viable striatal dopaminergic terminals.[129] In PD, the FD uptake in the putamen correlates inversely with the degree of motor impairment[130,131] and with numbers of remaining nigral DA neurons at autopsy, as well as with striatal DA levels.[132] Using a ligand for postsynaptic D2-receptors such as ^{11}C-raclopride, it has been possible to monitor upregulation of striatal D2 DA-receptors in PD patients[133,134] reflecting compensatory hypersensitivity due to the loss of striatal DA.

Histological studies have shown that the postoperative increase in FD-uptake in PET studies reflects survival of dopaminergic grafts reinnervating the striatum.[76,79,115,120] The FD-PET signal has been found to be significantly elevated at around 4-6 months after transplantation in several patients[77,79,117] and in some cases it shows a continued progressive increase for 2-3 years,[77] while in others it does not continue to increase between 6.5 and 18.3 months.[78]

In one cohort of patients, movement-related activation in the supplementary motor cortical area and dorsolateral prefrontal cortex were studied. These cortical areas are activated in

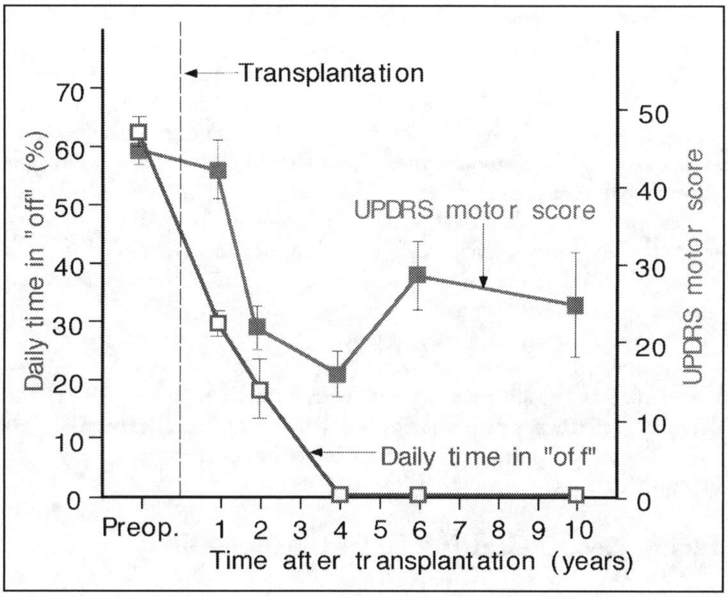

Figure 2. Percentage of the daily time spent in "off" by one patient and the changes in the UPDRS motor score (measured in the defined"off", when the patient had not received any anti-parkinson medication for 12 hours). First effects are clearly seen one year after grafting, but symptoms continue to improve up to 4 years after transplantation. The diagram is based on data from.[77]

normal subjects in conjunction with movement, whereas unmedicated PD patients do not exhibit the same activation pattern. Grafted PD patients exhibited restoration of this cortical activation at 18.3 months after surgery, but not at 6.5 months post-grafting.[78] Interestingly, the same patients displayed a partial and significant restoration of striatal FD-PET signals at the first time-point with no further increase by the second later time-point. Nevertheless the clinical improvement, monitored as a decrease in a motor impairment scale, continued between the two time-points.[78] These data suggest that improvement of clinical symptoms continues gradually even after the dopaminergic neurons have begun to produce DA, and that it may depend on protracted development of a functional integration between the graft and host brain, or be related to plastic changes in the host brain as the newly developed dopaminergic innervation exerts its effect.

Post-Mortem Histological Examination of Embryonic Dopaminergic Transplants

At least 14 PD patients who have received a neural transplant and subsequently died (all of them of unrelated causes), have come to autopsy (for review see Ref. 79). The first published case described the histological findings in a patient operated at the Yale University surgical center, 4 months after implantation of tissue from one embryo and with immunosuppressive treatment lasting for 7 weeks. No surviving TH-positive graft-derived cells were identified at autopsy.[135] In a series of five patients who had each been grafted with tissue from one donor (no immunosuppressive therapy) in Birmingham, United Kingdom, only few cells were found in three out of five cases 18-40 month after surgery.[90,136,137] Another three post-mortem cases operated by the Denver surgical center were given no immunosuppressive treatment and were each grafted with tissue from 2 donors into each putamen.[138] The numbers of surviving

TH- positive neurons were between 6.840-38.392 per putamen.[76] The survival was better in the cases operated in Tampa and was reported in great detail by[120] (for review see Ref. 90). These latter two patients died 18 and 19 months after grafting and had received tissue from 3-4 donors per putamen. Immunosuppression was continued for six month after surgery. Following cell counting, 81.905 and 126.162 TH-positive neurons were found in the left and right putamen, respectively, of the first patient, whereas the second patient had 138.000 surviving TH-positive cells in one putamen (other side not analysed).[119,139] In both cases extensive reinnervation of the host striatum was described, without any evidence for sprouting of host-derived TH-positive fibers.[119] Numerous macrophages and lymphocytes were also found infiltrating the graft area,[121] suggestive of an ongoing immune response. Both of these patients also exhibited marked increases in striatal FD uptake[115] (for review see Ref. 79).

Available autopsy and clinical data suggest that a graft survival has to exceed a threshold number of TH-positive neurons for functional effects to develop. This view is supported by the lack of documented clinical improvement in cases with virtually no graft survival (for review see Ref. 79). Based on animal experiments and available clinical data, it has been suggested that approximately 100.000 surviving grafted DA neurons per putamen are necessary to elicit significant clinical improvement (for review see Ref. 79).

Problems Facing Neural Grafting in Parkinson's Disease

Dyskinesias and Dystonia

In March 2001, the first double-blind placebo-controlled trial of neural transplantation was published by the Denver/New York group, and stimulated widespread media interest and scientific debate.[76]

This attention was partly due to the brief description of severe disabling and uncontrollable dyskinesias and dystonia (abnormal involuntary movements and postures) in 15% (5/33) of the grafted patients. The involuntary movements appeared more than one year post-grafting, even though the parkinsonian symptoms were reported to have undergone improvement during the first year after transplantation in these patients, allowing substantial reductions in L-Dopa medication.[76] The authors attributed this adverse effect to a 'relative excess of DA' released from the grafts and in order to avoid dyskinesias in future operations, it was suggested to graft less tissue.[76] However, this interpretation has been discussed and challenged repeatedly.[31,79,140-142] Most critics had the impression that patients from other centers using open-labeled trials had larger grafts, but still did not display disabling dyskinesias/dystonia of the same frequency or magnitude as those reported by the Denver/New York team. In 14 patients operated in Lund, who had been followed for up for up to 11 years after surgery, dyskinesias increased during "defined off" after grafting. However, the involuntary movements were generally mild or moderate. The relationship between graft size and dyskinesias in "defined off" is potentially important. In the Lund retrospective study, there was no correlation between the severity of dyskinesias in "defined off" and the extent of graft-induced symptomatic relief.[143] The development of dyskinesias in "defined off" seems to follow the clinical improvement with a delay of more than one year, suggesting a different mechanism to that underlying clinical improvement.[143] Furthermore, the severity of postoperative off-phase dyskinesias tended to correlate negatively with the preoperative FD-uptake in the putamen, which indicates that the severity of the dyskinesias in "defined off" may be related to the extent of striatal dopaminergic denervation.[143]

The Tampa/Mount Sinai surgical trial was the second randomized, double blind, placebo-controlled study to be performed. In this study, solid pieces of embryonic VM from one or four donors were grafted and patients were immunosuppressed with cyclosporine A for 6 months. No significant improvement in the UPDRS was found after two years. Patients who had received cells from four donors showed progressive improvement up to 6 months that was lost at the same time the immunosuppression was discontinued. This could suggest an immunresponse against the graft and maybe a late rejection. Striatal ^{18}F-DOPA uptake was significantly increased after transplantation and good survival of dopamine neurons was also observed at postmortem examination. However, in this study, fifty-six percent of the transplanted patients developed off-dyskinesias that persisted upon drug withdrawl.[143a] Four cases have come to autopsy and show activated microglia in and around the graft deposits. It has been suggested that the observed dyskinesias may reflect partial, but inadequate graft survival. The graft produces, stores and releases low levels of dopamine, but may not be sufficient to induce an antiparkinsonian response.[143a]

The Lund data do not support the suggestion by the Denver/New York group, that off dyskinesias may be a result of excess DA being produced by the grafts. However, differences in placement of graft tissue within the putamen, and the lack of immunosuppression and longer preoperative tissue storage time in the Denver/New York study may account for discrepancies in outcome between the two surgical centers.

In summary, there is no evidence that severe off phase dyskinesias are characteristic for DA cell replacement therapy per se. However, they can occur, and when they do, there is no evidence that they are the result of an excess of DA. The underlying mechanism is simply not well understood yet. Paradoxically, grafts have actually been described to ameliorate on-phase dyskinesias in some patients[71,127] and in rats with experimental parkinsonism.[144] In these cases, it has been suggested that the grafts provide the denervated striatum with a renewed capacity to buffer fluctuations in dopa levels. The sham controlled double blind studies illustrate that the present cell replacement procedures are far from optimal and that further research is necessary to optimize selection of patients and transplantation procedures.

Cell Death

There is a positive relationship between the number of surviving DA neurons and the extent of behavioral recovery. This is true for a certain window of graft sizes, beyond which there is a saturation effect such that additional surviving neurons do not give rise to further clinical relief (see also "Technical Aspects of Graft Preparation: Donor Age and Cell Number" section). In the unilateral 6-OHDA rat model of PD, around 2000 TH-positive neurons need to survive to reverse amphetamine-induced rotational behavior about 6 weeks after surgery (see review by Ref. 50). Whilst achieving large grafts may seem like a trivial surgical problem, most studies in rodents and humans report survival rates of transplanted DA neurons as low as 1-20%.[79,120,139] It has been suggested that nigral tissue from at least 4 human embryos has to be grafted to each side of the brain to obtain reliable functional effects in PD, even when using optimized transplantation protocols.

One major practical limitations of neural transplantation as a therapy for PD patients is therefore the need for multiple donors per patients. The use of tissue from very large number of embryos per patient can also be argued to raise specific ethical concerns.

During the transplantation procedure, DA neurons are subjected to several types of insults causing both apoptotic, necrotic and intermediate forms of cell death. The cell death in grafts is most likely triggered by one or more of the following mechanisms: hypoxia, hypoglycemia, mechanical trauma, free radicals and lack of appropriate growth factor stimulation. The cell death appears to begin already during retrieval of the donor tissue and has subsided after a few days in the new host brain (for reviews see refs. 145,146). Hypoxia and glucose deprivation during the removal of the embryo can result in oxidative stress[147] and may cause death in as many as 20% of the cells even prior to mechanical dissociation of the tissue.[145,148] During the tissue preparation, donor tissue is inevitably subjected to severe mechanical trauma, resulting in the death of another 30% of the neurons.[149,150] Cell death continues during the transplant injection and immediate period (around 7 days) thereafter.[150,151] However, during the graft maturation, i.e., beyond the first postoperative week up to several months, the total number of surviving cells does not change[151] (see for review Ref. 145).

In summary, cell death occurs during the entire transplantation procedure up to one week after grafting. Therefore neuroprotective strategies should primarily focus on the tissue preparation steps and the first few days after graft surgery.

Methods to Improve Graft Survival

Several pharmacological approaches have been used to promote survival of transplanted neurons in animal experiments. In this chapter we focus on those that can be applied clinically.

Flunarizine, an antagonist of voltage-gated calcium channels and also potent antioxidant,[152] has been shown to improve survival of grafted embryonic rat DA neurons.[150,153] To reduce oxidative stress in nigral grafts, lazaroids, a heterogeneous group of compounds that inhibit radical-evoked lipid peroxidation[154] are effective. Treatment with different experimental lazaroids improves survival of grafted DA neurons in rats,[91,155] and the clinically approved tirilazad mesylate has been used in an attempt to enhance graft survival in the clinical transplantation program in Lund.[74]

Apoptotic cell death occurs at least in some cells starting from the step of tissue preparation until 4 days after grafting.[148,156] Caspases, a family of proteases, play a central role in the apoptotic process[157] and caspase inhibitors have been shown to effectively block apoptosis in cultures of DA neurons and in nigral grafts[156] (for review see Ref. 145). The effects of lazaroids and caspase inhibition are at least partially additive[158] suggesting that different populations of dopaminergic neurons in the graft are killed by different mechanisms. Although caspase inhibitors have been found to reduce DNA fragmentation and caspase 3 activity in nigral cell suspensions, it is possible that they also inhibit other proteases important for neuronal death, such as cathepsin,[159] indicating that the high concentrations of caspase inhibitors employed in neural grafting experiments may act through multiple mechanisms. Finally, treatment of nigral cell suspensions with immunophilin ligands that inhibit calcineurin, namely cyclosporin A and FK506, can also improve the survival of DA neurons following implantation into an adult striatum.[160]

Different families of growth factors have also been investigated regarding their ability to improve survival of nigral grafts. Treatment with basic fibroblast growth factor (bFGF) significantly increases the survival of grafted neurons.[161-163] Glial cell line-derived neurotrophic factor (GDNF) was first described to have a survival-promoting effect on nigral grafts in 1996[164,165] and several subsequent studies have substantiated these findings (for review see Ref. 145). Of particular relevance to clinical trials, human GDNF can improve both survival and fiber outgrowth from transplanted human dopaminergic neurons in hemiparkinsonian rats.[166] Recently, the focus has been on how to improve graft survival by modifying the transplant procedure during the first few hours after implantation into the adult brain. Thus, we have found that treating the graft recipient with systemic injections of the lazaroid tirilazad mesylate increases survival of DA neurons.[167] Similarly treatment of the host with synthetic fibronectin peptide V improved the survival of grafted DA neurons.[168] Reducing the body temperature of the host to 32-33°C, causing hypothermia in the brain that receives the graft tissue also improves cell survival.[169] When combining pretreatment of rat embryonic graft tissue with two drugs—one lazaroid and one caspase inhibitor—and simultaneously cooling the body temperature of the graft recipient, it is possible to attain a survival rate around 55%.[170] While this particular combination of treatments may be difficult to apply clinically, the results demonstrate that around half of the grafted neurons may be possible to rescue. Should this be achieved in clinical trials, it would most probably be sufficient to graft mesencephalic tissue from one human embryo to each side of the patient's brain (Fig. 3).

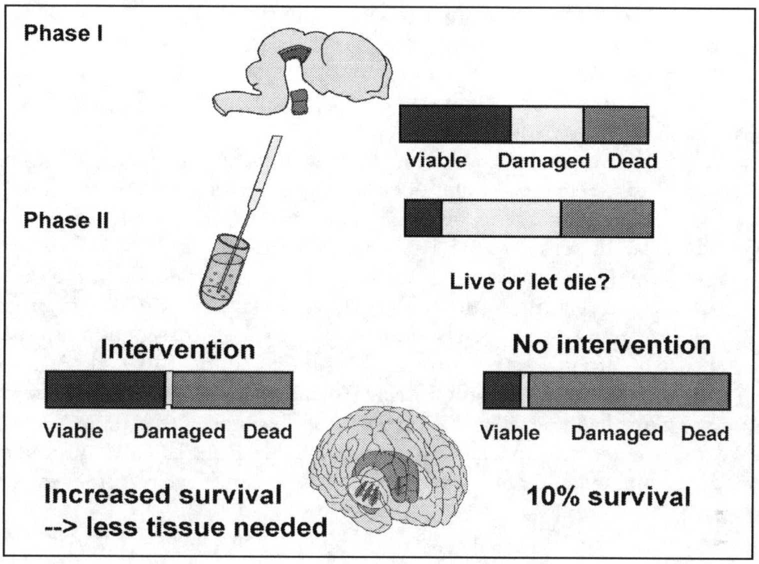

Figure 3. Already at the time-point of dissection of the donor tissue (phase I), some of the neurons are damaged (yellow) or dead (red), but a significant proportion are still viable (blue). During the dissociation step (phase II), there is a progressive increase in the number of damaged and dead neurons. If no intervention is made, with e.g., administration of neuroprotective drugs, there will only be around 10% surviving dopaminergic neurons in the grafts (right part of diagram). On the other hand, if neuroprotective measures are taken, a large proportion of damaged neurons can be rescued and a survival rate of around 50% can be attained (left part of diagram).

Alternative Sources of Donor Tissue

There is no doubt that the cell replacement strategy can give rise to symptomatic relief in PD. However, due to logistic and ethical problems connected with the use of human embryonic tissue the search for alternative cells for transplantation in PD is essential. In this section we will discuss cells that could eventually replace human embryonic DA neurons as a source for grafting in PD. We focus on cells that have either entered clinical trials or show promise that they could be developed into part of a clinical protocol in the foreseeable future.

Almost every existing type of DA-producing cell has at some point been considered for neural grafting in PD. However, it may not be sufficient for a cell to produce and secrete DA. Dopaminergic neurons also have the ability to reinnervate the host striatum over long distances, form synapses and possibly respond to host signals. The ideal cell for grafting in PD should fulfill those criteria. In addition, it should be available in unlimited amounts. Ideally, the supply of cells should also be possible to standardize, i.e., with large batches of identical cells being available at surgeries separated in time, and it should be possible to screen for infections before transplantation.

Neural Xenotransplants

Neural xenotranplantation, the grafting of embryonic neuroblasts from the nervous system of a different mammalian species, has been considered early on.[171] The use of porcine tissue to replace human embryonic tissue as opposed to embryonic nonhuman primate tissue has some advantages; pigs are easy to bread and produce large litters, they are not an endangered species. Therefore, xenotransplantation could potentially circumvent many of the logistic problems associated with the use of fresh embryonic tissue.[108,172,173] Porcine tissue has already been considered for other organ transplants before, donor tissue can be genetically modified to e.g., increase cell survival[174] or to knock out some of the dominating antigens responsible for the acute immunorejection (saccaride epitopes of the α-Gal on pig tissue) by naturally occurring xenoreactive antibodies.

Embryonic porcine neural tissue transplanted into hemiparkinsonian immunosuppressed rats not only grows anatomical correct connections,[175] but also restores behavioral deficits.[176,177] Although these are encouraging results, before xenotransplantation of neural tissue ought to enter large scale clinical trials, certain critical issues need to be clarified in animal experiments.[173] First, there is a potential risk of infection by zoonotic organisms, especially in the context of immunosuppression.[178] In xenotransplantation, however, donor tissue can be screened extensively for pathogens and animals can even be bread under pathogen-free conditions. The main concern is for the risk of infection by porcine endogenous retrovirus (PERV). However, several patients who have been transplanted with porcine pancreatic islets[179] or had been otherwise exposed to porcine tissue had no evidence of infections.[180] Possibly, protection against PERV can be reached by vaccination of recipients of porcine tissue against PERV, treatment with anti-retroviral drugs or elimination of the virus by knocking out the gene (for review see refs. 108,181).

Secondly, the major problem associated with the use of xenogeneic tissue is the immune host response. An immunosuppressive regimen needs to be established which ensures long-term graft survival while minimizing the risk of side effects associated with such therapy. Currently, the immunosuppressive regime may have to be administered lifelong. Several immunoprotection strategies have been devised, such as masking xenogeneic MHC class I antigens or altering the donor tissue, but none of them have proven very effective (for review see Ref. 173).

Clinical studies with porcine neural xenografts have already started. Twelve patients with PD were transplanted in a phase I study, testing the safety of the approach and a technique called "immunomasking", which was promoted by the company Diacrin in collaboration with an academic center. In the clinical study, 50% of the patients received the compound, a cell with anti-pig class I MHC F(ab) fragments to reduce the antigenicity of the tissue,[182] whereas the others received a low dose of cyclosporin A treatment.[183] Two or 3 patients experienced some benefit 1-2 years after grafting, however, this did not correspond to any positive changes in FD-PET.[183] One patient in this trial died from unrelated causes about 8 month after he had received a porcine transplant. Post-mortem analysis showed only about 600 cells surviving, even though each patient had received 12 million pig cells at the time of implantation. Lymphocytes were found around the remaining cells, indicating a chronic rejection.[184] A second clinical study was initiated in 1999 (Genzyme/Diacrine.com, phase II-study) using another immunosuppressive schedule and no immunomasking. In this study, 18 patients were randomly assigned to be either grafted or to receive sham surgery. Each of the treated patients in this trial received approximately 48 million cells (from 11 embryos) transplanted bilaterally. At 18 months after surgery, there was no significant difference between the transplanted and the placebo group with respect to the primary endpoint measure, the "UPDRS-off-score". There were no disabling dyskinesias reported in these patients.[185] Even though the clinical data for porcine neural xenografts are disappointing so far, porcine tissue still remains a promising alternative cell source for neurotranspantation in PD.

Stem Cells

Perhaps the most attractive alternative to embryonic tissue grafts would be a stem cell that could be proliferated in an unlimited fashion and then differentiated into a DA-producing cell with a full repertoire of neuronal features. Stem cells are cells characterized by the capacity of self-renewal and multipotentiality. There are several types of stem cells—the term being used liberally here, since some of these cells are committed precursors rather than true stem cells—that have been investigated regarding their prospective use for neurotransplantation.

1. Embryonic stem (ES) cells
2. Adult-derived stem cells
3. Neural progenitor cells

ES Cells

ES cells are derived from the inner cell mass of the blastocyst and can develop into virtually any cell type. Therefore this cell type is perhaps the ideal source to produce a specific cell type. ES cells spontaneously differentiate into all cell types derived from the three embryonic germ layers. Efforts are currently being made to isolate epigenetic factors, which stimulate the development of a certain cell type such as dopaminergic cells for transplantation in PD. To control cell growth and differentiation of these cells is one of the major challenges science is currently facing.

Several authors have reported a derivation of significant amounts of dopaminergic cells gained from ES cells in vitro: Different in vitro induction procedures have been most successful, so far using either mouse[186,187] or primate ES cells.[188] Both methods have led to the induction of cells expressing a dopaminergic phenotype in vitro varying between 23%[186] or 16% dopaminergic neurons out of the total number of cells.[187] A third group has established an ES cell line transfected with a rat nuclear receptor related-1 (Nurr1) gene, coding for a transcription factor that has a role in the differentiation of midbrain precursors into DA neurons.[189] Using stable Nurr1-ES cells in combination with defined culture conditions, as many as 78% of the neurons were generated under in vitro condition were TH-positive. These cells were also elegantly shown to posses several characteristics of differentiated DA neurons and to be functional in vitro and in vivo. Most importantly when grafted to the striatum, they could reverse both drug-induced and spontaneous motor deficits in the unilateral 6-OHDA-lesion rat model of PD.[189] Importantly, there was no teratoma formation in contrast to grafting undifferentiated mouse ES cells in another study.[190] In this latter study, undifferentiated mouse ES cells were transplanted in low cell numbers into hemiparkinsonian rats. In some cases the implants developed into relatively large grafts with many cells with a dopaminergic phenotype. These data were intriguing because there was no special protocol to induce differentiation of dopaminergic neurons from the ES cells. However, the study reported a troubling high 25% frequency of teratomas in animals with a surviving graft, and a further 16% of the grafted animals contained nonneuronal cells that were positive for mesodermal markers.[190] Dopaminergic differentiation of human ES cells has now also been shown in vitro. Cells can be differentiated using a co-culture system with mouse stromal cell lines[191,192] or other epigenetic factors such as transforming growth factor α (TGFα)[193] or a more complex sequence of different factors.[191] However, studies showing that human ES cell-derived dopaminergic neurons can survive grafting, express TH in vivo, integrate into the host brain without forming tumors, release DA in a regulated fashion and reverse behavioral deficits in animal models of PD have to be awaited.

Clearly, several safety aspects need to be addressed before grafts from ES cells can be considered for clinical trials.

Furthermore, there are ethical concerns associated with the use of human ES cells. Human ES cells are derived from the inner cell mass of human embryos about 4 days after fertilization, most of which are generated in in vitro fertilization clinics.[194] In August 2001, US president George Bush allowed govermental funding to be used for research involving the already established human ES cell lines (71 reported worldwide to the NIH so far) (for review see Ref. 195), but not for the derivation of new human cell lines.

Fetal Derived Neural Progenitors

In contrast to ES cells, neural progenitor cells are isolated from the developing embryo at later stages. They have a limited proliferative potential and a restricted differentiation capacity. They are characteristic for the tissue they derive from and can probably only give rise (as far as we currently know) to cell types present in the respective tissue. Thus, neural precursors have the ability to differentiate into neurons, oligodendrocytes and astrocytes.

Neural progenitor cells have been investigated regarding their use for cell replacement in PD. The goal is to proliferate neural progenitors in culture and subsequently differentiate these cells along the dopaminergic line.

Several groups have tried to optimize dopaminergic differentiation of mouse neural precursors of different brain regions[196,197] mouse striatum,[198] of precursor cells from rat and human cortex,[199] and rat mesencephalic precursors,[200-203] using different in vitro protocols and achieving up to 40% TH-positive cells out of all cells,[203] or even up to 73-98% using a rat mesencephalic dopaminergic progenitor clone.[204]

Mesencephalic precursor cells transplanted into a PD rat model have led to partial recovery of behavior.[200,204] However, the main problem connected with the use of progenitor cells for transplantation purposes is currently the poor survival of these cells.[205] The relative rate of survival was only about 3-5%,[200]or even lower.[204]

Human midbrain presursor cells can be expanded under conditions using low oxygen, using similar protocols.[206] About 0.93% of the cells expressed markers of a DA phenotype if cultured under certain conditions. Upon transplantation human neural progenitor cells gave rise to mainly astrocytes and few cells expressing neuronal markers. Cells that were TH-positive, and therefore probably dopaminergic neurons, were only rarely found.[207]

Adult-Derived Stem Cells

Stem cells have also been identified in a variety of adult tissues such as bone marrow, blood as well as within the skin, liver, muscle and even the adult brain. These adult derived stem cells (ASC) are pluripotent and have the ability of self-renewal. Furthermore, in comparison to ES cells, ASCs are easier to access and devoid of serious ethical issues because they can, e.g., be harvested from the patient, which would also make immunosuppressive treatment after transplantation unnecessary. In this section, we discuss the possible differentiation potential of ASC's and highlight some recent advances on ASC research.

Adult-derived neural stem cells could be isolated from various brain regions such as the subgranular zone of the dentate gyrus of the hippocampus, the ependymal/subventricular zone and the spinal cord.[208] Even the adult rat substantia nigra as been discussed as a site of ongoing neurogenesis and continous formation of newly differentiated DA neurons,[208a] that even further increases in a lesion model of PD. Carefully conducted studies,[209,209a] question this findings, so that the occurence of neurogenesis in the adult substantia nigra is currently debated. In other studies, only a glial response in the SNc was observed after lesioning the nigrostriatal system.[210] Adult-derived neural stem cells can be grown as neurospheres in culture and differentiate into all three major CNS cell types (neurons, astrocytes and oligodendrocytes). They have also been shown to differentiate into multiple nonneural cell types,[210a-214] (for review see Ref. 196). In humans, it has been possible to isolate stem cells with proliferative capacity from a variety of post-mortem brain regions, including the hippocampus, ventricular zone, motor cortex, temporal cortex and corpus callosum.[215]

Given their expected capacity to self-renew and differentiate efficiently into a desired cell type, clonal populations of neural stem cells promise to produce high numbers of DA neurons if they can be propagated, enriched and manipulated to differentiate along the DA lineage. However, it is not clear whether and to which extent this is possible.

The ability of adult neural stem cells to adopt a certain phenotype in vitro does still not necessarily guarantee that cells, once transplanted, incorporate into the host tissue and function in the expected way. So far it seems that site-specific differentiation in the embryonic as well as the adult brain only occurs in neurogenic regions. This could be especially true since a study showed that astrocytes from the hippocampus and the neonatal spinal cord have the ability to instruct neurogenesis in adult neural stem cells, whereas astryocytes from other regions than neurogenic regions fail to do so.[216] So the presence of endogenous neurogenic cues seems to be a prerequisite for neuronal differentiation and integration of donor cells (for review see Ref. 217). Neurogenic activity persists in the ventricular/subventricular zone, the olfactory bulb and the dentate gyrus of the hippocampus in the adult brain.[208,217]

Efforts to induce dopaminergic phenotypes in neurons derived from adult neural stem cells which have been propagated in vitro are currently made and transplantation experiments need to be awaited.

Evidence suggests that adult bone marrow may contain stem cells that can be pluripotent under appropriate conditions. After systemic infusion of bone marrow stem cells, low numbers of neuronal marker-positive (such as NeuN, b-III tubulin or MAP2) cells were detected in different regions of the brain.[218-220] In addition, BMSCs transplanted into lesioned spinal cord or cortical ischemic brain could differentiate into myelin-forming cells and repair the demyelinated CNS[221] or become neuronal marker-positive and improve functional recovery of the ischemic brain.[222] Most interestingly, a very recent study showed that cells copurified with mesenchymal stem cell from adult mouse or rat bone marrow differentiate into not only mesenchymal cells, but also cells with visceral mesoderm, neuroectoderm and endoderm characteristics in vitro and in vivo.[223] Remarkably, in vitro as many as 30% of these cells can differentiate into DA (TH-positive) neurons. When a single cell was introduced into a blastocyst, 33% of the mice formed chimaeras, and exhibited stem cell-derived cells in brain, lung, myocardium, skeletal muscle, liver, intestine, kindey, spleen, bone marrow, blood and skin. In addition, after intravenous systemic infusion, these cells differentiate into various tissue-specific cells, including bone marrow stem cells, but not neuronal cells. Importantly, human bone marrow stromal cells have been shown to generate a neural stem cell-like population that differentiates into neurons and glial cells. Some of these cells express TH and seem to release dopamine upon stimulation with KCl.[223a] These exciting findings suggest that mesenchymal stem cells are pluripotent and could serve as an ideal cell source for neurotransplantation in PD.

Besides bone marrow-derived stem cells, other stem cells derived from skin[224,224a] and fat tissue[225] e.g., have been discussed as a possible source for neurons. However, none of the data gives evidence for the derivation of TH-positive neurons (Fig. 4).

Concluding Remarks

Neural transplantation using nigral embryonic tissue has proven successful in animal experiments and clinical studies in PD patients. However, the relative shortage of suitable donor tissue, as well as ethical issues have lead to the demand of improved survival of grafted tissue or new tissue sources for neural transplantation. Several measures regarding a better survival of transplanted cells have been investigated and partially translated into the clinical situation. However, the use of xenogeneic tissue or, alternatively, dopaminergic cells derived from stem cells may solve problems with donor tissue in the foreseeable future. Overall, the fast developing field of neural transplantation gives the hope, that a standardized transplantation strategy for a larger number of patients can be expected in the not to distant future.

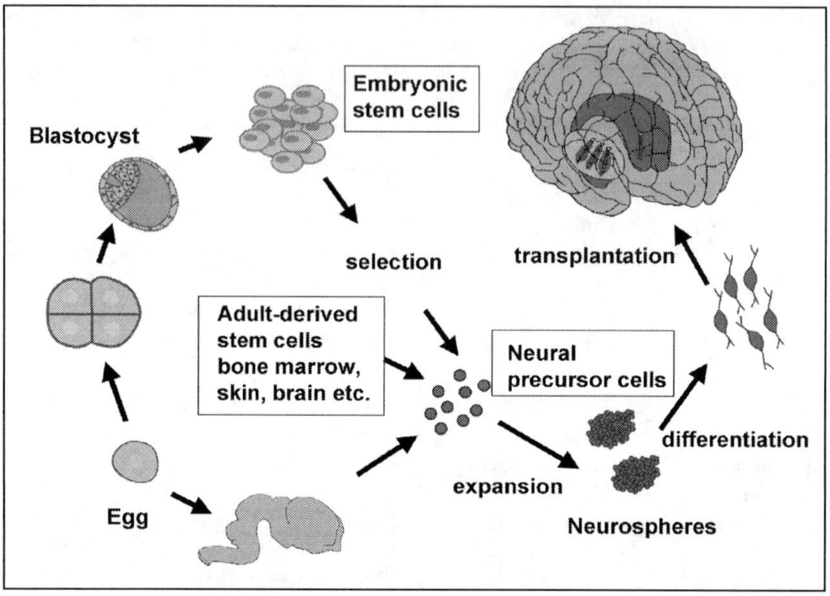

Figure 4. Dopaminergic neurons can be derived from different types of stem cells/progenitor cells. Embryonic stem (ES) cells are derived from the inner cell mass of the blastocyst stage of an embryo; and, after selection, can be differentiated into neural precursors. Neural precursors can also be derived from later stages of embryonic development or from stem cells residing in adult tissues, such as bone marrow stem cells.

Acknowledgments

We would like to thank Håkan Widner for valuable comments on the manuscript. Our own work mentioned in this review was supported by grants from the Swedish Research Council and the Segerfalk Foundation. Gesine Paul holds an Individual Marie Curie Fellowship from the European Union.

References

1. Lang AE, Lozano AM. Parkinson's disease. Second of two parts. N Engl J Med 1998a; 339:1130-1143.
2. Lang AE, Lozano AM. Parkinson's disease. First of two parts. N Engl J Med 1998b; 339:1044-1053.
3. Oertel WH, Hoglinger GU, Caraceni T et al. Depression in Parkinson's disease. An update. Adv Neurol 2001; 86:373-383.
4. Mayeux R, Chen J, Mirabello E et al. An estimate of the incidence of dementia in idiopathic Parkinson's disease. Neurology 1990; 40:1513-1517.
5. McKeith IG, Ballard CG, Perry RH et al. Prospective validation of consensus criteria for the diagnosis of dementia with Lewy bodies. Neurology 2000; 54:1050-1058.
6. Aarsland D, Tandberg E, Larsen JP et al. Frequency of dementia in Parkinson disease. Arch Neurol 1996; 53:538-542.
7. Kish SJ, Shannak K, Hornykiewicz O. Uneven pattern of dopamine loss in the striatum of patients with idiopathic Parkinson's disease. Pathophysiologic and clinical implications. N Engl J Med 1988; 318:876-880.
8. Agid Y. Parkinson's disease: Pathophysiology. Lancet 1991; 337:1321-1324.
9. Benamer TS, Patterson J, Grosset DG et al. Accurate differentiation of parkinsonism and essential tremor using visual assessment of [123I]-FP-CIT SPECT imaging: The [123I]-FP-CIT study group. Mov Disord 2000; 15:503-510.

10. Ichise M, Kim YJ, Ballinger JR et al. SPECT imaging of pre and postsynaptic dopaminergic alterations in L- dopa-untreated PD. Neurology 1999; 52:1206-1214.
11. Kruger R, Schols L, Muller T et al. Evaluation of the gamma-synuclein gene in German Parkinson's disease patients. Neurosci Lett 2001; 310:191-193.
12. Hirsch EC, Faucheux BA. Iron metabolism and Parkinson's disease. Mov Disord 1998; 13:39-45.
13. Jenner P. Oxidative mechanisms in nigral cell death in Parkinson's disease. Mov Disord 1998; 13:24-34.
14. Schapira AH. Pathogenesis of Parkinson's disease. Baillieres Clin Neurol 1997; 6:15-36.
15. Mouradian MM. Recent advances in the genetics and pathogenesis of Parkinson disease. Neurology 2002; 58:179-185.
16. Polymeropoulos MH, Lavedan C, Leroy E et al. Mutation in the alpha-synuclein gene identified in families with Parkinson's disease. Science 1997; 276:2045-2047.
17. Spillantini MG, Schmidt ML, Lee VM et al. Alpha-synuclein in Lewy bodies. Nature 1997; 388:839-840.
18. Goedert M. Alpha-synuclein and neurodegenerative diseases. Nat Rev Neurosci 2001; 2:492-501.
19. Lotharius J, Brundin P. Impaired dopamine storage resulting from alpha-synuclein mutations may contribute to the pathogenesis of Parkinson's disease. Hum Mol Genet 2002; 11(20):2395-2407.
20. Leroy E, Boyer R, Auburger G et al. The ubiquitin pathway in Parkinson's disease. Nature 1998; 395:451-452.
21. Lorick KL, Jensen JP, Fang S et al. RING fingers mediate ubiquitin-conjugating enzyme (E2)-dependent ubiquitination. Proc Natl Acad Sci USA 1999; 96:11364-11369.
22. Sherman MY, Goldberg AL. Cellular defenses against unfolded proteins: A cell biologist thinks about neurodegenerative diseases. Neuron 2001; 29:15-32.
23. Shastry BS. Parkinson disease: Etiology, pathogenesis and future of gene therapy. Neurosci Res 2001; 41:5-12.
24. Rascol O, Goetz C, Koller W et al. Treatment interventions for Parkinson's disease: An evidence based assessment. Lancet 2002; 359:1589-1598.
25. Koller W, Pahwa R, Busenbark K et al. High-frequency unilateral thalamic stimulation in the treatment of essential and parkinsonian tremor. Ann Neurol 1997; 42:292-299.
26. Lang AE. Surgery for Parkinson disease: A critical evaluation of the state of the art. Arch Neurol 2000; 57:1118-1125.
27. Pahwa R, Wilkinson S, Smith D et al. High-frequency stimulation of the globus pallidus for the treatment of Parkinson's disease. Neurology 1997; 49:249-253.
28. Bjorklund A. Cell replacement strategies for neurodegenerative disorders. Novartis Found Symp 2000; 231:7-15.
29. Alexander GE, Crutcher MD. Functional architecture of basal ganglia circuits: Neural substrates of parallel processing. Trends Neurosci 1990; 13:266-271.
30. DeLong MR. Primate models of movement disorders of basal ganglia origin. Trends Neurosci 1990; 13:281-285.
31. Dunnett SB, Bjorklund A, Lindvall O. Cell therapy in Parkinson's disease - stop or go? Nat Rev Neurosci 2001; 2:365-369.
32. Olson L, Malmfors T. Growth characteristics of adrenergic nerves in the adult rat. Fluorescence histochemical and 3H-noradrenaline uptake studies using tissue transplantations to the anterior chamber of the eye. Acta Physiol Scand Suppl 1970; 348:1-112.
33. Olson L, Seiger A. Brain tissue transplanted to the anterior chamber of the eye. 1. Fluorescence histochemistry of immature catecholamine and 5- hydroxytryptamine neurons reinnervating the rat iris. Z Zellforsch Mikrosk Anat 1972; 135:175-194.
34. Stenevi U, Bjorklund A, Svendgaard NA. Transplantation of central and peripheral monoamine neurons to the adult rat brain: Techniques and conditions for survival. Brain Res 1976; 114:1-20.
35. Bjorklund A, Stenevi U, Svendgaard N. Growth of transplanted monoaminergic neurones into the adult hippocampus along the perforant path. Nature 1976; 262:787-790.
36. Perlow MJ, Freed WJ, Hoffer BJ et al. Brain grafts reduce motor abnormalities produced by destruction of nigrostriatal dopamine system. Science 1979; 204:643-647.
37. Bjorklund A, Stenevi U. Reconstruction of the nigrostriatal dopamine pathway by intracerebral nigral transplants. Brain Res 1979; 177:555-560.

38. Dunnett SB, Bjorklund A, Schmidt RH et al. Intracerebral grafting of neuronal cell suspensions. V. Behavioural recovery in rats with bilateral 6-OHDA lesions following implantation of nigral cell suspensions. Acta Physiol Scand Suppl 1983a; 522:39-47.
39. Dunnett SB, Bjorklund A, Schmidt RH et al. Intracerebral grafting of neuronal cell suspensions. IV. Behavioural recovery in rats with unilateral 6-OHDA lesions following implantation of nigral cell suspensions in different forebrain sites. Acta Physiol Scand Suppl 1983b; 522:29-37.
40. Schmidt RH, Bjorklund A, Stenevi U et al. Intracerebral grafting of neuronal cell suspensions. III. Activity of intrastriatal nigral suspension implants as assessed by measurements of dopamine synthesis and metabolism. Acta Physiol Scand Suppl 1983; 522:19-28.
41. Dunnett SB, Bjorklund A, Stenevi U et al. Grafts of embryonic substantia nigra reinnervating the ventrolateral striatum ameliorate sensorimotor impairments and akinesia in rats with 6-OHDA lesions of the nigrostriatal pathway. Brain Res 1981; 229:209-217.
42. Dunnett SB. Anatomical and behavioral consequences of cholinergic-rich grafts to the neocortex of rats with lesions of the nucleus basalis magnocellularis. Ann NY Acad Sci 1987; 495:415-430.
43. Mandel RJ, Brundin P, Bjorklund A. The importance of graft placement and task complexity for transplant- induced recovery of simple and complex sensorimotor deficits in dopamine denervated rats. Eur J Neurosci 1990; 2:888-894.
44. Mahalik TJ, Finger TE, Stromberg I et al. Substantia nigra transplants into denervated striatum of the rat: Ultrastructure of graft and host interconnections. J Comp Neurol 1985; 240:60-70.
45. Freund TF, Bolam JP, Bjorklund A et al. Efferent synaptic connections of grafted dopaminergic neurons reinnervating the host neostriatum: A tyrosine hydroxylase immunocytochemical study. J Neurosci 1985; 5:603-616.
46. Zetterstrom T, Brundin P, Gage FH et al. In vivo measurement of spontaneous release and metabolism of dopamine from intrastriatal nigral grafts using intracerebral dialysis. Brain Res 1986; 362:344-349.
47. Strecker RE, Sharp T, Brundin P et al. Autoregulation of dopamine release and metabolism by intrastriatal nigral grafts as revealed by intracerebral dialysis. Neuroscience 1987; 22:169-178.
48. Arbuthnott G, Dunnett S, MacLeod N. Electrophysiological properties of single units in dopamine-rich mesencephalic transplants in rat brain. Neurosci Lett 1985; 57:205-210.
49. Doucet G, Murata Y, Brundin P et al. Host afferents into intrastriatal transplants of fetal ventral mesencephalon. Exp Neurol 1989; 106:1-19.
50. Brundin P, Duan W, Sauer H. Functional effects of mesencephalic dopamine neurons and adrenal chromaffin cells grafted to the rodent striatum. In: Dunnett SaBA, ed. Functional neural transplantation. New York: Raven Press, 1994:9-46.
51. Lindvall O. Clinical application of neuronal grafts in Parkinson's disease. J Neurol 1994; 242:S54-56.
52. Bjorklund A, Kirik D, Rosenblad C et al. Towards a neuroprotective gene therapy for Parkinson's disease: Use of adenovirus, AAV and lentivirus vectors for gene transfer of GDNF to the nigrostriatal system in the rat Parkinson model. Brain Res 2000; 886:82-98.
53. Lindvall O, Rehncrona S, Gustavii B et al. Fetal dopamine-rich mesencephalic grafts in Parkinson's disease. Lancet 1988; 2:1483-1484.
54. Lindvall O, Rehncrona S, Brundin P et al. Human fetal dopamine neurons grafted into the striatum in two patients with severe Parkinson's disease. A detailed account of methodology and a 6-month follow-up. Arch Neurol 1989; 46:615-631.
55. Madrazo I, Leon V, Torres C et al. Transplantation of fetal substantia nigra and adrenal medulla to the caudate nucleus in two patients with Parkinson's disease. N Engl J Med 1988; 318:51.
56. Madrazo I, FrancoBourland R, Ostrosky-Solis F et al. Fetal homotransplants (ventral mesencephalon and adrenal tissue) to the striatum of parkinsonian subjects. Arch Neurol 1990b; 47:1281-1285.
57. Madrazo I, FrancoBourland R, Ostrosky-Solis F et al. Neural transplantation (auto-adrenal, fetal nigral and fetal adrenal) in Parkinson's disease: The Mexican experience. Prog Brain Res 1990a; 82:593-602.
58. Backlund EO, Granberg PO, Hamberger B et al. Transplantation of adrenal medullary tissue to striatum in parkinsonism. First clinical trials. J Neurosurg 1985; 62:169-173.
59. Lindvall O, Backlund EO, Farde L et al. Transplantation in Parkinson's disease: Two cases of adrenal medullary grafts to the putamen. Ann Neurol 1987; 22:457-468.

60. Madrazo I, Drucker-Colin R, Diaz V et al. Open microsurgical autograft of adrenal medulla to the right caudate nucleus in two patients with intractable Parkinson's disease. N Engl J Med 1987; 316:831-834.
61. Goetz CG, Stebbins IIIrd GT, Klawans HL et al. United parkinson foundation neurotransplantation registry on adrenal medullary transplants: Presurgical, and 1- and 2-year follow-up. Neurology 1991; 41:1719-1722.
62. Quinn N. The modern management of Parkinson's disease. J Neurol Neurosurg Psychiatry 1990; 53:93-95.
63. Espejo EF, Montoro RJ, Armengol JA et al. Cellular and functional recovery of Parkinsonian rats after intrastriatal transplantation of carotid body cell aggregates. Neuron 1998; 20:197-206.
64. Luquin MR, Montoro RJ, Guillen J et al. Recovery of chronic parkinsonian monkeys by autotransplants of carotid body cell aggregates into putamen. Neuron 1999; 22:743-750.
65. Itakura T, Komai N, Ryujin Y et al. Autologous transplantation of the cervical sympathetic ganglion into the parkinsonian brain: Case report. Neurosurgery 1994; 35:155-157, (discussion 157-158).
66. Itakura T, Uematsu Y, Nakao N et al. Transplantation of autologous sympathetic ganglion into the brain with Parkinson's disease. Long-term follow-up of 35 cases. Stereotact Funct Neurosurg 1997; 69:112-115.
67. Subramanian T. Cell transplantation for the treatment of Parkinson's disease. Semin Neurol 2001; 21:103-115.
68. Lindvall O, Rehncrona S, Brundin P et al. Neural transplantation in Parkinson's disease: The Swedish experience. Prog Brain Res 1990b; 82:729-734.
69. Brundin P, Hagell, Peter. The neurobiology of cell transplantation in Parkinsons's disease. Clinical Neuroscience Research 2001; 1:507-520.
70. Defer GL, Geny C, Ricolfi F et al. Long-term outcome of unilaterally transplanted parkinsonian patients. I. Clinical approach. Brain 1996; 119:41-50.
71. Hauser RA, Freeman TB, Snow BJ et al. Long-term evaluation of bilateral fetal nigral transplantation in Parkinson disease. Arch Neurol 1999; 56:179-187.
72. Mendez I, Dagher A, Hong M et al. Enhancement of survival of stored dopaminergic cells and promotion of graft survival by exposure of human fetal nigral tissue to glial cell line—derived neurotrophic factor in patients with Parkinson's disease. Report of two cases and technical considerations. J Neurosurg 2000; 92:863-869.
73. Olanow CW, Kordower JH, Freeman TB. Fetal nigral transplantation as a therapy for Parkinson's disease. Trends Neurosci 1996; 19:102-109.
74. Brundin P, Pogarell O, Hagell P et al. Bilateral caudate and putamen grafts of embryonic mesencephalic tissue treated with lazaroids in Parkinson's disease. Brain 2000b; 123:1380-1390.
75. Hagell P, Schrag A, Piccini P et al. Sequential bilateral transplantation in Parkinson's disease: Effects of the second graft. Brain 1999; 122:1121-1132.
76. Freed CR, Greene PE, Breeze RE et al. Transplantation of embryonic dopamine neurons for severe Parkinson's disease. N Engl J Med 2001; 344:710-719.
77. Piccini P, Brooks DJ, Bjorklund A et al. Dopamine release from nigral transplants visualized in vivo in a Parkinson's patient. Nat Neurosci 1999; 2:1137-1140.
78. Piccini P, Lindvall O, Bjorklund A et al. Delayed recovery of movement-related cortical function in Parkinson's disease after striatal dopaminergic grafts. Ann Neurol 2000; 48:689-695.
79. Hagell P, Brundin P. Cell survival and clinical outcome following intrastriatal transplantation in Parkinson disease. J Neuropathol Exp Neurol 2001; 60:741-752.
80. Brundin P. Dissection, preparation and implantation of human embryonic brain tissue. In: SB, Dunnett AB, eds. Neural Transplantation, A Practical approach. Oxford: University Press, 1992:139-160.
81. Freeman TB, Spence MS, Boss BD et al. Development of dopaminergic neurons in the human substantia nigra. Exp Neurol 1991; 113:344-353.
82. Boer GJ. Ethical guidelines for the use of human embryonic or fetal tissue for experimental and clinical neurotransplantation and research. Network of European CNS Transplantation and Restoration (NECTAR). J Neurol 1994; 242:1-13.

83. Freed CR, Breeze RE, Rosenberg NL et al. Transplantation of human fetal dopamine cells for Parkinson's disease. Results at 1 year. Arch Neurol 1990a; 47:505-512.
84. Petersen A, Hansson O, Emgard M et al. Grafting of nigral tissue hibernated with tirilazad mesylate and glial cell line-derived neurotrophic factor. Cell Transplant 2000; 9:577-584.
85. Freeman TB, Sanberg PR, Nauert GM et al. The influence of donor age on the survival of solid and suspension intraparenchymal human embryonic nigral grafts. Cell Transplant 1995a; 4:141-154.
86. Folkerth RD, Durso R. Survival and proliferation of nonneural tissues, with obstruction of cerebral ventricles, in a parkinsonian patient treated with fetal allografts. Neurology 1996; 46:1219-1225.
87. Mamelak AN, Eggerding FA, Oh DS et al. Fatal cyst formation after fetal mesencephalic allograft transplant for Parkinson's disease. J Neurosurg 1998; 89:592-598.
88. Pakkenberg B, Moller A, Gundersen HJ et al. The absolute number of nerve cells in substantia nigra in normal subjects and in patients with Parkinson's disease estimated with an unbiased stereological method. J Neurol Neurosurg Psychiatry 1991; 54:30-33.
89. Frodl EM, Duan WM, Sauer H et al. Human embryonic dopamine neurons xenografted to the rat: Effects of cryopreservation and varying regional source of donor cells on transplant survival, morphology and function. Brain Res 1994; 647:286-298.
90. Kordower J, Habury H, Bankiewicz KS. Neuropathology of dopaminergic transplants in patient's with Parkinson's disease. In: FTaW H, ed. Cell transplantation for neurological disorders: Towards reconstruction of the human central nervous system. Totowa, NJ: Humana Press, 1998:51-75.
91. Nakao N, Frodl EM, Duan WM et al. Lazaroids improve the survival of grafted rat embryonic dopamine neurons. Proc Natl Acad Sci USA 1994; 91:12408-12412.
92. Clarkson ED, Freed CR. Development of fetal neural transplantation as a treatment for Parkinson's disease. Life Sci 1999; 65:2427-2437.
93. Freeman TB, Olanow CW, Hauser RA et al. Bilateral fetal nigral transplantation into the postcommissural putamen in Parkinson's disease. Ann Neurol 1995b; 38:379-388.
94. Freed CR, Breeze RE, Rosenberg NL et al. Therapeutic effects of human fetal dopamine cells transplanted in a patient with Parkinson's disease. Prog Brain Res 1990b; 82:715-721.
95. Sauer H, Frodl EM, Kupsch A et al. Cryopreservation, survival and function of intrastriatal fetal mesencephalic grafts in a rat model of Parkinson's disease. Exp Brain Res 1992; 90:54-62.
96. Zawada WM, Zastrow DJ, Clarkson ED et al. Growth factors improve immediate survival of embryonic dopamine neurons after transplantation into rats. Brain Res 1998; 786:96-103.
97. Hagell P, Piccini P, Bjorklund A et al. Dyskinesias following neural transplantation in Parkinson's disease. Nat Neurosci 2002b; 5:627-628.
98. Brundin P, Strecker RE, Widner H et al. Human fetal dopamine neurons grafted in a rat model of Parkinson's disease: Immunological aspects, spontaneous and drug-induced behaviour, and dopamine release. Exp Brain Res 1988; 70:192-208.
99. Forni C, Brundin P, Strecker RE et al. Time-course of recovery of dopamine neuron activity during reinnervation of the denervated striatum by fetal mesencephalic grafts as assessed by in vivo voltammetry. Exp Brain Res 1989; 76:75-87.
100. Brundin P, Strecker RE, Lindvall O et al. Intracerebral grafting of dopamine neurons. Experimental basis for clinical trials in patients with Parkinson's disease. Ann NY Acad Sci 1987; 495:473-496.
101. Lindvall O, Brundin P, Widner H et al. Grafts of fetal dopamine neurons survive and improve motor function in Parkinson's disease. Science 1990a; 247:574-577.
102. Peschanski M, Defer G, N'Guyen JP et al. Bilateral motor improvement and alteration of L-dopa effect in two patients with Parkinson's disease following intrastriatal transplantation of foetal ventral mesencephalon. Brain 1994; 117:487-499.
103. Bhatia KP, Marsden CD. The behavioural and motor consequences of focal lesions of the basal ganglia in man. Brain 1994; 117:859-876.
104. Holthoff-Detto VA, Kessler J, Herholz K et al. Functional effects of striatal dysfunction in Parkinson disease. Arch Neurol 1997; 54:145-150.
105. Annett LE, Torres EM, Ridley RM et al. A comparison of the behavioural effects of embryonic nigral grafts in the caudate nucleus and in the putamen of marmosets with unilateral 6- OHDA lesions. Exp Brain Res 1995; 103:355-371.
106. Widner H, Brundin P. Immunological aspects of grafting in the mammalian central nervous system. A review and speculative synthesis. Brain Res 1988; 472:287-324.

107. Brundin P, Widner H, Nilsson OG et al. Intracerebral xenografts of dopamine neurons: The role of immunosuppression and the blood-brain barrier. Exp Brain Res 1989; 75:195-207.
108. Brevig T, Holgersson J, Widner H. Xenotransplantation for CNS repair: Immunological barriers and strategies to overcome them. Trends Neurosci 2000; 23:337-344.
109. Widner H. Transplantation of neuronal and nonneuronal cells into the brain. Immune Response in the Nervous System. 2 Edition. In: Rothwell N, ed. Bios Scientific Publishers, 2002:189-217.
110. Duan WM, Cameron RM, Brundin P et al. Rat intrastriatal neural allografts challenged with skin allografts at different time points. Exp Neurol 1997b; 148:334-347.
111. Duan WM, Brundin P, Widner H. Addition of allogeneic spleen cells causes rejection of intrastriatal embryonic mesencephalic allografts in the rat. Neuroscience 1997a; 77:599-609.
112. Kordower JH, Goetz CG, Freeman TB et al. Dopaminergic transplants in patients with Parkinson's disease: Neuroanatomical correlates of clinical recovery. Exp Neurol 1997a; 144:41-46.
113. Widner H. The lund transplant program for parkinson's disease and patients with MPTP-Induced parkinsonism. In: FTaW H, ed. Cell transplantation for Neurological Disorders: Toward Reconstruction of the Human Central Nervous System. Totowa, NJ: Humana Press Inc.,1998:1-17.
114. Widner H. Immunological issues in rodent and primate transplants (Allografts). In: FTaW H, ed. Cell transplantation for Neurological Disorders: Toward Reconstruction of the Human Central Nervous System. Totowa, NJ: Humana Press Inc.,1998a:171-187.
115. Kordower JH, Freeman TB, Snow BJ et al. Neuropathological evidence of graft survival and striatal reinnervation after the transplantation of fetal mesencephalic tissue in a patient with Parkinson's disease. N Engl J Med 1995; 332:1118-1124.
116. Wenning GK, Odin P, Morrish P et al. Short- and long-term survival and function of unilateral intrastriatal dopaminergic grafts in Parkinson's disease. Ann Neurol 1997; 42:95-107.
117. Lindvall O, Hagell P. Clinical observations after neural transplantation in Parkinson's disease. Prog Brain Res 2000; 127:299-320.
118. Freed C, Breeze RE, Fahn S et al. Preoperative response to Levadopa is the best predictor of transplant outcome. Ann Neurology 2004; 6:846.
119. Kordower JH, Rosenstein JM, Collier TJ et al. Functional fetal nigral grafts in a patient with Parkinson's disease: Chemoanatomic, ultrastructural, and metabolic studies. J Comp Neurol 1996; 370:203-230.
120. Kordower JH, Freeman TB, Chen EY et al. Fetal nigral grafts survive and mediate clinical benefit in a patient with Parkinson's disease. Mov Disord 1998b; 13:383-393.
121. Kordower JH, Styren S, Clarke M et al. Fetal grafting for Parkinson's disease: Expression of immune markers in two patients with functional fetal nigral implants. Cell Transplant 1997b; 6:213-219.
122. Defer GL, Widner H, Marie RM et al. Core assessment program for surgical interventional therapies in Parkinson's disease (CAPSIT-PD). Mov Disord 1999; 14:572-584.
123. Langston JW, Widner H, Goetz CG et al. Core assessment program for intracerebral transplantations (CAPIT). Mov Disord 1992; 7:2-13.
124. van Hilten JJ, van der Zwan AD, Zwinderman AH et al. Rating impairment and disability in Parkinson's disease: Evaluation of the Unified Parkinson's Disease Rating Scale. Mov Disord 1994; 9:84-88.
125. Hagell P, Crabb L, Pogarell O et al. Health-related quality of life following bilateral intrastriatal transplantation in Parkinson's disease. Mov Disord 2000; 15:224-229.
126. Lindvall O, Sawle G, Widner H et al. Evidence for long-term survival and function of dopaminergic grafts in progressive Parkinson's disease. Ann Neurol 1994; 35:172-180.
127. Widner H, Tetrud J, Rehncrona S et al. Bilateral fetal mesencephalic grafting in two patients with parkinsonism induced by 1-methyl-4-phenyl-1,2,3,6-tetrahydropyridine (MPTP). N Engl J Med 1992; 327:1556-1563.
128. Baker KK, Ramig LO, Johnson AB et al. Preliminary voice and speech analysis following fetal dopamine transplants in 5 individuals with Parkinson disease. J Speech Lang Hear Res 1997; 40:615-626.
129. Brooks DJ. Monitoring neuroprotection and restorative therapies in Parkinson's disease with PET. J Neural Transm Suppl 2000; 60:125-137.

130. Morrish PK, Sawle GV, Brooks DJ. An [18F]dopa-PET and clinical study of the rate of progression in Parkinson's disease. Brain 1996; 119:585-591.
131. Remy P, Samson Y, Hantraye P et al. Clinical correlates of [18F]fluorodopa uptake in five grafted parkinsonian patients. Ann Neurol 1995; 38:580-588.
132. Snow BJ, Tooyama I, McGeer EG et al. Human positron emission tomographic [18F]fluorodopa studies correlate with dopamine cell counts and levels. Ann Neurol 1993; 34:324-330.
133. Rinne UK, Laihinen A, Rinne JO et al. Positron emission tomography demonstrates dopamine D2 receptor supersensitivity in the striatum of patients with early Parkinson's disease. Mov Disord 1990; 5:55-59.
134. Brooks DJ, Ibanez V, Sawle GV et al. Striatal D2 receptor status in patients with Parkinson's disease, striatonigral degeneration, and progressive supranuclear palsy, measured with 11C-raclopride and positron emission tomography. Ann Neurol 1992; 31:184-192.
135. Redmond Jr DE, Leranth C, Spencer DD et al. Fetal neural graft survival. Lancet 1990; 336:820-822.
136. Henderson BT, Clough CG, Hughes RC et al. Implantation of human fetal ventral mesencephalon to the right caudate nucleus in advanced Parkinson's disease. Arch Neurol 1991; 48:822-827.
137. Hitchcock E. Stereotactic neural transplantation. Stereotact Funct Neurosurg 1994; 62:120-133.
138. Freed CR, Breeze RE, Fahn S. Placebo surgery in trials of therapy for Parkinson's disease. N Engl J Med 2000; 342:353-354, (discussion 354-355).
139. Kordower JH, Freeman TB, Olanow CW. Neuropathology of fetal nigral grafts in patients with Parkinson's disease. Mov Disord 1998a; 13:88-95.
140. Brundin P, Dunnett S, Bjorklund A et al. Transplanted dopaminergic neurons: More or less? Nat Med 2001; 7:512-513.
141. Olanow CW, Freeman T, Kordower J. Transplantation of embryonic dopamine neurons for severe Parkinson's disease. N Engl J Med 2001; 345:146, (discussion 147).
142. Redmond Jr DE, Sladek JR, Spencer DD. Transplantation of embryonic dopamine neurons for severe Parkinson's disease. N Engl J Med 2001; 345:146-147.
143. Hagell P, Piccini P, Bjorklund A et al. Dyskinesias following neural transplantation in Parkinson's disease. Nat Neurosci 2002a; 3:3.
143a. Olanow CW, Goetz CG, Kordower JH et al. A double-blind controlled trial of bilateral fetal nigral transplantation in Parkinson's disease. Ann Neurol 2003; 54:403-414.
144. Lee CS, Cenci MA, Schulzer M et al. Embryonic ventral mesencephalic grafts improve levodopa-induced dyskinesia in a rat model of Parkinson's disease. Brain 2000a; 123:1365-1379.
145. Castilho RF, Hansson O, Brundin P. Improving the survival of grafted embryonic dopamine neurons in rodent models of Parkinson's disease. Prog Brain Res 2000b; 127:203-231.
146. Brundin P, Karlsson J, Emgard M et al. Improving the survival of grafted dopaminergic neurons: A review over current approaches. Cell Transplant 2000a; 9:179-195.
147. Glozman S, Yavin E. Lipid peroxides are generated by the fetal rat brain after episodes of global ischemia in utero. Neurochem Res 1997; 22:201-208.
148. Emgård M, Blomgren K, Brundin P. Characterisation of cell damage and death in embryonic mesencephalic tissue: a study on ultrastructure, vital stains and protease activity. Neuroscience 2002; 115(4):1177-1187.
149. Fawcett JW, Barker RA, Dunnett SB. Dopaminergic neuronal survival and the effects of bFGF in explant, three dimensional and monolayer cultures of embryonic rat ventral mesencephalon. Exp Brain Res 1995; 106:275-282.
150. Schierle GS, Brundin P. Excitotoxicity plays a role in the death of tyrosine hydroxylase-immunopositive nigral neurons cultured in serum-free medium. Exp Neurol 1999; 157:338-348.
151. Emgard M, Karlsson J, Hansson O et al. Patterns of cell death and dopaminergic neuron survival in intrastriatal nigral grafts. Exp Neurol 1999; 160:279-288.
152. Takei M, Hiramatsu M, Mori A. Inhibitory effects of calcium antagonists on mitochondrial swelling induced by lipid peroxidation or arachidonic acid in the rat brain in vitro. Neurochem Res 1994; 19:1199-1206.
153. Kaminski Schierle GS, Hansson O, Brundin P. Flunarizine improves the survival of grafted dopaminergic neurons. Neuroscience 1999; 94:17-20.

154. Braughler JM, Pregenzer JF. The 21-aminosteroid inhibitors of lipid peroxidation: Reactions with lipid peroxyl and phenoxy radicals. Free Radic Biol Med 1989; 7:125-130.
155. Karlsson J, Love RM, Clarke DJ et al. Effects of anaesthetics and lazaroid U-83836E on survival of transplanted rat dopaminergic neurones. Brain Res 1999; 821:546-550.
156. Schierle GS, Hansson O, Leist M et al. Caspase inhibition reduces apoptosis and increases survival of nigral transplants. Nat Med 1999; 5:97-100.
157. Pettmann B, Henderson CE. Neuronal cell death. Neuron 1998; 20:633-647.
158. Hansson O, Castilho RF, Kaminski Schierle GS et al. Additive effects of caspase inhibitor and lazaroid on the survival of transplanted rat and human embryonic dopamine neurons. Exp Neurol 2000; 164:102-111.
159. Gray J, Haran MM, Schneider K et al. Evidence that inhibition of cathepsin-B contributes to the neuroprotective properties of caspase inhibitor Tyr-Val-Ala-Asp- chloromethyl ketone. J Biol Chem 2001; 276:32750-32755.
160. Castilho RF, Hansson O, Brundin P. FK506 and cyclosporin A enhance the survival of cultured and grafted rat embryonic dopamine neurons. Exp Neurol 2000a; 164:94-101.
161. Mayer E, Dunnett SB, Pellitteri R et al. Basic fibroblast growth factor promotes the survival of embryonic ventral mesencephalic dopaminergic neurons—I. Effects in vitro. Neuroscience 1993; 56:379-388.
162. Takayama H, Ray J, Raymon HK et al. Basic fibroblast growth factor increases dopaminergic graft survival and function in a rat model of Parkinson's disease. Nat Med 1995; 1:53-58.
163. Zeng BY, Jenner P, Marsden CD. Altered motor function and graft survival produced by basic fibroblast growth factor in rats with 6-OHDA lesions and fetal ventral mesencephalic grafts are associated with glial proliferation. Exp Neurol 1996; 139:214-226.
164. Rosenblad C, Martinez-Serrano A, Bjorklund A. Glial cell line-derived neurotrophic factor increases survival, growth and function of intrastriatal fetal nigral dopaminergic grafts. Neuroscience 1996; 75:979-985.
165. Sinclair SR, Svendsen CN, Torres EM et al. GDNF enhances dopaminergic cell survival and fibre outgrowth in embryonic nigral grafts. Neuroreport 1996; 7:2547-2552.
166. Ahn Y-H, Bensadoun J-C, Aebischer P. Increased fiber outgrowth from xeno-transplanted human embryonic dopaminergic neurons with coimplants of polymer-encapsulated genetically modified cells releasing glial-cell line derived neurotrophic factor. Brain Res Bull; 66(2):135-42.
167. Karlsson J, Emgard M, Brundin P. Comparison between survival of lazaroid-treated embryonic nigral neurons in cell suspension, cultures and transplants. Brain Res 2002; 955(1-2):268-280.
168. Duan WM, Zhao LR, Westerman M et al. Enhancement of nigral graft survival in rat brain with the systemic administration of synthetic fibronectin peptide V. Neuroscience 2000; 100:521-530.
169. Karlsson J, Emgard M, Gido G et al. Increased survival of embryonic nigral neurons when grafted to hypothermic rats. Neuroreport 2000; 11:1665-1668.
170. Karlsson J, Petersen A, Gidö G et al. Combined caspase inhibitor and lazaroid pretreatment of embryonic donor tissue with mild hypothermia of the graft recipient-effects on the survival of the transplanted dopaminergic neurons and nondopaminergic cells. Submitted 2002b.
171. Freeman TB, Wojak JC, Brandeis L et al. Cross-species intracerebral grafting of embryonic swine dopaminergic neurons. Prog Brain Res 1988; 78:473-477.
172. Edge AS, Gosse ME, Dinsmore J. Xenogeneic cell therapy: Current progress and future developments in porcine cell transplantation. Cell Transplant 1998; 7:525-539.
173. Barker RA, Kendall AL, Widner H. Neural tissue xenotransplantation: What is needed prior to clinical trials in Parkinson's disease? Neural Tissue Xenografting Project. Cell Transplant 2000; 9:235-246.
174. Cozzi E, White DJ. The generation of transgenic pigs as potential organ donors for humans. Nat Med 1995; 1:964-966.
175. Isacson O, Deacon TW, Pakzaban P et al. Transplanted xenogeneic neural cells in neurodegenerative disease models exhibit remarkable axonal target specificity and distinct growth patterns of glial and axonal fibres. Nat Med 1995; 1:1189-1194.
176. Galpern WR, Burns LH, Deacon TW et al. Xenotransplantation of porcine fetal ventral mesencephalon in a rat model of Parkinson's disease: Functional recovery and graft morphology. Exp Neurol 1996; 140:1-13.

177. Larsson LC, Czech KA, Brundin P et al. Intrastriatal ventral mesencephalic xenografts of porcine tissue in rats: Immune responses and functional effects. Cell Transplant 2000; 9:261-272.
178. Michaels MG, McMichael JP, Brasky K et al. Screening donors for xenotransplantation. The potential for xenozoonoses. Transplantation 1994; 57:1462-1465.
179. Heneine W, Tibell A, Switzer WM et al. No evidence of infection with porcine endogenous retrovirus in recipients of porcine islet-cell xenografts. Lancet 1998; 352:695-699.
180. Paradis K, Langford G, Long Z et al. Search for cross-species transmission of porcine endogenous retrovirus in patients treated with living pig tissue. The XEN 111 Study Group. Science 1999; 285:1236-1241.
181. Dobson R. Cloning of pigs bring xenotransplants closer. Bmj 2000; 320:826.
182. Pakzaban P, Deacon TW, Burns LH et al. A novel mode of immunoprotection of neural xenotransplants: Masking of donor major histocompatibility complex class I enhances transplant survival in the central nervous system. Neuroscience 1995; 65:983-996.
183. Schumacher JM, Ellias SA, Palmer EP et al. Transplantation of embryonic porcine mesencephalic tissue in patients with PD. Neurology 2000; 54:1042-1050.
184. Deacon T, Schumacher J, Dinsmore J et al. Histological evidence of fetal pig neural cell survival after transplantation into a patient with Parkinson's disease. Nat Med 1997; 3:350-353.
185. Freeman TBW, Hauser RL, Bakay RA et al. A Prospective, randomized, double-blind, surgical placebo-controlled trial of intrastriatal transplantation of fetal porcine ventral mesencephalic tissue (Neurocell-PD) in subjects with parkinson's disease. Experimental Neurology 2002; 83(175):426, (Abstract).
186. Lee SH, Lumelsky N, Studer L et al. Efficient generation of midbrain and hindbrain neurons from mouse embryonic stem cells. Nat Biotechnol 2000b; 18:675-679.
187. Kawasaki H, Mizuseki K, Nishikawa S et al. Induction of midbrain dopaminergic neurons from ES cells by stromal cell-derived inducing activity. Neuron 2000; 28:31-40.
188. Kawasaki H, Suemori H, Mizuseki K et al. Generation of dopaminergic neurons and pigmented epithelia from primate ES cells by stromal cell-derived inducing activity. Proc Natl Acad Sci USA 2002; 99:1580-1585.
189. Kim JH, Auerbach JM, Rodriguez-Gomez JA et al. Dopamine neurons derived from embryonic stem cells function in an animal model of Parkinson's disease. Nature 2002; 20:20.
190. Bjorklund LM, Sanchez-Pernaute R, Chung S et al. Embryonic stem cells develop into functional dopaminergic neurons after transplantation in a Parkinson rat model. Proc Natl Acad Sci USA 2002; 99:2344-2349.
191. Perrier AL, Tabar V, Barberi T et al. Derivation of midbrain dopamine neurons from human embryonic stem cells. Proc Natl Acad Sci USA 2004; 101:12543-12548.
192. Zeng X, Cai J, Chen J et al. Dopaminergic differentiation of human embryonic stem cells. Stem Cells 2004; 22:925-940.
193. Park S, Lee KS, Lee YJ et al. Generation of dopaminergic neurons in vitro from human embryonic stem cells treated with neurotrophic factors. Neurosci Lett 2004; 359:99-103.
194. Thomson JA, Itskovitz-Eldor J, Shapiro SS et al. Embryonic stem cell lines derived from human blastocysts. Science 1998; 282:1145-1147.
195. Paul G, Li JY, Brundin P. Stem cells: Hype or hope? Drug Discov Today 2002; 7:295-302.
196. Daadi MM, Weiss S. Generation of tyrosine hydroxylase-producing neurons from precursors of the embryonic and adult forebrain. J Neurosci 1999; 19:4484-4497.
197. Daadi M, Arcellana-Panlilio MY, Weiss S. Activin cooperates with fibroblast growth factor 2 to regulate tyrosine hydroxylase expression in the basal forebrain ventricular zone progenitors. Neuroscience 1998; 86:867-880.
198. Stull ND, Iacovitti L. Sonic hedgehog and fgf8: Inadequate signals for the differentiation of a dopamine phenotype in mouse and human neurons in culture. Exp Neurol 2001; 169:36-43.
199. Theofilopoulos S, Goggi J, Riaz SS et al. Parallel induction of the formation of dopamine and its metabolites with induction of tyrosine hydroxylase expression in foetal rat and human cerebral cortical cells by brain-derived neurotrophic factor and glial-cell derived neurotrophic factor. Brain Res Dev Brain Res 2001; 127:111-122.
200. Studer L, Tabar V, McKay RD. Transplantation of expanded mesencephalic precursors leads to recovery in parkinsonian rats. Nat Neurosci 1998; 1:290-295.

201. Ling ZD, Potter ED, Lipton JW et al. Differentiation of mesencephalic progenitor cells into dopaminergic neurons by cytokines. Exp Neurol 1998; 149:411-423.
202. Potter ED, Ling ZD, Carvey PM. Cytokine-induced conversion of mesencephalic-derived progenitor cells into dopamine neurons. Cell Tissue Res 1999; 296:235-246.
203. Studer L, Csete M, Lee SH et al. Enhanced proliferation, survival, and dopaminergic differentiation of CNS precursors in lowered oxygen. J Neurosci 2000; 20:7377-7383.
204. Carvey PM, Ling ZD, Sortwell CE et al. A clonal line of mesencephalic progenitor cells converted to dopamine neurons by hematopoietic cytokines: A source of cells for transplantation in Parkinson's disease. Exp Neurol 2001; 171:98-108.
205. Brundin P, Bjorklund A. Survival of expanded dopaminergic precursors is critical for clinical trials. Nat Neurosci 1998; 1:537.
206. Storch A, Paul G, Csete M et al. Long-term proliferation and dopaminergic differentiation of human mesencephalic neural precursor cells. Exp Neurol 2001; 170:317-325.
207. Svendsen CN, Caldwell MA, Shen J et al. Long-term survival of human central nervous system progenitor cells transplanted into a rat model of Parkinson's disease. Exp Neurol 1997; 148:135-146.
208. Gage FH. Mammalian neural stem cells. Science 2000; 287:1433-1438.
208a. Zhao M, Momma S, Delfani K et al. Evidence for neurogenesis in the adult mammalian substantia nigra. Proc Natl Acad Sci USA 2003; 100:7925-7930.
209. Frielingsdorf H, Schwarz K, Brundin P et al. No evidence for new dopaminergic neurons in the adult mammalian substantia nigra. Proc Natl Acad Sci USA 2004; 101:10177-10182.
209a. Cooper O, Isacson O. Intrastriatal transforming growth factor alpha delivery to a model of Parkinson's disease induces proliferation and migration of endogenous adult neural progenitor cells without differentiation into dopaminergic neurons. J Neurosci 2004; 24:8924-8931
210. Lie DC, Dziewczapolski G, Willhoite AR et al. The adult substantia nigra contains progenitor cells with neurogenic potential. J Neurosci 2002; 22:6639-6649.
210a. Bjornson CR, Rietze RL, Reynolds BA et al. Turning brain into blood: A hematopoietic fate adopted by adult neural stem cells in vivo. Science 1999; 283:534-537.
211. Morshead CM, Benveniste P, Iscove NN et al. Hematopoietic competence is a rare property of neural stem cells that may depend on genetic and epigenetic alterations. Nat Med 2002; 8:268-273.
212. Galli R, Borello U, Gritti A et al. Skeletal myogenic potential of human and mouse neural stem cells [In Process Citation]. Nat Neurosci 2000; 3:986-991.
213. Rietze RL, Valcanis H, Brooker GF et al. Purification of a pluripotent neural stem cell from the adult mouse brain. Nature 2001; 412:736-739.
214. Clarke DL, Johansson CB, Wilbertz J et al. Generalized potential of adult neural stem cells. Science 2000; 288:1660-1663.
215. Palmer TD, Schwartz PH, Taupin P et al. Cell culture. Progenitor cells from human brain after death. Nature 2001; 411:42-43.
216. Song H, Stevens CF, Gage FH. Astroglia induce neurogenesis from adult neural stem cells. Nature 2002; 417:39-44.
217. Rossi F, Cattaneo E. Opinion: Neural stem cell therapy for neurological diseases: Dreams and reality. Nat Rev Neurosci 2002; 3:401-409.
218. Brazelton TR, Rossi FM, Keshet GI et al. From marrow to brain: Expression of neuronal phenotypes in adult mice. Science 2000; 290:1775-1779.
219. Mezey E, Chandross KJ, Harta G et al. Turning blood into brain: Cells bearing neuronal antigens generated in vivo from bone marrow. Science 2000; 290:1779-1782.
220. Priller J, Persons DA, Klett FF et al. Neogenesis of cerebellar Purkinje neurons from gene-marked bone marrow cells in vivo. J Cell Biol 2001; 155:733-738.
221. Sasaki M, Honmou O, Akiyama Y et al. Transplantation of an acutely isolated bone marrow fraction repairs demyelinated adult rat spinal cord axons. Glia 2001; 35:26-34.
222. Zhao LR, Duan WM, Reyes M et al. Human bone marrow stem cells exhibit neural phenotypes and ameliorate neurological deficits after grafting into the ischemic brain of rats. Exp Neurol 2002; 174:11-20.
223. Jiang Y, Jahagirdar BN, Reinhardt RL et al. Pluripotency of mesenchymal stem cells derived from adult marrow. Nature 2002; 20:20.

223a. Hermann A, Gastl R, Liebau S et al. Efficient generation of neural stem cell-like cells from adult human bone marrow stromal cells. J Cell Sci 2004; 117:4411-4422.
224. Toma JG, Akhavan M, Fernandes KJ et al. Isolation of multipotent adult stem cells from the dermis of mammalian skin. Nat Cell Biol 2001; 3:778-784.
224a. Joannides A, Gaughwin P, Schwiening C et al. Efficient generation of neural precursors from adult human skin: astrocytes promote neurogenesis from skin-derived stem cells. Lancet 2004; 364:172-178
225. Zuk PA, Zhu M, Mizuno H et al. Multilineage cells from human adipose tissue: Implications for cell- based therapies. Tissue Eng 2001; 7:211-228.

Index

A

B

C

D

E

F

G

H

I

L

M

N

O

P

R

S

T

V

W

X